FOUNDATIONS OF NANOTECHNOLOGY

VOLUME 2

NANOELEMENTS FORMATION AND INTERACTION

FOUNDATIONS OF NANOTECHNOLOGY

VOLUME 2

NANOELEMENTS FORMATION AND INTERACTION

Sabu Thomas, PhD, Saeedeh Rafiei,
Shima Maghsoodlou, and Arezo Afzali

Apple Academic Press

TORONTO   NEW JERSEY

Apple Academic Press Inc. | Apple Academic Press Inc.
3333 Mistwell Crescent | 9 Spinnaker Way
Oakville, ON L6L 0A2 | Waretown, NJ 08758
Canada | USA

First issued in paperback 2021

Exclusive worldwide distribution by CRC Press, a member of Taylor & Francis Group
No claim to original U.S. Government works

ISBN 13: 978-1-77463-105-8 (pbk)
ISBN 13: 978-1-77188-028-2 (hbk)

Library of Congress Control Number: 2014948525

Library and Archives Canada Cataloguing in Publication

Foundations of nanotechnology.
(AAP research notes on nanoscience & nanotechnology book series)
Contents: Volume 2. Nanoelements formation and interaction/Sabu Thomas, PhD, Saeedeh Rafiei, Shima Maghsoodlou, and Arezo Afzali.

Includes bibliographical references and index.
ISBN 978-1-77188-028-2 (v. 2 : bound)
1. Nanotechnology. I. Series: AAP research notes on nanoscience & nanotechnology book series

T174.7.F69 2014   620'.5   C2014-905376-2

ABOUT AAP RESEARCH NOTES ON NANOSCIENCE & NANOTECHNOLOGY

AAP Research Notes on Nanoscience & Nanotechnology reports on research development in the field of nanoscience and nanotechnology for academic institutes and industrial sectors interested in advanced research.

BOOKS IN THE AAP RESEARCH NOTES ON NANOSCIENCE & NANOTECHNOLOGY BOOK SERIES

**Nanostructure, Nanosystems and Nanostructured Materials:
Theory, Production, and Development**
Editors: P. M. Sivakumar, PhD, Vladimir I. Kodolov, DSc,
Gennady E. Zaikov, DSc, and A. K. Haghi, PhD

Nanostructures, Nanomaterials, and Nanotechnologies to Nanoindustry
Editors: Vladimir I. Kodolov, DSc, Gennady E. Zaikov, DSc, and
A. K. Haghi, PhD

**Foundations of Nanotechnology:
Volume 1: Pore Size in Carbon-Based Nano-Adsorbents**
A. K. Haghi, PhD, Sabu Thomas, PhD, and Moein MehdiPour MirMahaleh

**Foundations of Nanotechnology:
Volume 2: Nanoelements Formation and Interaction**
Sabu Thomas, PhD, Saeedeh Rafiei, Shima Maghsoodlou, and Arezo Afzali

**Foundations of Nanotechnology:
Volume 3: Mechanics of Carbon Nanotubes**
Saeedeh Rafiei

ABOUT THE AUTHORS

Sabu Thomas, PhD

Sabu Thomas, PhD, is a Professor of Polymer Science and Engineering at the School of Chemical Sciences and Director of the International and Inter University Centre for Nanoscience and Nanotechnology at Mahatma Gandhi University, Kottayam, Kerala, India. He received his BSc degree (1980) in Chemistry from the University of Kerala, BTech. (1983) in Polymer Science and Rubber Technology from the Cochin University of Science and Technology, and PhD (1987) in Polymer Engineering from the Indian Institute of Technology, Kharagpur. The research activities of Professor Thomas include surfaces and interfaces in multiphase polymer blend and composite systems, phase separation in polymer blends, compatibilization of immiscible polymer blends, thermoplastic elastomers, phase transitions in polymers, nanostructured polymer blends, macro-, micro- and nanocomposites, polymer rheology, recycling, reactive extrusion, processing–morphology–property relationships in multiphase polymer systems, double networking of elastomers, natural fibers and green composites, rubber vulcanization, interpenetrating polymer networks, diffusion and transport and polymer scaffolds for tissue engineering. He has supervised 65 PhD theses, 30 MPhil theses, and 40 Masters theses. He has three patents to his credit. He also received the coveted Sukumar Maithy Award for the best polymer researcher in the country for the year 2008. Very recently Professor Thomas received the MRSI and CRSI medals for his excellent work. With over 600 publications to his credit and over 17,500 citations, with an h-index of 67, Dr. Thomas has been ranked fifth in India as one of the most productive scientists.

Saeedeh Rafiei

Saeedeh Rafiei is a professional textile engineer and a Research Fellow at Technopark, Kerala, India's first technology park and among the three largest IT parks in India today. Saeed Rafiei earned a BSc in Textile Engineering, an MSc on Textile Engineering, and has published several papers in journals and conferences.

Shima Maghsoodlou

Shima Maghsoodlou is a professional textile engineer and a Research Fellow at Technopark, Kerala, India's first technology park and among the three largest IT parks in India today. Shima Maghsoodlou received a BSc in Textile Engineering, an MSc in Textile Engineering, and has published several papers in journals and conferences.

Arezoo Afzali

Arezoo Afzali is a professional textile engineer and a Research Fellow at Technopark, Kerala, India's first technology park and among the three largest IT parks in India today. She has a BSc in Textile Engineering, an MSc in Textile Engineering, and published several papers in journals and conferences.

CONTENTS

List of Abbreviations.. *xi*

List of Symbols.. *xiii*

Preface..*xvii*

1. Nanoscale Science and Technology: An Overview 1

2. Nanoelement Manufacturing: Self-Assembly 33

3. Nanomaterials: Properties and Application 55

4. Modeling and Simulation 101

5. Molecular Simulation for Nanomaterials 131

6. Numerical Simulation of Nanoelements 167

7. Numerical Study of Axial and Coaxial Electrospinning Process 197

Index *391*

LIST OF ABBREVIATIONS

A	Attractive Segment
AF	Attractive Force
AFM	Atomic Force Microscope
AP	Asymmetric Packing Segment
APP	Asymmetric Packing Process
BAMs	Bulk-Amorphous Metals
BD	Brownian Dynamics
BSA	Bovine Serum Albumin
CNT	Carbon Nanotube
D	Directional Segment
DC	Direct Current
DF	Directional Force
DOD	Dynamic Oblique Deposition
DPD	Dissipative Particle Dynamics
DSMC	Direct Monte Carlo Simulation
E	Energy of the System
ED	External Force-Induced Directional Factor
EF-F	External Force-Specific Functional Segment
F-BU	Fabrication Building Unit
FET	Field Effect Transistors
LEDs	Light Emitting Diodes
M&S	Modeling and Simulation
MC	Monte Carlo
MCP	Mechanochemical Processing
MD	Molecular Dynamics
MEMS	Micro Electromechanical Systems
MMT	Montmorillonite
MRI	Magnetic Resonance Imaging
MSA	Molecular Self- Assembly
MWCNT	Multi Wall Carbon Nanotube
N-CE	Nano-Communication Element
NEMS	Nano Electromechanical Systems
N-ME	Nano-Mechanical Element

N-PE	Nano-Property Element
NSC	National Science Foundation
N-SE	Nano-Structural Element
NSET	Nanoscale Science, Engineering, and Technology
PEM	Proton Exchange Ma
PMMA	Polymethyl Methacrylate
PSL	Polystyrene Latex
PZT	Plumbum Zirconate Titanate
R	Repulsive Segment
R-BU	Reactive Building Unit
RF	Repulsive Force
RISC	Reduced Instruction Set Computer
RVE	Representative Volume Element
SA-BU	Self-Assembly Building Unit
SAMs	Self-Assembled Monolayers
SOFC	Solid Oxide Fuel Cells
SWCNT	Single Wall Carbon Nanotubes
TBT	Tributyltin
UV	Ultraviolet
VLS	Vapor-Liquid-Solid

LIST OF SYMBOLS

A	an arbitrary physical quantity
A_L	new liquid surface area
A_S	surface area of the solid destroyed
A_{SL}	new liquid/solid interfacial area
D	particle diameter
D	characteristic size of the nanoelement
d	diameter of the nanoparticles
E	tensile strain
$E\gg(f)$	second derivative of the energy density surface
E_{Ani}	anisotropic energy
E_{App}	energy associated with an applied magnetic field
E_{Dem}	demagnetization energy
E_{Exc}	exchange energy
$e^{-i\omega t}$	monochromatic excitation
F	force
$\vec{F}_i(t)$	a random set of forces at a given temperature
$F_{X_1}^{ke}, F_{X_2}^{ke}\ F_{X_3}^{ke},$	external forces acting on nanoelements
$F_{X_1}^{kj}, F_{X_2}^{kj}\ F_{X_3}^{kj},$	the interaction forces of nanoelements
F_{ij}^C	conservative force
F_{ij}^D	dissipative force
F_{ij}^R	random force
F_{max}	the maximal force of the interaction of the nanoparticles
G	acceleration of gravity
H	Planck's constant
$\hbar$	reduced Planck constant
H_0	an incident electromagnetic wave
H_a	anisotropy field
H_t	transverse oscillating field
H_{demag}	demagnetizing field
H_{ext}	external field
H_{int}	internal field

H_{res}	residual field
I	area in the direction of the bending motion
J	the magnetisation oscillation
$J_{z_1}^k, J_{z_2}^k, J_{z_3}^k$	moments of inertia of a nanoelement
K	bulk modulus
k'	wave number
L	width of the potential well
l	length of the pendulum rod
L_0	latent heat of melting
M	magnetization vector
m	mass of the particle
m_i	the mass of the -th atom
$M_{z_1}^{ke}, M_{z_2}^{ke}, M_{z_3}^{ke},$	external moments acting on nanoelements
$M_{z_1}^{kj}, M_{z_2}^{kj}, M_{z_3}^{kj},$	the moment of forces of the nanoelement interaction
M^k	a mass of a nanoelements
M_s	saturation magnetization
N	anisotropy
n	the number of interatomic interaction types
$\hat{n}$	normal vector
N_e	the number of nanoelements
N_k	the number of atoms forming each nanoparticle
P	hydrostatic pressure
ρ	density
$\vec{P}_{ij}$	radius-vector
R	nanosphere radius
r_1	the position vector of the center of mass of particle
S	entropy
S_c	distance between the centers of mass of the nanostructure nanoelements
T	melting point of the extended system
Γ	shear strain
T_0	melting temperature of the bulk material
T_i	torque
U	anisotropy energy
$\cup$	total interaction energy
$u(r,t)$	macroscopic velocity
$V(\vec{r},t)$	varying potential influencing the particle's motion

V_i	the velocity vector of the center of mass of particle
V	the volume of the particle
v	velocity
V_L	volume of liquid
v_i	the translational velocity
$W_D (r_{ij})$ and $W_R (r_{ij})$	weighting functions
w_i	the angular velocity vector
$\bar{V}_{i0}, \bar{V}_i$	initial and current velocities of the -th atom
$\bar{x}_{i0}, \bar{x}_i$	original and current coordinates of the -th atom
X_{max}	arbitrary amplitude

GREEK SYMBOLS

$\psi(\vec{r})$	a function of space
Δ	dilatation
α	Gilbert damping constant
λ	longitudinal displacement of the nanospring
l'	mean free path
φ	mean magnetization direction
γ	surface tension
σ	tensile stress
DA	increment in surface area
DG_{Bulk}	free energy of the bulk material
S_L	surface energy of the liquid per unit area
γ_S	solid surface energy per unit area
γ_{SL}	solid/liquid interfacial energy per unit area
$\Theta_1, \Theta_2, \Theta_3,$	mutual orientation of the nanoelements
Ω_k	nanoelement area
α_i	the "friction" coefficient in the atomic structure
δ_{ij}	Kronecker delta
ζ_{ij}	stochastic variable inducing the random motion of particles
θ_{ij}	stochastic variable
ρ_j and ρ_i	density functions for microscopic states j and
σ^2	variance
Δn_i^{B}	rotational displacement
$\varepsilon(w)$	dielectric function
ε_0	dielectric constant of free space
τ	shear stress

τ the relaxation time (dimensionless)

$\Phi\left(\vec{\rho}_{ij}\right)$ the potential depending on the mutual positions of all
                                       the atoms

δ arbitrary initial phase angle

η viscosity

ξ friction coefficient

0-D   zero-dimensional

1-D   one-dimensional

2-D   two-dimensional

3-D   three-dimensional

PREFACE

One of the main tasks in making nanocomposites is building the dependence of the structure and shape of the nanoelements forming the basis for the composite of their sizes. This is because with an increase or a decrease in the specific size of nanoelements, their physical-mechanical properties, such as the coefficient of elasticity, strength, deformation parameter, etc., vary by over one order. The calculations show that this is primarily due to a significant rearrangement of the atomic structure and the shape of the nanoelement. The investigation of the above parameters of the nanoelements is technically complicated and laborious because of their small sizes. When the characteristics of powder nanocomposites are calculated, it is also very important to take into account the interaction of the nanoelements since the changes in their original shapes and sizes in the interaction process and during the formation of the nanocomposite can lead to a significant change in its properties and a cardinal structural rearrangement. In addition, the studies show the appearance of the processes of the ordering and self-assembling leading to a more organized form of a nanosystem. The above phenomena play an important role in nanotechnological processes. They allow nanotechnologies to be developed for the formation of nanostructures by the self-assembling method (which is based on self-organizing processes) and building up complex spatial nanostructures consisting of different nanoelements.

The study of the above dependences based on the mathematical modeling methods requires the solution of the aforementioned problem at the atomic level. This requires large computational aids and computational time, which makes the development of economical calculation methods urgent. The objective of this volume is the development of such a technique in various nanosystems.

— **Sabu Thomas, Saeedeh Rafiei, Shima Maghsoodlou, and Arezo Afzali**

CHAPTER 1

NANOSCALE SCIENCE AND TECHNOLOGY: AN OVERVIEW

CONTENTS

Abstract 2

1.1 Nanoscale Science and Technology 2

1.2 Review of Definitions 3

1.3 The Relationship Between Nanoscience and Mechanic Quantum 6

1.4 The Vision for Nanomaterials Technology 9

1.5 Surface to Volume 15

1.6 Production of Nanomaterials 21

1.7 Combined Top-Down and Bottom-Up Nanomanufacturing..................... 26

1.8 Conclusion 28

Keywords 28

References... 29

ABSTRACT

Nanoscience is the science, which study fundamental interactions between atoms and molecules. Modern properties in various systems can be atmostly obtained at dimensions between 1 nm to 100 nm. Nanoelement categories consist of atom clusters/assemblies or structures possessing at least one dimension between 1 and 100 nm. Quantum physics gives a completely different version of the world on the nanometric scale than that given by traditional physics. The most typical way of classifying nanomaterials is to identify them according to their dimensions. There are two approaches for synthesis of nanomaterials and the fabrication of nanostructures, Top down and Bottom up approaches. Combined top-down and bottom-up nanomanufacturing is also used for producing nanoelements such as various methods of self-assembly. In this chapter, the concepts of nanoscience and nanotechnology are reviewed in different views.

1.1   NANOSCALE SCIENCE AND TECHNOLOGY

1.1.1   KNOWLEDGE AND SCIENTIFIC UNDERSTANDING

Nanoscience is the science measured on a nanometer scale (10^{-9} meters). At this level everything is attenuated to fundamental interactions between atoms and molecules. A nanoelement compares to a basketball, like a basketball to the size of the earth. The aim of nonscientists is to manipulate and control the infinitesimal particles to create novel structures with unique properties. The science of atoms and simple molecules and the science of matter from microstructures to larger scales are generally established, in parallel. The remaining size related challenge is at the nanoscale where the fundamental properties of materials are determined and can be engineered. A revolution has been occurring in science and technology, based on the ability to measure, manipulate and organize matter on this scale. These properties are incorporated into useful and functional devices. Therefore, nanoscience will be transformed into nanotechnology. Through a basic understanding of ways to control and manipulate matter at the nanometer scale and through the incorporation of nanostructures and nanoprocesses into technological innovations, nanotechnology will provide the capacity to create affordable products with dramatically improved performance. Nanotechnology involves the ability to manipulate, measure, and model physical, chemical, and biological systems at nanometer dimensions, in order to exploit nanoscale phenomena [1].

Novel properties in biological, chemical, and physical systems can be approximately obtained at dimensions between 1 nanometer to 100 nanometers. These properties can differ in fundamental ways from the properties of individual atoms and molecules and those of bulk materials [1].

Nowadays, advances in nanoscience and nanotechnology indicate to have major implications for health, wealth, and peace. Knowledge in this field due to fundamental scientific advances, will lead to dramatic changes in the ways that materials, devices, and systems are understood and created. Nanoscience will redirect the scientific approach toward more generic and interdisciplinary research [1, 2].

1.2 REVIEW OF DEFINITIONS

In order to provide the necessary perspective, the following working definitions of nanotechnology, and its distinction from nanoscience, are listed below.

1.2.1 NANOMETER

The prefix 'nano' is derived from the Greek word "nanos" which means, "dwarf" in English [3–5]. 1 nanometer (nm) is 1 billionth of a meter (10^{-9} meter). For comparison purposes, the width of an average hair is 100.000 nanometers. Human blood cells are 2.000 to 5.000 nm long, a strand of DNA has a diameter of 2.5 nm, and a line of ten hydrogen atoms is one nm. The last three statistics are especially enlightening [3, 6–11].

1.2.2 NANOSCALE

The nanoscale, based on the nanometer (nm) or one-billionth of a meter, exists specifically between 1 and 100 nm. In the general sense, materials with at least one dimension below one micron but greater than one nanometer can be considered as nanoscale materials [12–14].

1.2.3 NANOELEMENTS

Nanoelement categories consist of atom clusters/assemblies or structures possessing at least one dimension between 1 and 100 nm, containing 10^3–10^9 atoms with masses of 10^4–10^{10} Daltons. Nanoelements are homogenous, uniform nanoparticles exhibiting well-defined (a) sizes, (b) shapes, (c) surface chemistries, and (d) flexibilities (i.e., polarizability). Typical nanoelement categories exhibit certain nanoscale atom mimicry features such as (a) core-shell architectures, (b) predominately (0-D) zero dimensionality (i.e, 1-D in some cases), (c) react and behave as discrete, quantized modules in their manifestation of nanoscale physico-chemical properties, and (d) display discrete valencies, stoichiometries, and mass combining ratios as a consequence of active atoms or reactive/passive functional groups presented in the outer valence shells of their core-shell architectures. Nanoelements must be accessible by synthesis or fractionation/separation methodologies with typical monodispersities [90% (i.e., uniformity) [15] as a function of mass, size, shape, and valency]. Wilcoxon et al. have shown that hard nanoparticle Au nanoclusters are as monodisperse as 99.9% pure (C) [15]. Soft nanoparticle dendrimers are routinely produced as high as generation = 6–8 with polydispersities ranging from 1.011 to 1.201 [16–18]. Nanoelement categories must be robust enough to allow reproducible analytical measurements to confirm size, mass, shape, surface chemistries, and flexibility/polarizability parameters under reasonable experimental conditions.

1.2.4   NANOMETROLOGY

Nanometrology (nanomanufacturing) is the science of measurement at the nanoscale level. Nanometrology has an important role to produce nanomaterials and devices with a high degree of accuracy and reliability [19–24]. Nanometrology includes length or size measurements (where dimensions are typically given in nanometers and the measurement uncertainty is often less than 1 nm) as well as measurement of force, mass, electrical and other properties [25–26].

1.2.5   NANOMATERIAL

These materials are revolutionizing the functionality of material systems. Due to the materials very small size, they have some remarkable, and in some cases, novel properties. Significant enhancement of optical, mechanical, electrical, structural and magnetic properties are commonly found with these materials [27–28]. Some key attributes include:
  • Grain size on the order of 10^{-9} m (1–100 nm).
  • Extremely large specific surface area.
  • Manifest fascinating and useful properties.
  • Structural and nonstructural applications.

- Stronger, more ductile materials.
- Chemically very active materials.

Two principal factors cause the properties of nanomaterials to differ significantly from other materials: increased relative surface area, and quantum effects. These factors can change or enhance properties such as reactivity, strength and electrical characteristics. As a particle decreases in size, a greater proportion of atoms are found at the surface compared to those inside.

Thus nanoparticles have a much greater surface area per unit mass compared with larger particles. As growth and catalytic chemical reactions occur at surfaces, this means that a given mass of material in nanoparticulate form will be much more reactive than the same mass of material made up of larger particles [29, 30].

In coupling with surface-area effects, quantum effects can begin to dominate the properties of matter as size is reduced to the nanoscale. These can affect the optical, electrical and magnetic behavior of materials, particularly as the structure or particle size approaches the smaller end of the nanoscale. Materials that exploit these effects include quantum dots, and quantum well lasers for optoelectronics [31–33].

1.2.6   NANOSCIENCE

Nanoscience is the study of phenomena and manipulation of materials at atomic, molecular and macromolecular scales, where properties differ significantly from those at a larger scale. Nanoscience is the study of nanoscale materials-materials that exhibit remarkable properties, functionality, and phenomena due to the influence of small dimensions. A couple of major distinctions between the size and biology are existed [34, 35].

1.2.7   NANOTECHNOLOGY

The National Science and Technology Council, Committee on Technology, Subcommittee on Nanoscale Science, Engineering, and Technology (NSET) formally established the following boundaries of nanotechnology in 2000.

The US national Nanotechnology Initiative defines Nanotechnology as: The science, engineering, and technology related to the understanding and control of matter at the length scale of approximately 1 to 100 nanometers and a joint report by the British Royal Society and the Royal Academy of Engineering similarly defined nanotechnology as "The design, characterization, production, and application of structures, devices and systems by controlling shape and size at nanometer scale."

There are many different forms of the above definitions in the media and scientific literature, but essentially all the definitions, after distillation and purification, crystallize into a few key forms-in particular that nanotechnology consists of materials with small dimensions, remarkable properties, and great potential.

So it can be said that nanotechnology research and development includes manipulation under control of the nanoscale structures and their integration into larger material components, systems and architectures. The novel and differentiating properties and functions are developed at a critical length scale of matter typically under 100 nm. In some particular cases, the critical length scale for novel properties and phenomena may be under 1 nm (manipulation of atoms at 0.1 nm) or be larger than 100 nm (nanoparticle reinforced polymers have the unique feature at 200–300 nm as a function of the local bridges or bonds between the nanoparticles and the polymer) [34].

The potential of nanominerals, as just one sector of nanomaterials technology have some very real and useful outcomes:

- Production of materials and products with new properties.
- Contribution to solutions of environmental problems.
- Improvement of existing technologies and development of new applications.
- Optimization of primary conditions for practical applications

1.3   THE RELATIONSHIP BETWEEN NANOSCIENCE AND MECHANIC QUANTUM

Atom (element)-based chemistry discipline before the advent of quantum mechanics and electronic theory, Dalton's atom/molecular theory is:

1. Each element consists of picoscale particles called atoms.
2. The atoms of a given element are identical; the atoms of different elements are different in some fundamental way(s).
3. Chemical compounds are formed when atoms of different elements combine with each other. A given compound always has the same relative number in types of atoms.
4. Chemical reactions involve reorganization of atoms (i.e., changes in the way they are bound).

Critical parameters that allowed this important progress evolved around discrete, reproducible features exhibited by each atomic element such as well-defined (a) atomic masses, (b) reactivities, (c) valency, (d) stoichiometries, (e) mass-combining ratios, and (f) bonding directionalities. These intrinsic elemental properties, inherent in all atom-based elemental structures.

Isaac Newton created, more than 300 years ago, classical mechanics by finding the laws of motion for solids and of gravitation between masses. This theory was so successful for the deterministic description of motions. At the beginning of the twentieth century, then, experimental results accumulated which contributed essentially to the emergence of a new physics, quantum physics. Also known as quantum or wave mechanics, this branch of physics was created by Max Planck who showed that the exchange of energy between matter and radiation occurred in discontinuous quantities (quanta). The quantum mechanics is presented as one of the most important and successful theories to solve physical problems. This is totally in the sense of most physicists, who applied, until the 1970s of the twentieth century, in a first quantum revolution quantum mechanics with overwhelming success not only to atom and particle physics but also to nearly all other science branches as chemistry, solid state physics, biology or astrophysics. Because of the success in answering essential questions in these fields fundamental open problems concerning the theory itself were approached only in rare cases. This situation has changed since the last decade of the twentieth century [36].

The "second quantum revolution" as this continuing further development of quantum physical thinking is called by Alain Aspect, one of the pioneers in this field, one expects a deeper understanding of quantum physics Fig. 1.1 itself but also applications in engineering. There is already the term "quantum engineering" which describes scientific activities to apply particle wave duality or entanglement for practical purposes, for example, nano-machines, quantum computers, etc. [37–38].

The nanoworld is part of our world, but in order to understand this, concepts other than the normal ones, such as force, speed, weight, etc., must be taken into consideration. The nanoworld is subject to the laws of quantum physics, yet evolution has conditioned us to adapt to this ever-changing world. This observation has led to further investigate theories based on the laws of physics that deal with macroscopic phenomena. In the macroworld, sizes are continuous, however, this is not the case in the nanoworld. When we investigate and try to understand what is happening on this scale, the way we look at things must be changed. New concepts of quantum physics can only come directly from our surroundings. However, our world is fundamentally quantum. Our common sense in this world has no value in the nanoworld [39].

Quantum physics gives a completely different version of the world on the nanometric scale than that given by traditional physics. A cloud of probability describes a molecule with the presence of electrons at discrete energy levels. This can only be represented as a simulation. All measurable sizes are subject to the laws of quantum physics, which condition every organism in the world, from the atom to the different states of matter. The nanoworld must therefore be addressed with quantum concepts. Chemistry is quantum. The chemistry of living organisms is quantum. Is the functioning of our brain closer to the concept of a quantum computer or to the most sophisticated microprocessors? All properties of matter are explicable only by quantum physics [39].

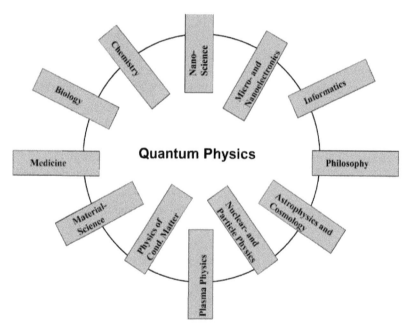

FIGURE 1.1 The relationship between important science branches and the field of quantum physics.

Traditional physics, which is certainly efficient and sufficient in the macroscopic domain, only deals with large objects (remember that there are nearly 10^{23} atoms per cm^3 in a solid), while quantum physics only deals with small discrete objects. However, the evolution of techniques and the use of larger and larger objects stemming from scientific discoveries make us aware of the quantum nature of the world in all its domains. Everything starts with the atom, the building block of the nanoworld, and also of our world. In mechanic quantum view, Particles can behave like waves. This property, particularly for electrons, is used in different investigation. On the other hand, waves can also act like particles: the photoelectric effect shows the corpuscular properties of light [39].

1.4 THE VISION FOR NANOMATERIALS TECHNOLOGY

Nanomaterials will deliver new functionality and options of material types. A diverse range of nanomaterial building blocks with well-defined properties and stable compositions will enable the design of nanomaterials that provide levels of functionality and performance which are not available in conventional materials.

Manufacturers will combine the benefits of traditional materials and nanomaterials to create new generations of nanomaterial-enhanced products that can be seamlessly integrated into complex systems. In some occasions, nanomaterials will serve as stand-alone devices, providing incomparable functionality. Nanomaterials show a prodigious opportunity for industry to introduce a host of new products that would energize the economy, solve major societal problems, revitalize existing industries, and create entirely new businesses. The race to research, develop, and commercialize nanomaterials is obviously global.

The nanoscience concept proposed the following: (a) creation of a nanomaterials roadmap (Fig. 1.2) focused solely on well-defined (i.e., >90% monodisperse), (0-D) and (1-D) nanoscale materials; (b) these well defined materials were divided into hard and soft nanoparticles, broadly following compositional/architectural criteria for traditional inorganic and organic materials; (c) a preliminary table of hard and soft nanoelement categories consisting of six (6) hard matter and six (6) soft matter particles was proposed. Elemental category selections were based on "atom mimicry" features and the ability to chemically combine or self assemble like atoms; (d) these hard and soft nanoelement categories produce a wide range of stoichiometric nanostructures by chemical bonding or nonbonding assembly. An abundance of literature examples provides the basis for a combinatorial library of hard-hard, hard-soft and soft-soft nano-compounds, many of which have already been characterized and reported. However, many such predicted constructs remain to be synthesized and characterized, (e) based on the presumed conservation of critical module design parameters, many new emerging nano-periodic property patterns have been reported in the literature for both the hard and soft nanoelement categories and their compounds [40].

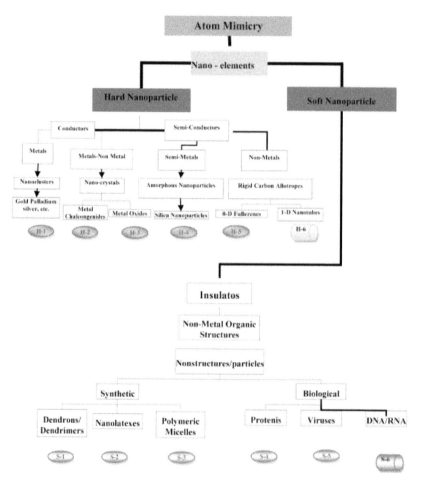

FIGURE 1.2 Nanomaterials Classification Roadmap.

1.4.1 CLASSIFICATION OF NANOMATERIALS

To properly understand and appreciate the diversity of nanomaterials, some form of categorization is required. Currently, the most typical way of classifying nanomaterials is to identify them according to their dimensions.

When one or more of the dimensions of a solid are reduced sufficiently in size, its physico-chemical behavior departs significantly from the bulk state. Low-dimensional structures exhibit properties that are different from those of the miniature versions of bulk structures. Nanostructures constitute a bridge between molecules and bulk materials. Suitable control of the properties and response of nanostructures can lead to new devices and technologies. Accordingly, nanoscience and nanotechnology primarily deal with the synthesis, characterization, exploration and exploitation of nanostructured materials. With reduction in size, different and often new electrical, mechanical, chemical, magnetic and optical properties emerge. The resulting structure is a low-dimensional structure. The typical dimensions are usually in the range of a few nanometers. The confinement of particles in a low-dimensional structure actuates to an imperial change in their behavior and to the manifestation of novel size-dependent effects, which usually fall into the category of quantum size effects. Nanostructures are low-dimensional structures. Quantum size effects appear in their electrical, thermal, magnetic and optical properties, depending on the dimensionality, and offer a rich palette of phenomena to be technological exploited. Several old techniques have been improved and several new techniques have been devised to fabricate and characterize nanostructures.

As sizes approach the atomic scale, the relevant physical laws change from the classical to the quantum -mechanical laws of physics. The quantum nature of nanostates and specific statistical laws dominates in the nanoworld. The basic principles are the superposition principle as an attribute of the quantum microworld decoherence of a wave function as an attribute of the classical macroworld. The Physical behavior at the nanometer scale is accurately predicted by quantum mechanics, as represented by the Schrödinger equation, which therefore provides a quantitative understanding of the properties of low-dimensional structures. In the Schrödinger description of quantum mechanics, a particle (electron, hole, exciton, etc.) or physical system (i.e., an atom) is described by a wave function $\psi(\vec{r}, t)$, which depends on the variables describing the degrees of freedom of the system, and is interpreted as the probability amplitude of finding a particle at spatial location $\vec{r} = (x, y, z)$ and time t. Thus, while the state of motion of a particle in classical mechanics is specified by the particle's position and velocity, in quantum mechanics the state of motion is specified by the particle's wave function, which contains all the information that may be obtained about the particle. Thus, in quantum mechanics, the idea of a trajectory must be eliminated in favor of a more subtle description in terms of quantum states and wave functions [28, 41, 42].

The wave function of an uncharged particle with no spin satisfies the Schrödinger equation:

$$-\frac{\hbar^2}{2m}\nabla^2 + V(\vec{r},t)\psi(\vec{r},t) = \hbar^2 \frac{\partial \psi(\vec{r},t)}{\partial t} \tag{1.1}$$

where $\nabla^2 \equiv \dfrac{\partial^2}{\partial x^2} + \dfrac{\partial^2}{\partial y^2} + \dfrac{\partial^2}{\partial z^2}$ is the Laplacian operator?

The Hamiltonian $H(\vec{r},t) = -\dfrac{\hbar^2}{2m}\nabla^2 + V(\vec{r},t)$ is implicit in this nonrelativistic equation. The first part of the Hamiltonian corresponds to the kinetic energy, the second to potential energy. The Hamiltonian thus describes the total energy of the system. As with the force in Newton's second law, the exact form of the Hamiltonian is not provided by the Schrödinger equation and must be independently formulated from the physical properties of a given system [28, 43].

For many real-world systems, the potential does not depend on time $V(\vec{r},t) = V(\vec{r})$. Then, the dependence on time and spatial coordinates of $\psi(\vec{r},t)$ can be separated as:

$$\psi(\vec{r},t) = e^{-iEt/\hbar}\psi(\vec{r}) \tag{1.2}$$

On using the representation of the wave function in the Schrödinger Eq. (1), the time- independent Schrödinger equation is [28]:

$$-\frac{\hbar^2}{2m}\nabla^2 + V(\vec{r})\psi(\vec{r}) = E\psi(\vec{r}) \tag{1.3}$$

Low-dimensional systems are usually classified according to the number of reduced dimensions. More precisely, the dimensionality refers to the number of degrees of freedom in the momentum. Accordingly, depending on the dimensions of the system, different cases can be distinguished, as shown in Fig. 1.3 [44]:

- Zero-dimensional (0-D)
- One-dimensional (1-D)
- Two-dimensional (2-D)
- Three-dimensional (3-D)

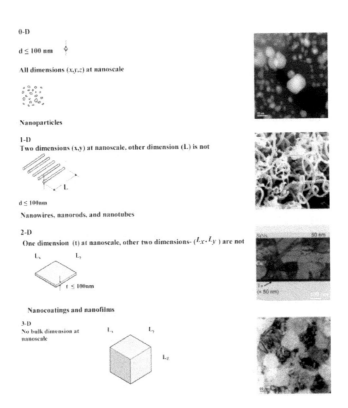

FIGURE 1.3   Classification of nanomaterials according to 0-D, 1-D, 2-D and 3-D.

This classification is based on the number of dimensions, which are not confined to the nanoscale range (<100 nm). As we will see in what follows, as these categories of nanomaterials move from the 0-D to the 3-D configuration, categorization becomes more and more difficult to define as well [45]

1.4.1.1   ZERO-DIMENSIONAL (0-D)

Beginning with the most clearly defined category, zero-dimensional nanomaterials are materials wherein all the dimensions are measured within the nanoscale (no dimensions, or 0-D, are larger than 100 nm). The most common representation of zero dimensional nanomaterials is nanoparticles. These nanoparticles can [7, 46]:

- be amorphous or crystalline.
- be single crystalline or polycrystalline.
- be composed of single or multichemical elements.
- exhibit various shapes and forms.

- exist individually or incorporated in a matrix.
- be metallic, ceramic, or polymeric.

1.4.1.2 ONE-DIMENSIONAL (1-D)

On the other hand, 1-D nanomaterials differ from 0-D nanomaterials in that the former have one dimension that is outside the nanoscale. This difference in material dimensions leads to needlelike-shaped nanomaterials. One-dimensional nanomaterials include nanotubes, nanorods, and nanowires, these 1-D nanomaterials can be [7, 46–48]:

- amorphous or crystalline;
- single crystalline or polycrystalline;
- chemically pure or impure;
- standalone materials or embedded in within another medium;
- metallic, ceramic, or polymeric.

1.4.1.3 TWO-DIMENSIONAL (2-D)

Two-dimensional nanomaterials are more difficult to classify. However, assuming for the time being the aforementioned definitions for 0-D and 1-D nanomaterials, 2-D nanomaterials are materials in which two of the dimensions are not bounded to the nanoscale. As a result, 2-D nanomaterials exhibit plate-like shapes. Two-dimensional nanomaterials include nanofilms, nanolayers, and nanocoatings. These 2-D nanomaterials can be [7, 46–47]:

- amorphous or crystalline;
- made up of various chemical compositions;
- used as a single layer or as multilayer structures;
- deposited on a substrate;
- integrated in a surrounding matrix material;
- metallic, ceramic, or polymeric.

1.4.1.4 THREE-DIMENSIONAL (3-D)

Three-dimensional nanomaterials, also known as bulk nanomaterials, are relatively difficult to classify. However, in keeping with the dimensional parameters we've established so far, it is true to say that bulk nanomaterials are materials that are not confined to the nanoscale in any dimension. These materials are thus characterized by having three arbitrarily dimensions above 100 nm. With the introduction of these arbitrary dimensions, it could be asked why these materials are called nanomaterials. The reason for the continued classification of these materials as nanomaterials is: Despite their nanoscale dimensions, these materials possess a nanocrystalline structure or involve the presence of features at the nanoscale. In terms of nanocrystalline structure, bulk nanomaterials can be composed of a multiple arrangement of nanosize crystals, most typically in different orientations [7, 46, 47].

With respect to the presence of features at the nanoscale, 3-D nanomaterials can contain dispersions of nanoparticles, bundles of nanowires, and nanotubes as well as multinanolayers.

Three-dimensional nanomaterials can be:

- amorphous or crystalline;
- chemically pure or impure;
- composite materials;
- composed of multinanolayers;
- metallic, ceramic, or polymeric.

This procedure of classification by dimensions allows nanomaterials to be identified and classified in a 3-D space. The x, y and z represent dimensions below 100 nm (Fig. 1.4).

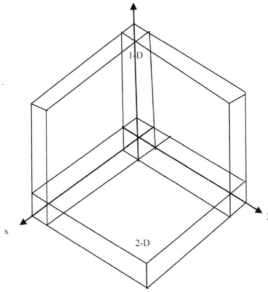

FIGURE 1.4 Three-dimensional space showing the relationships among 0-D, 1-D, 2-D, and 3-D nanomaterials.

1.5   SURFACE TO VOLUME

Nanomaterials have an increased surface-to-volume ratio compared to bulk materials. This means that for a given total volume of material, the external surface is greater if it is made of an ensemble of nanomaterial subunits rather than of bulk material (Fig. 1.5). The increased surface-to-volume of nanomaterials affects onto the material physical properties such as its melting and boiling points, as well as its chemical reactivity like reactions which occur at the material surface (such as catalysis reactions, detection reactions, and reactions that require the physical adsorption of certain species at the material's surface to initiate). Finally, the higher surface-to-volume of nanomaterials allows using less material, which has environmental and economic benefits, as well as fabricating highly miniaturized devices, which can be portable and could use less power to operate [7].

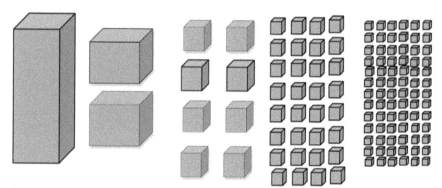

FIGURE 1.5 Schematic drawing showing how surface-to-volume increases when moving from a bulk material (far left) to nano-sized particles (far right).

Individual nanoclusters, nanostructures, and matrix nanosystems possess unique properties as compared to those of individual macromolecules and bulk solids. This can be explained by the specific features of their structure. Moreover, it seems likely that the nanoworld various types of geometries of constant curvature, namely, the Euclidean, Lobachevski, Riemannian geometries and certain constructions of projective geometry [49].

FIGURE 1.6 The structural diversity of the nanoworld: zero-dimensional (point), one-dimensional (linear), fractal, two-dimensional, and three-dimensional nanoparticle fragments.

1.5.1 THE FUNDAMENTAL IMPORTANCE OF SIZE

Some of the technologies deal with systems on the micrometer range and not on the nanometer range (1–100 nm). In fact, distance scales used in science go too much smaller than nanometers and much larger than meters. All the experience in the macroscopic world suggests that matter is continuous. This, however, leads to a paradox because if matter were a continuum, it could be cut into smaller and smaller pieces without end. If one were able to keep cutting a piece of matter in two, each of those pieces into two, and so on ad infinitum, one could, at least in principle, cut it out of existence into pieces of nothing that could not be reassembled.

Nowadays it can be studied pieces of matter of smaller and smaller size right down to the atom. The important result is that the properties of the pieces start to change at sizes much bigger than a single atom. When the size of the material crosses into the nanoworld, its fundamental properties start to change and become dependent on the size of the piece. It is an important issue to know how the behavior of a piece of material can become critically dependent on its size [44].

1.5.2 NANOTECHNOLOGY AND APPROACHES TO FABRICATING

There are two approaches for synthesis of nanomaterials and the fabrication of nanostructures. Top down approach refers to slicing or successive cutting of a bulk material to get nanosized particle, which enables to control the manufacture of smaller, more complex objects, as illustrated by micro and nanoelectronics. Bottom up approach refers to the buildup of a material from the bottom: atom by atom, molecule by molecule or cluster by cluster which enables to control the manufacture of atoms and molecules, as illustrated by supramolecular chemistry.

Both approaches play very important role in modern industry and most likely in nanotechnology as well. There are advantages and disadvantages in both approaches.

Attrition or Milling is a typical top down method in making nanoparticles, where as the colloidal dispersion is a good example of bottom up approach in the synthesis of nanoparticles (Fig. 1.7).

The biggest problem with top down approach is the imperfection of surface structure and significant crystallographic damage to the processed patterns. These imperfections, which in turn lead to extra challenges in the device design and fabrication. But this approach navigates to the bulk production of nanomaterial. Regardless of the defects produced by top down approach, it will be continued to play an important role in the synthesis of nanostructures.

Though the bottom up approach frequently referred in nanotechnology, it is not a newer concept. All the living beings in nature observe growth by this approach only and also it has been in industrial use for over a century. Examples include the production of salt and nitrate in chemical industry. Although the bottom up approach is nothing new, it plays an important role in the fabrication and processing of nanostructures. There are several reasons for this and explained as below.

When structures fall into a nanometer scale, there is a little chance for top down approach. All the tools we have possessed are too big to deal with such tiny subjects. Bottom up approach also promises a better chance to obtain nanostructures with less defects, more homogeneous chemical composition. On the contrary, top down approach most likely introduces internal stress, in addition to surface defects and contaminations [50, 51].

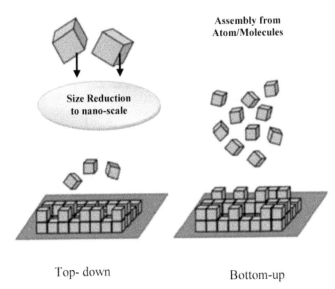

Assembly from Atom/Molecules

Size Reduction to nano-scale

Top- down   Bottom-up

FIGURE 1.7   Top-down and bottom-up approaches in nanotechnology.

1.5.2.1   TOP-DOWN APPROACH

The idea behind the top-down approach is the following: An operator first designs and controls a macroscale machine shop to produce an exact copy of itself, but smaller in size. Subsequently, this downscaled machine shop will make a replica of itself, but also a few times smaller in size. This process of reducing the scale of the machine shop continues until a nanosize machine shop is produced and is capable of manipulating nanostructures. One of the emerging fields based on this top-down approach is the field of Nano and Micro-Electromechanical Systems (NEMS and MEMS, respectively). MEMS research has already produced various micromechanical devices, smaller than 1 mm^2, which are able to incorporate microsensors, cantilevers, microvalves, and micro pumps [7, 52].

1.5.2.2 BOTTOM-UP APPROACH

The concept of the bottom-up approach is that one starts with atoms or molecules, which build up to form larger structures (Fig. 1.8). In this context, there are three important enabling bottom-up technologies, namely:
- supramolecular and molecular chemistry;
- scanning probes;
- biotechnology [52].

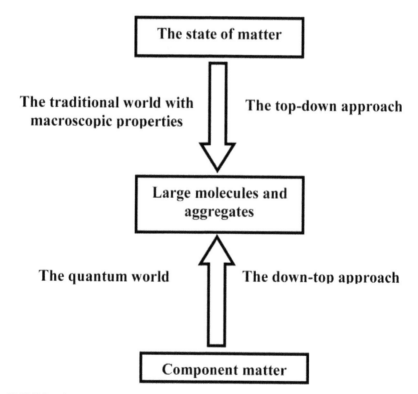

FIGURE 1.8   The schematic of nanomaterial fabrication.

1.6 PRODUCTION OF NANOMATERIALS

There are various widely known methods to produce nanomaterials other than by direct atom manipulation. In plasma arcing, the very high temperatures associated with the formation of an arc or plasma is used to effectively separate the atomic species of feedstock, which quickly recombine outside the plasma to form nanosized particles, which may have novel compositions [53–55]. In the case of chemical vapor deposition, feed gasses are reacted in a chamber and the resulting species attracted to a substrate. Once again the reaction products can be controlled and not only in terms of composition but also in terms of how they are deposited. The substrate effectively provides a template from where the deposited coating can grow in a very well controlled manner. Electro-deposition involves a similar process; however, the controlled coating is deposited from solution by the application of an electric field. Sol-gel synthesis uses chemical means to produce intimately mixed compounds that are hydrolyzed into gels. The gels can be deposited on any surface and shape at well controlled thicknesses and on subsequent heating, decompose to leave a thin layer of the desired coating. This technique is well suited to coating large surface areas with very well defined nanometer scale compounds. In high intensity ball milling, as the name suggested, high impact collisions are used to reduce macrocrystalline materials down into nano-crystalline structures without chemical change. A relatively new technique termed Mechanochemical Processing (MCP) technology, being developed by Advanced Nanotechnologies based in Perth, is a novel, solid-state process for the manufacture of a wide range of nanopowders. Dry milling is used to induce chemical reactions through ball-powder collisions that result in nanoparticles formed within a salt matrix. Particle size is defined by the chemistry of the reactant mix, milling and heat treatment conditions. Particle agglomeration is minimized by the salt matrix, which is then removed by a simple washing procedure [56].

1.6.1 TOP-DOWN PROCESSES

Top-down processes are best described this way: Take the material and drug, beat and freeze it. Drug it by alloying. Beat it by severe plastic deformation. Freeze it to stop the fine-scale structure, once formed, from coarsening.

1.6.1.1   RAPID SOLIDIFICATION

Materials would rather be crystals than glasses. If ordered, all atoms sit at exactly the distance from their neighbors that best satisfies their interatomic bonds. Disorder disrupts this comfortable seating arrangement, stretching some bonds and squeezing others. Liquids are disordered because heat shakes the atoms so violently that they are sprung from their low-energy, crystalline arrangement. The melting point is the temperature at which this disruption occurs; below it the thermal shaking is too weak to disrupt bonds, and the liquid crystallizes.

Cooled at normal rates, most liquids solidify to give solids with large crystals, or "grains." The trick in making nanocrystalline or amorphous materials by casting them is to deprive the material of the time or the means to transform from liquid to solid. Some materials, of which window glass is one, are easily duped into retaining their glassy structure-their high viscosity when liquid slows the rearrangement of the molecules to form crystals. Many polymers, too, are "glassy" because their tangled molecules cannot reorganize in any normal timeframe to form the crystal they would like to be. Metals and ceramics, by contrast, crystallize at the drop of a hat. It takes extreme measures to make them retain their liquid like structure or adopt a structure with exceedingly small grains.

The first step is to mix in elements with different-sized atoms, each preferring a different atomic spacing and crystal structure, making crystallization difficult. The second step is to cool quickly, leaving little or no time for crystallization. Early alloys required precipitous cooling rates exceeding 1,000,000°C/sec, limiting the form to thin wires and ribbons from which heat can be conducted quickly [7].

Figure 1.9 shows one way of achieving such cooling rates. A jet of liquid alloy-one designed to be hard to crystallize-is squirted onto a spinning, water-cooled copper drum. The process is called melt spinning. The liquid layer cools fast enough to become amorphous or, if not that, then nanocrystalline. The laminated cores of many transformers and the read/write heads of magnetic tape and disc recorders are made that way, exploiting the special magnetic properties of the amorphous state. Newer bulk-amorphous metals (BAMs) remain glassy even at relatively slow rates of cooling (10°C/sec), allowing thick sections (up to 20 mm) to be cast [7].

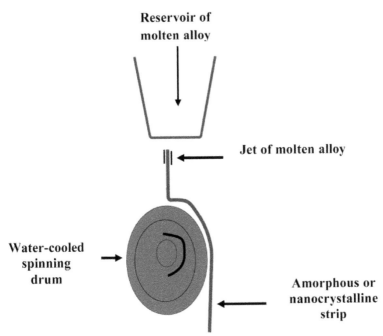

Reservoir of molten alloy

Jet of molten alloy

Water-cooled spinning drum

Amorphous or nanocrystalline strip

FIGURE 1.9 Melt spinning allows cooling rates up to 1 million degrees Centigrade per second.

Figure 1.10 shows a second way of cooling a liquid fast: Zap it with a laser beam, scanning the beam fast enough that it melts only a very thin surface layer. The layer is already stuck to the cold material beneath. Conduction of heat into this cold substrate is fast enough to trap the amorphous structure. This laser surface melting is the way amorphous silicon for cheap solar cells is made. The same technique is used to make nanocrystalline surface layers [7].

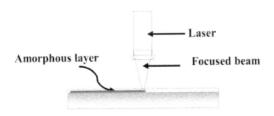

Amorphous layer

Laser

Focused beam

FIGURE 1.10 Laser surface hardening.

1.6.1.2 EQUIANGLE EXTRUSION

The second approach is that of extreme plastic deformation. Figure 1.11 shows equiangle extrusion-a way of generating extreme deformation while suppressing fracture. A rod, forced through the die, is savagely sheared and extruded, emerging as a rod with the same diameter as that with which it started but with a much-refined structure. It can then be reinserted into the die and tortured further until the structure is sufficiently refined. The process is now a standard one for making metals with grain sizes in the 100–500 nm range [7].

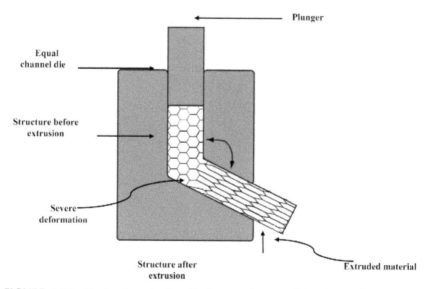

FIGURE 1.11   Equiangle extrusion. Each pass shears and breaks up the structure; repeated, the process reduces it to the nanoscale.

1.6.2 BOTTOM-UP PROCESSES: CONSOLIDATION OF NANOCLUSTERS AND MILLED POWDERS

Most of the processes described thus far do not make solid objects; they make clusters, powders, or chips. The long-established way to consolidate powder is to press in a die that has the desired form, then heat to a temperature at which diffusional bonding takes place. This powder-pressing and sintering route to manufacturing products (Fig. 1.12) is widely used to make engine parts for cars and components for household appliances such as washing machines. The difficulty in using it to consolidate nanostructured particles is that sintering takes time, and at the sintering temperature, the structure coarsens. It is not an easy problem to solve; during consolidation there will always be some coarsening. The question is how to minimize it. One way is to compact the powder in such a clean environment that the particles bond, even at room temperature, but that is seldom possible. An alternative is to sinter so fast that there is little time for coarsening. The powder (or swarf) is compressed in a die through which a bank of capacitors is discharged. The blast of heat, generated by the resistance of the packed powder, is enough to create good bonding without leaving enough time for serious coarsening [7].

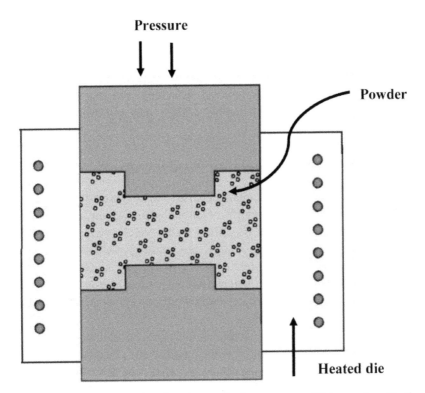

FIGURE 1.12 Pressure sintering, the standard way to consolidate powders. Heating is slow, allowing time for the structure to coarsen.

1.7   COMBINED TOP-DOWN AND BOTTOM-UP NANOMANUFACTURING

Current nanotechnology research focuses on surface modification, matching molecules and "sockets" at the level of manipulating few to several-hundred particles or molecules to be assembled into desirable configurations. Commercial scale-up will not be realized unless one can perform high-rate/high-volume assembly of nanoelements economically and using environmentally benign processes. High-rate/high-volume directed self-assembly will accelerate the creation of highly anticipated commercial products and enable the creation of an entirely new generation of applications yet to be imagined, because they are developed with scalability and integration as a requirement. This includes understanding what is essential for a rapid multistep or reel-to-reel process, as well as for accelerated-life testing of nanoelements and defect-tolerance. For example, a fundamental understanding of the interfacial behavior and forces required assembling, detaching, and transfer nanoelements, required for guided self-assembly at high-rates and over large areas is needed.

1.7.1   DIRECTED SELF-ASSEMBLY OF NANOELEMENTS

Assembly techniques such microchannnels [57, 58], and electric fields [59], have been explored for local assembly of carbon nanotubes for interconnects and electromechanical probe [60, 61]. These techniques, however, do not provide precise large-scale assembly at high-rates and high-volumes. The electrostatically addressable nanotemplate offers a simple means for controlling the placement and positioning of nanoelements for transfer using conductive nanowires. Gold nanowires have been used initially, and other conductors will be developed for use in templates.

To demonstrate how the large-scale assembly process will work, the electrostatically addressable nanotem plate that controls the placement and positioning of carbon nanotubes, nanoparticles, or other nanoelements is chosen. The nanotubes align on the charged wires of the nanotemplate. The nanotemplate and nanoelements (Step 2) can form a device or can function as a template to transfer patterned arrays of nanoelements onto another substrate as shown in Steps 3 and 4 [62] (Fig. 1.13).

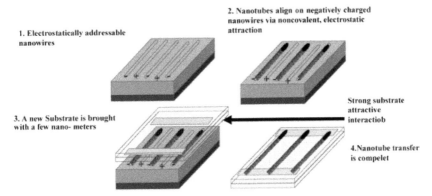

FIGURE 1.13   Steps of 2-D molecular assembly.

1.7.2   DIRECTED SELF-ASSEMBLY OF NANOELEMENTS USING NANOTEMPLATES

Nanotemplates can be used to enable precise assembly and orientation of various nanoelements such as nanoparticles and nanotubes. The directed assembly of colloidal nanoparticles into non-uniform 2D nanoscale features has been demonstrated via template-assisted electrophoretic deposition. The assembly process is controlled by adjusting the applied voltage, assembly time, or the geometric design of templates. The control of the assembly process to produce monolayers or multilayers is as well as full or partial assembly of nanoparticles. This approach offers a simple, fast means of nanoscale directed self-assembly of nanoparticles and other nanoelements over a large scale.

Electrostatically addressable nanotemplate could also be used to directly assemble carbon nanotubes.

1.7.3   HIGH RATE ASSEMBLY AND TRANSFER OF NANOELEMENTS

Nanoelements such as carbon nanotubes, conducting polymers and carbon black are of great interest in the researcher's community due to their high mechanical, thermal and electrical properties. They are often combined with polymers to enhance their properties. However, it is required for some of the specific applications that the particles to be in patterned form over the polymeric surface. The electrophoretic deposition of carbon nanotubes (Single wall (SWCNT) as well as multi wall (MWCNT)) and conducting polyaniline onto the circuits followed by transfer to a polyurethane film by thermoforming are studied. Two circuit designs can be used. Nanoelements were deposited onto the Cu and Au wires using the electrophoresis method with direct current (DC) voltage. A novel mold design can be chosen for the thermoforming process has a removable insert to keep the patterned nanoelements circuit inside the mold. The high rate assembly could be the best method in such situations [63].

1.8   CONCLUSION

In this chapter, the principles of nanoscience and nanotechnology were reviewed. Nanomaterials have an increased surface-to-volume ratio compared to bulk materials. Beginning with the most clearly defined category, 0-D nanomaterials are materials wherein all the dimensions are measured within the nanoscale. On the other hand, 1-D nanomaterials differ from 0-D nanomaterials in that the former have one dimension that is outside the nanoscale. This difference in material dimensions leads to needlelike-shaped nanomaterials. Two-dimensional nanomaterials are more difficult to classify. Three-dimensional nanomaterials, also known as bulk nanomaterials, are relatively difficult to classify. Nowadays it can be studied pieces of matter of smaller and smaller size right down to the atom. The important result is that the properties of the pieces start to change at sizes much bigger than a single atom. Top down approach refers to slicing or successive cutting of a bulk material to get nanosized particle that enables to control the manufacture of smaller, more complex objects, as illustrated by micro and nanoelectronics. Bottom up approach refers to the buildup of a material from the bottom: atom by atom, molecule by molecule or cluster by cluster which enables to control the manufacture of atoms and molecules, as illustrated by supramolecular chemistry. Both approaches play very important role in modern industry and most likely in nanotechnology as well. Directed and high rat self-assembly are the efficient methods for nanoelements production, which classified into combined top-Down and bottom-Up nanomanufacturing.

KEYWORDS

- **Nanoelements**
- **Nanomanufacturing**
- **Nanoscience**

REFERENCES

1. Bainbridge, W. S. (2001). *Societal Implications of Nanoscience and Nanotechnology* Springer.
2. Shelton, R. D., et al., *Societal Implications of Nanoscience and Nanotechnology.*
3. Keiper, A. (2003). *The Nanotechnology Revolution.* The New Atlantis. 2, 17–34.
4. Ramakrishna, S. (2005). *An Introduction to Electrospinning and Nanofibers*: World Scientific.
5. Johansson, M. (2003). *'Plenty of Room at the Bottom': Towards an Anthropology of Nanoscience.* Anthropology Today. 19(6), 3–6.
6. Jha, Z., et al. (2011). *Nanotechnology: Prospects of Agricultural Advancement.* Nano Vision. 1(2), 88–100.
7. Schodek, D. L., Ferreira, P., Ashby, M. F. (2009). *Nanomaterials, Nanotechnologies and Design: An Introduction for Engineers and Architects*: Butterworth-Heinemann.
8. Kennedy, J. (2008). *Nanotechnology: The Future is Coming Sooner Than You Think*, in *The Yearbook of Nanotechnology in Society, Volume I: Presenting Futures*, Springer. 1–21.
9. Glenn, L. M. D., Boyce, J. S. (2008). *Nanotechnology: Considering the Complex Ethical, Legal, and Societal Issues with the Parameters of Human Performance.* Nano Ethics. 2(3), 265–275.
10. Glenn, L. M. D., Boyce, J. S. (2012). *Regenerative Nanomedicine: Ethical, Legal, and Social Issues*, in *Nanotechnology in Regenerative Medicine*, Springer. 303–316.
11. Jha, Z., et al. (2011). *Nanotechnology: Prospects of Agricultural Advancement.* Nano Vision. 1(2), 88–100.
12. Cortie, M. B. (2004). *The Weird World of Nanoscale Gold.* Gold Bulletin. 37(1–2), 12–19.
13. Lieber, C. M. (2003). *Nanoscale Science and Technology: Building a Big Future from Small Things.* Mrs Bulletin. 28(07), 486–491.
14. Kelsall, R. W., et al. (2005). *Nanoscale Science and Technology*: Wiley Online Library. 472.
15. Park, S. Y. et al., (2008). *DNA-Programmable Nanoparticle Crystallization.* Nature. 451(7178). : 553–556.
16. Islam, M. T., et al. (2005). *HPLC Separation of Different Generations of Poly (amido-amine) Dendrimers Modified with Various Terminal Groups.* Analytical chemistry. 77(7), 2063–2070.
17. Islam, M. T., Majoros I. J., Baker, J. R. (2005). *HPLC Analysis of PAMAM Dendrimer Based Multifunctional Devices.* Journal of Chromatography B. 822(1), 21–26.
18. Desai, A., Shi, X., Baker, J. R. (2008). *CE of Poly (amidoamine) Succinamic Acid Dendrimers Using a Poly (vinyl alcohol) Coated Capillary.* Electrophoresis. 29(2), 510–515.
19. Wang, Z., et al. (2011). *Robotic Nanoassembly: Current Developments and Challenges.* International Journal of Computer Applications in Technology. 41(3), 185–194.
20. Olyaee, S., Hamedi, S., Dashtban, Z. (2012). *Efficient Performance of Neural Networks for Nonlinearity Error Modeling of Three-Longitudinal-Mode Interferometer in Nano-Metrology System.* Precision Engineering. 36(3), 379–387.
21. Korzenowski, A. L., *Inference and Classification in Overlapped Nanoparticles.*
22. Herrera-Basurto, R., Simonet, B. *Nanometrology.* Encyclopedia of Analytical Chemistry.
23. Stadnyk, B., Yatsyshyn, S., Seheda, O. (2012). *Research in Nanothermometry. Part 6. Metrology of Raman Thermometer with Universal Calibration Artifacts.* Research in Nanothermometry. 142(7), 1726–5479.

24. Stadnyk, B., Yatsyshyn, S., Seheda, O. (2012). *Research in Nanothermometry. Part 6. Metrology of Raman Thermometer with Universal Calibration Artifacts.* Sensors & Transducers. 142(7), 1726–5479.

25. Newell, D. B., et al. (2003). *The NIST Microforce Realization and Measurement Project.* Instrumentation and Measurement, IEEE Transactions on. 52(2), 508–511.

26. Teague, E. C. (1991). *Nanometrology.* in *AIP Conference Proceedings.*

27. Lines, M. G. (2008). *Nanomaterials for Practical Functional Uses.* Journal of Alloys and Compounds. 449(1), 242–245.

28. Edelstein, A. S., Cammaratra, R. C. (1998). *Nanomaterials: Synthesis, Properties and Applications*: CRC Press. 616.

29. Hsiao, I. L., Huang, Y. J. (2011). *Effects of Various Physico-chemical Characteristics on the Oxicities of ZnO and TiO2 Nanoparticles Toward Human Lung Epithelial Cells.* Science of The Total Environment. 409(7), 1219–1228.

30. Lanone, S., Boczkowski, J. (2006). *Biomedical Applications and Potential Health Risks of Nanomaterials: Molecular Mechanisms.* Current molecular medicine. 6(6), 651–663.

31. Singh, a. k. (2007). *Science & Technology For Upsc*: McGraw-Hill Education (India) Pvt Limited.

32. Joshi, D. R. (2010). *Engineering Physics*: McGraw-Hill Education (India) Pvt Limited.

33. Edelstein, A. S., Cammaratra, R. C. (1998). *Nanomaterials: Synthesis, Properties and Applications, Second Edition*: Taylor & Francis.

34. Hornyak, G. L., Moore, J. J., Tibbales, H. F. (2009). *Fundamentals of Nanotechnology*: CRC press Boca Raton.

35. Joachim, C. (2005). *To be Nano or not to be Nano?* Nature materials. 4(2), 107–109.

36. Tegmark, M., Wheeler, J. A. (2001). *100 Years of the Quantum.* arXiv preprint quant-ph/0101077. 284, 68–75.

37. Luth, H. (2013). *Quantum Physics in the Nanoworld*: Springer.

38. Aspect, A. (2007). *Quantum Mechanics: to be or not to be Local.* Nature. 446(7138). : 866–867.

39. Schodek, D. L., Ferreira, P., Ashby, M. F. (2009). *Nanomaterials, Nanotechnologies and Design: An Introduction for Engineers and Architects*: Elsevier Science. 560.

40. Tomalia, D. A. (2010). *Dendrons/dendrimers: Quantized, Nano-element like Building Blocks for Soft-soft and Soft-hard Nano-Compound Synthesis.* Soft Matter. 6(3), 456–474.

41. Nelson, E. (1966). *Derivation of the Schrödinger equation from Newtonian mechanics.* Physical Review. 150(4), 1079.

42. Feit, M. D., Fleck Jr, J., Steiger, A. A. (1982). *Solution of the Schrödinger Equation by a Spectral Method.* Journal of Computational Physics. 47(3), 412–433.

43. Shirley, J. H. (1965). *Solution of the Schrödinger equation with a Hamiltonian periodic in time.* Physical Review. 138(4B): B979.

44. Binns, C. (2010). *Introduction to Nanoscience and Nanotechnology.* Vol. 14, John Wiley & Sons. 301.

45. Tiwari, J. N., Tiwari, R. N., Kim, K. S. (2012). *Zero-Dimensional, One-Dimensional, Two-Dimensional and Three-Dimensional Nanostructured Materials for Advanced Electrochemical Energy Devices.* Progress in Materials Science. 57(4), 724–803.

46. Koch, C. (2002). *Nanostructured Materials: Processing, Properties and Applications*: Taylor & Francis.

47. Nalwa, H. S. (2001). *Nanostructured Materials and Nanotechnology: Concise Edition*: Gulf Professional Publishing.

48. Tang, Z., Kotov, N. A. (2005). *One- Dimensional Assemblies of Nanoparticles: Preparation, Properties, and Promise.* Advanced Materials. 17(8), 951–962.

49. Shevchenko, V. Y., Madison, A. E., Shudegov, V. E. (2003). *The Structural Diversity of the Nanoworld*. Glass Physics and Chemistry. 29(6), 577–582.

50. Kelsall, R. W., et al. (2005). *Nanoscale Science and Technology*: Wiley Online Library.

51. Cao, G., Wang, Y. (2011). *Nanostructures and Nanomaterials: Synthesis, Properties, and Applications*. Vol. 2, World Scientific. 581.

52. Ozin, G. A. (2009). *Nanochemistry: A Chemical Approach to Nanomaterials*: Royal Society of Chemistry. 820.

53. Wilson, M., et al. (2002). *Nanotechnology: Basic Science and Emerging Technologies*: Taylor & Francis.

54. Gogotsi, Y. (2006). *Nanomaterials Handbook*: Taylor & Francis.

55. Zhong, W. H. (2012). *Nanoscience and Nanomaterials: Synthesis, Manufacturing and Industry Impacts*: Destech Publications. 303.

56. Lines, M. (2008). *Nanomaterials for Practical Functional Uses*. Journal of Alloys and Compounds. 449(1), 242–245.

57. Messer, B., Song., J. H., Yang, P. (2000). *Microchannel Networks for Nanowire Patterning*. Journal of the American Chemical Society. 122(41), 10232–10233.

58. Huang, Y., et al. (2001). *Directed Assembly of Ane-Dimensional Nanostructures into Functional Networks*. Science. 291 5504, 630–633.

59. Duan, X., et al. (2001). *Indium Phosphide Nanowires as Building Blocks for Nanoscale Electronic and Optoelectronic Devices*. Nature. 409(6816), 66–69.

60. Li, J., et al. (2003). *Bottom-up Approach for Carbon Nanotube Interconnects*. Applied Physics Letters. 82(15), 2491–2493.

61. Guillorn, M. A., et al. (2002). *Individually Addressable Vertically Aligned Carbon Nanofiber-Based Electrochemical Probes*. Journal of Applied Physics. 91(6), 3824–3828.

62. Busnaina, A., Mehta, M. (2010). *Introduction to Nanoscale Manufacturing and the State of the Nanomanufacturing Industry in the United States*, in *Nanomanufacturing Handbook*. 432.

63. Gultepe, E., et al. (2008). *Large Scale 3D Vertical Assembly of Single-wall Carbon Nanotubes at Ambient Temperatures*. Nanotechnology. 19(45), 455309.

CHAPTER 2

NANOELEMENT MANUFACTURING: SELF-ASSEMBLY

CONTENTS

Abstract 34
2.1　Nanoelement Manufacturing 34
2.2　Self-Assembly 35
2.3　Molecular Self-Assembly 39
2.4　General Trends 46
2.5　Investigation Into Some New Studies...................................... 47
2.6　Conclusion 50
Keywords 51
References.. 51

ABSTRACT

Self-assembly is applied to the formation processes that involve nanoparticles as pre existing components are reversible, and can be controlled by proper component design. Development in the self-assembly approach makes functional material or device can be constructed using self-assembly of simple constituent nanoscale elements into an ordered nanocrystalline array, which can actuate to important performances in nanotechnology. Nanoassembled systems can have a variant range of structures and physical/chemical properties and diverse functional properties. Most of the nanoproperty elements originate because the nanostructural elements are in the nanoscale. And the changes in nanoproperty elements can be feasible because the changes in nanostructural elements are practical through nanoassembly. In this chapter, this method is investigated in detail.

2.1 NANOELEMENT MANUFACTURING

2.1.1 MANUFACTURING AT NANOSCALE

The physico-chemical properties of nano-sized materials are really unprecedented, exquisite and sometimes even adjustable in contrast to the bulk phase. For instance, quantum confinement phenomena allow semiconductor nanoparticles to sustain a dilating of their band gap energy as the particle size becomes smaller. Thereby it causes the blue shifts in the optical spectra and a change in their energy density from continuous to discrete energy levels as the transition moves from the bulk to the nanoscale quantum dot state [1–4]. In addition, interesting electrical properties including resonance tunneling and Coulomb blockade effects are observed with metallic and semiconducting nanoparticles, and endohedral fullerenes and carbon nanotubes can be processed to exhibit a tunable band gap of either metallic or semiconducting properties [5, 6]. These very different phenomena are mainly due to larger surface area-to-volume ratio at the nanoscale compared to the bulk. Thus, the surface forces become more important when the nano-sized materials exhibit unique optical or electrical properties. The surface (or molecular) forces can be generally categorized as electro- static, hydration (hydrophobic, and hydrophilic), Van der Waals, capillary forces, and direct chemical interactions [7, 8]. Based on these forces, the synthesis and processing techniques of these interesting nano-sized materials have been well established as capable of producing high-quality mono- disperse nanocrystals of numerous semiconducting and metallic materials, fullerenes of varying properties, single- and multiwall carbon nanotubes, conducting polymers, and other nano-sized systems [9]. The next key step in the application of these materials to device fabrication is undoubtedly the formation of sub-nanoelements into functional and desired nanostructures without mutual aggregation. To achieve the goal of innovative developments in the areas of microelectronic, optoelectronic and photonic devices with unique physical and chemical characteristics of the nano-sized materials, it may be necessary to immobilize these materials on surfaces and/or assemble them into an organized network [8].

Many significant advances in one- to three-dimensional arrangements in nanoscale have been achieved using the 'bottom-up' approach. Unlike typical top-down photolithographic approaches, the bottom-up process offers numerous attractive advantages, including the substantiation of molecular-scale feature sizes, the potential of three- dimensional assembly and an economical mass fabrication process [10]. Self-assembly is one of the few vital techniques available for controlling the orchestration of nanostructures via this bottom-up technology. The self-assembly process is defined as the autonomous organization of components into well-organized structures. It can be characterized by its numerous advantages such as cost-effective, versatile, facile, and the process seeks the thermodynamic minima of a system, resulting in stable and robust structures [11]. As the description suggests, it is a process in which defects are not energetically favored, thus the degree of perfect organization is relatively high. As described earlier, there are various types of interaction forces by which the self-assembly of molecules and nanoparticles can be accomplished [12, 13].

2.2 SELF-ASSEMBLY

The term "self-assembly" is understood and used differently in many fields of science from living cell biology to the evolution of galaxies. We use "self-assembly" as applied to the formation processes that involve nanoparticles as pre existing components are reversible, and can be controlled by proper component design [14]. We will consider the formation processes, which lead to the ordered periodic structures. Self-assembly is now a practical method of nanotechnology to design ensembles of nanoparticles for many applications [15]. The objective of this chapter is to make available a set of parameters that could be changed easily to enable the researcher to manipulate the self-assembling nanoelements into nanomaterials.

The enormous potential of nanotechnology for industry can be explained by the size dependence of many important physical and chemical properties at some scale because the properties of individual atoms cannot be directly correlated with the properties of bulk materials. Different regimes of understanding of physical and chemical properties could be suggested: bulk, nanoparticles/ nanostructures and atomic clusters. The given in systematic of size-dependent properties includes optical [16, 17], magnetic [18], catalytic [19], thermodynamic [20], electrochemical [21] and electric transport properties [22].

The opportunities of self-assembly approach to the design of three-dimensional (3D) photonic crystals are especially important [23]. The idea on periodic 3D dielectric structures with electromagnetic band gap due to refractive index modulation, which can permit distinct laser modes, was formulated by Yablonovitch [24] and John [25]. Yablonovitch has also shown that this concept is observed in such experimental systems as hexagonally close-packed glass or polystyrene spheres [24]. As a result of further studies, the use of photonic crystals for direction, control, splitting, guiding, and localization of light was suggested. Photonic crystals have a dominant role in recent investigations toward elaboration of all-optical integrated circuits for future communication and computing systems. The resolution that can be routinely achieved by lithographic techniques is at the level of several micrometers, and new emerging methods require expensive installations. Although some limitations exist, the self-assembling approach remains the most practically important method of fabrication of periodic structures for photonic applications. Electron-beam and scanning probe lithography pattern the substrate sequentially and cause the method becomes very time consuming [26].

Many examples show that nanoconfinement makes new opportunities in organizing, ordering or assembling nanoparticles. This can lead to materials with qualitatively new physical and chemical properties. For example, the traditional devices cannot be used to design sharp waveguide bends without notable signal loss. Two-dimensional photonic nanostructures consisting of periodic 2-D arrays of columns of dielectric material can serve as efficient waveguides [10, 27].

Development in the self-assembly approach makes functional material or device can be constructed using self-assembly of simple constituent nanoscale elements into an ordered nanocrystalline array, which can actuate to important performances in nanotechnology. While the fabrication techniques of commercial importance such as lithography belong in the "top-down" category, a "bottom-up" approach may offer a number of advantages. The main advantages of the bottom-up approach are experimental simplicity, possibility of 3-D assembly and potential for inexpensive mass production [10]. There are three purposes of the bottom-up approach [28] which are listed below:

1. Self-assembly of nanometer-scale structures from nanocrystals dispersed in solution;

2. Self-organization of these structures on technologically relevant substrates; and

3. Processing of organized arrays into structures convenient for practical applications.

The substantiation of self-assembly of nanoparticles into ordered arrays for design on nanodevices can be divided into the following steps:

1. Understanding the interaction between nanoparticles and the main physico-chemical variables governing formation of functional nanocrystalline arrays.

2. Understanding of function and functional requirements for the desired device.

3. Realization of the desired functional structure at the nanolevel, construction of the nanodevice.

Self-assembling molecules, macromolecules, molecular assemblies and supramolecules are outsides the scope of this chapter.

The self-assembly processes lead to the formation of periodically ordered nanocrystalline arrays, starting from ordered aggregates into one-dimensional (1-D) templates and with 2-D and 3-D ordering on planar or nonplanar substrates. The contributions of self-assembly phenomenon in nanodesign can be divided into three groups [29, 30]:

1. Self-assembly of particles into nanoaggregates.

2. Self-assembly of nanoparticles on quasi-1-D templates (such as nanotubes).

3. Self-assembly of nanoparticles on 2-D planar or not planar substrates.

The objective of this chapter is not to provide a comprehensive review of all available studies comprising self-assembly as a main or important step in the fabrication of new nanostructures or nanodevices. Instead, it will be tried to present the most characteristic investigations to provide the reader with the necessary tools to navigate in current publications on nanotechnology.

There are some difficulties systematizing the known approaches based on the self-assembly of nanoparticles. It was realized that the important part of the presented examples of the use of self-assembly for the formation of ordered periodic nanoarrays have simple evidences of more than one characteristic features.

2.2.1 SELF-ASSEMBLY: FROM NATURE TO THE LAB

In the 1980's, scientists discovered that alkanethiols spontaneously assembled on noble metals. This new area of science opened the doors to a simple way of creating surfaces of almostly any desired chemistry by placing a gold sub-strate into a millimolar solution of an alkanethiol in ethanol. This result in crystalline-like monolayers formed on the metal surface, called self-assem-bled monolayers (SAMs) [31].

Over the years, the mechanism of the self-assembly process has been well studied and elucidated. Researchers have found that a typical alkanethiol monolayer forms a structure on gold with the thiol chains tilted approximately 30 degrees from the surface normal. The exact structure of the monolayer depends on the chemistry of the chain [32].

Self-assembly forms the basis for many natural processes including pro-tein folding, DNA transcribing and hybridization, and the formation of cell membranes. The process of self- assembly in nature is governed by inter- and intra-molecular forces that drive the molecules into a stable, low energy state. These forces include hydrogen bonding, electrostatic interactions, hydropho-bic interactions, and Vander Waals forces [33, 34].

As with self-assembly in nature, there are several driving forces for the assembly of alkanethiols onto noble metal surfaces. The first is the affinity of sulfur for the gold surface. The next driving force for assembly is the hy-drophobic, Vander Waals interactions between the methylene carbons on the alkane chains. For alkanethiol monolayers, this interaction causes the thiol chains to tilt in order to maximize the interaction between the chains and low-er the overall surface energy. A well-ordered monolayer forms from an alkane chain of at least 10 carbons. With carbon chains of this length, hydrophobic interactions between the chains can overcome the molecules' rotational de-grees of freedom [35, 36].

A simple alkanethiol molecule is shown in Fig. 2.1. An alkanethiol can be thought of as containing three parts: a sulfur binding group for attachment to a noble metal surface, a spacer chain [typically made up of methylene groups, $(CH_2)_n$], and a functional head group. As mentioned above, the sulfur atom and the carbons in the methylene groups act as the main driving forces for assembly of the alkanethiols. The head group then provides a platform where any desired group can be used to produce surfaces of effectively any type of chemistry. By simply changing the head group, a surface can be created that is hydrophobic (methyl head group), hydrophilic (hydroxyl or carboxyl head group), protein resistant (ethlylene glycol head group), or allows chemical binding (NTA, azide, carboxyl, amine head groups). This enables a researcher to custom design a surface to serve any desired function [37, 38].

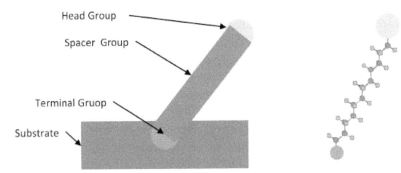

FIGURE 2.1   Schematic diagram of a thiol molecule.

2.2.2   SELF-ASSEMBLY: UNLIMITED OPPORTUNITIES

Due to self-assembly has applications in many areas, it is interested for various researches. A few examples include surface wetting, nonfouling property, electrochemistry, surface passivation, protein binding, DNA assembly, corrosion resistance, biological arrays, cell interactions and molecular electronics. These and other topics have been summarized in previous review articles. Self-assembly has truly opened the doors toward direct surface engineering [39–41].

2.3   MOLECULAR SELF-ASSEMBLY

Molecular self-assembly is the assembly of molecules without guidance or management from an outside source. Self- assembly can happen spontaneously in nature, for example, in cells such as the self-assembly of the lipid bilayer membrane. It usually results in an increase in internal organization of the system. Many biological systems use self-assembly to assemble various molecules and structures. Imitating these strategies and creating novel molecules with the ability to self-assemble into supramolecular assemblies is an important technique in nanotechnology [42–44].

In self-assembly, the final desired structure is 'encoded' in the shape and properties of the molecules that are used, as compared to traditional techniques, such as lithography, where the desired final structure must be carved out from a larger block of matter [45].

On a molecular scale, the accurate and controlled application of inter-molecular forces can lead to new and previously unachievable nanostructures. This is why molecular self- assembly (MSA) is a highly topical and promising field of research in nanotechnology today. With many complex examples all around in nature, MSA is a widely perceived phenomenon that has yet to be completely understood. Biomolecular assemblies are so-phisticated and often hard to isolate, making systematic and progressive analyzes of their fundamental science very difficult. What in fact are need-ed are simpler MSAs, the constituent molecules of which can be readily synthesized by chemists. These molecules would self-assemble into simpler constructs that can be easily assessed with current experimental techniques [46, 47].

Of the diverse approaches possible for Molecular Self-assembly, two strategies have received significant research attention, electrostatic Self-assembly (or layer- by-layer assembly) and "Self-assembled Monolayers (SAMs). Electrostatic self-assembly involves the alternate adsorption of an-ionic and cationic electrolytes onto a suitable substrate. Typically, only one of these is the active layer while the other enables the composite multilayered film to be bound by electrostatic attraction. The latter strategy of Self-assem-bled monolayers or SAMs based on constituent molecules, such as thiols and silanes [48, 49].

For SAMs, synthetic chemistry is used only to construct the basic building blocks (that is the constituent molecules), and weaker intermolecular bonds such as Vander Waals bonds are involved in arranging and binding the blocks together into a structure. This weak bonding makes solution, and hence re-versible, processing of SAMs (and in general, MSAs) possible. Thus, solu-tion processing and manufacturing of SAMs offer the enviable goal of mass production with the possibility of error correction at any stage of assembly. It is well recognized that, this method could prove to be the most cost-effective way for the semiconductor electronics industry to produce functional nanode-vices such as nanowires, nanotransistors, and nanosensors in large numbers [41, 50].

2.3.1   NANOSELF-ASSEMBLY

In the previous sections self-assembly was defined as assembly of its building units. All possible entities (atoms, molecules, colloidal particles, etc.) that can take part in this process are self-assembly building units. Building units for nanotechnology systems have more structural hierarchies. Nanotechnology systems can be built not only through self-assembly processes but through an external manipulation as well. All these efforts to create nanotechnology systems can be considered as the processes for assembling nanotechnology systems. We will define this as a nanoassembly, which can be stated as a "thermodynamic, kinetic, or manipulative assembly of nanoassembly building units." Spontaneous assembly of nanoassembly building units will be a great route for building nanotechnology systems [51, 52].

However, assembling them, for example, using an atomic force microscope through a one-by-one type of operation with any type of nanoassembly building units will also be a great alternative for creating nanotechnology systems. Figure 2.2 (left-hand side) shows that nanoassembled systems are assembled from three basic nanoassembly building units. They are a self-assembly building unit, a fabrication-building unit, and a reactive building unit. As will be described in the next section with more details, the structures of all three basic nanoassembly building units can be analyzed based on the concept of segmental analysis. In other words, the segmental analysis that was developed for self-assembly building units can be expanded for the two other types of building units. Figure 2.3 explains this. All three basic nanoassembly building units can be analyzed with the three fundamental and two additional segments. And all segments from the three basic nanoassembly-building units interact through the force balance with any possible combinations. The whole process resembles the self-assembly process. But it now occurs in a "quasi-three-dimensional" way, which is to imply that there are three different types of building units instead of just one (as for self-assembly). The concept of force balance is directly applied not only between self-assembly building units, between fabrication building units, or between reactive building units, but between all three different types of building units as well. This gives us an important insight for the third part of this book that there can be great possibilities for building nanoassembled systems once the three basic building units are well identified and the relationships between them are well controlled. The roles of the five segments during the assemblies of nanoassembled systems are the same as for the assemblies of self-assembled systems.

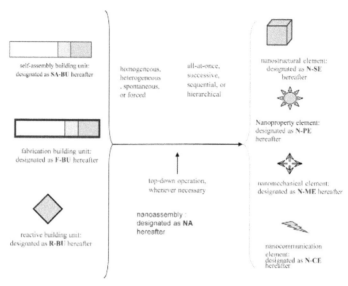

FIGURE 2.2   Three basic nanoassembly building units construct the four nanoelements. Force balance between the nanoassembly building units plays a key role.

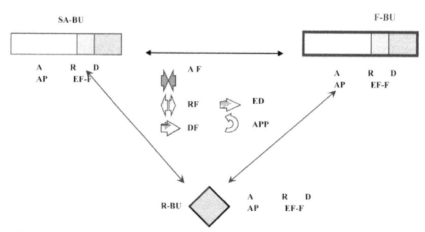

FIGURE 2.3   The fundamental and additional segments of self-assembly building unit (SA-BU), fabrication building unit (F-BU), and reactive building unit (R-BU) can interact through the force balance with any possible combinations. AF, RF, and DF represent attractive force between As, repulsive force between Rs, and directional force between Ds, respectively. A, R, D, AP, and EF-F refer to attractive, repulsive, directional, asymmetric packing, and external force–specific functional segments, respectively. ED is external force–induced directional factor. APP is asymmetric packing process.

2.3.2 NANOELEMENTS

Figure. 2.2 showed that nanoassembled systems are obtained through nanoassembly with three basic nanoassembly building units. Nanoassembled systems can have a variant range of structures and physical/chemical properties and diverse functional properties. For many nanoassembled systems, these general properties are those that are already known to other existing systems such as macroscopic counter parts. They can be straightly characterized. However, for many others, they can be novel properties that cannot be easily recognized and characterized. There are also nanoassembled systems whose general properties are overlapped by others. The concept of force balance for nanoassembly makes it possible for us to evaluate the specific properties that can be expected from certain nanoassembly building units. It can provide a nice insight when choosing a proper nanoassembly route for a specific nanoassembled system and help clarify intended nanoscale properties (or nanoproperties) with a reasonable degree of accuracy. Four elemental properties (which will be called nanoelements hereafter) for nanoassembled systems are proposed here in order to address these properties in a systematic manner. They are nanostructural element, nanoproperty element, nanomechanical element, and nanocommunication element [53].

The symbols for each nanoelement are also shown in Fig. 2.2. Table 2.1 shows representative examples of the four nanoelements. A nanostructural element is the structural features that are inherited or designed from nanoassembly itself.

As shown in the Table 2.1, most of the nanostructure-based nanoassembled systems belong to this. A nanoproperty element is the properties that are inherited, induced, or designed from nanoassembly and its framework. Some of them could be the same properties as macroscale counterparts but in the nanoscale while others are those that emerge only when the systems have nanoscale features. A nanomechanical element is the unit operations that are designed to express the motional aspects of nano-assembled systems. Finally, a nanocommunication element is a signal, energy, or work that is designed to communicate with the macroworld. This nanoelement is almost exclusively for nanofabricated systems, nanointegrated systems, nanodevices, and nanomachines [54].

TABLE 2.1 Representative Examples of the Four Nanoelements: N-SE, N-PE, N-ME, and N-CE (refer to nanostructural, nanoproperty, nanomechanical, and nanocommunication elements, respectively)

	Nanoparticle		Gating and Switching
N-SE	Nanopore	N-ME	Rotation and oscillation
	Nanofilm		Tweezering and fingering
	Nanotube		Rolling and bearing
	Nanorod		Self-directional movement
	Nano hollow sphere		Capture and release
	Nanofabricated surface		Sensing
N-PE	Surface Plasmon	N-CE	Any macroscale performance by na-nointegarated system and energy ex-change, which are performed by nano-machines
	Quantum size effect		
	Single electron tunneling		
	Surface catalytic activity		
	Mechanical strength		
	Energy conversion		
	Nano-confinement effect		

2.3.3   GENERAL ASSEMBLY DIAGRAM

The outcome of self-assembly is self-assembled aggregate. For nanoassembly, it goes one more step. The apparent initial outcome of nanoassembly is a nanoassembled system. But it is the nanoelements that make nanoassembled systems distinctive from self-assembled aggregates. Self-assembled aggregates have their own characteristic properties, which in many ways are effective, and many applications have been established using them over a wide range of scientific and technological fields. For nanoassembled systems, it is the nanoelements that define their characteristic properties, and with which we are seeking practical applications for nanotechnology systems [55].

Figure 2.4 presents the general rules of nanoassembly and their relationship with nanoelements. As a nanoassembly becomes more desired (moving toward the right-hand direction on the horizontal arrow of attractive interaction–repulsive interaction balance), the nanoelement that will be expressed is a nanostructural element. Typical nanopores, nanoparticles, nanocrystals, nanoemulsions, and nanocomposites are more likely to be obtained on this side of the arrow. On the other hand, if a nanoassembly moves toward the left-hand side, it is more likely to obtain nanoassembled systems that usually need an aid of external force for their assembly. Colloidal crystal is one good example, especially when the size of nanoassembly building unit (colloidal particle) is increased. Many top-down operation-based nanoelements are other examples.

When a nanoassembly is involved with a directional interaction, the most likely nanoelement will be a film or surface-based nanoscale operation. Examples include most of the nanostructured films regardless of their detailed morphology. Nanoporous film, nanolayered film, and nanopatterned film are among them. It also includes most of the nanoscale products that are obtained as a result of directional growth (from the spherical-shape) such as nanorods, nanoneedles, and nanotubes. A good deal of nanofabrication is basically the nanoscale process that is performed on the surface, and thus becomes one prominent example for the upward direction on the vertical arrow. The opposite direction produces nanoelements, too. Some nanoparticles and nanocrystals can be obtained at this end. Most of the nanoproperty elements come along with nanostructural elements. And they are coupled to each other in many ways [56, 57].

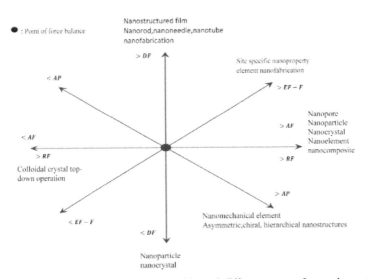

FIGURE 2.4   General rules of nanoassembly and different types of nanoelements (AF, RF, and DF refer to attractive, repulsive, and directional forces, respectively. AP and EF-F are short for asymmetric packing and external force–specific functional segments, respectively.

Most of the nanoproperty elements originate because the nanostructural elements are in the nanoscale. And the changes in nanoproperty elements can be feasible because the changes in nanostructural elements are practical through nanoassembly. The nanoassemblies that occur with external force, specific functional and asymmetric packing segments are critical for nanomechanical and nanocommunication elements. Electron tunneling and Coulomb blockade are good examples. Nanofabrication can take advantage of the unique features of external stimulus–specific nanoassembly, too. For a chiral nanoassembly, the chirality that is specific on each system can be used for the development of the nanostructures that can take advantage of the uniqueness, which includes highly asymmetric nanostructures, chiral nanoparticles, and some hierarchically constructed multiple-length-scale nanomaterials [58, 59].

It is also important for many unique types of nanomechanical elements. By coupling with the external stimulus–specific nanoassembly, the development of nanoelements on this side (right-hand side of both external stimulus-specific and chiral nanoassemblies) can be much more fruitful. As far as the application for nanotechnology systems goes, the other side (left-hand side) of both diagonal arrows does not have much use in the development of specific nanoelements [60].

2.4   GENERAL TRENDS

Each nanofabricated system is a unique product of each fabrication system. Each nanoelement of the nanofabricated system is a unique expression of its building units. They can be coupled locally or as a whole. They also can have a synergistic or an antagonistic outcome after the fabrication. All of these aspects have some degree of impact on the nanoelements of the nanofabricated systems. For some cases, different nanofabrication processes become the major reason for differentiating the nanoelements, even though the nanofabricated system might be the same [61].

Figure 2.5 shows a general trend of nanofabrication that covers these aspects from the three approaches. The mass assembling capability of nanofabrication becomes critically important when it goes to industrial scale. Generally, this capability is enhanced where fabrication is performed based on the bottom-up or the bottom-up/top-down hybrid approach. Because of the technical difficulties of top-down techniques and the limitations of the starting bulk materials that are comparable to them, the diversity of building units is much increased when the bottom-up or the hybrid approach is used. More diverse building units mean more diverse nanofabricated systems and more diverse nanoelements that can be explored. Structural diversity, hierarchy, and chirality are also important for widening the practicality of the nanofabricated systems [62].

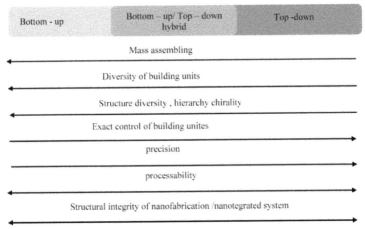

FIGURE 2.5 General trend of the three main approaches to nanofabrication.

It is easier to take advantage of two methods with a bottom-up or hybrid approach. Exact control of the building units is useful when the nanoelement is determined by the local control of a few main building units. As the top-down approach provides an advantage for this. Generally, the top-down approach has higher precision because of capability of manipulative assembly. Process ability means ease of the fabrication. This factor is important because it determines how practical a specific fabrication can be. It is, however, very much dependent on each nanofabrication system. Another factor is the structural integrity of the nanofabricated systems. It might appear that top-down processed systems would have better structural integrity because they are the products of the bulk materials. But structural integrity is measured not by the absolute strength of the nanofabricated systems but by their relative stability during actual use. As long as they perform the desired functions at given conditions, arguments about their absolute strength are less meaningful. Pure bottom-up fabrication in many cases provides enough, sometimes surprisingly strong, structural strength and resilience for nanofabricated systems to make them function properly even under harsh conditions [63, 64].

2.5 INVESTIGATION INTO SOME NEW STUDIES

The fields of nanoscience and nanotechnology generally concern the synthesis, fabrication and use of nanoelements and nanostructures at atomic, molecular and supramolecular levels. The nanosize of these elements and structures offers significant potential for research and applications across the scientific disciplines, including materials science, physics, chemistry, computer science, engineering and biology. Biological processes and methods, for example, are expected to be developed based entirely on nanoelements and their assembly into nanostructures. Other applications include developing nanodevices for use in semiconductors, electronics, photonics, optics, materials and medicine [65].

One class of nanoelements that has garnered considerable interest consists of carbon nanotubes. A carbon nanotube has a diameter on the order of nanometers and can be several micrometers in length. These nanoelements feature concentrically arranged carbon hexagons. Carbon nanotubes can behave as metals or semiconductors depending on their chirality and physical geometry. Other classes of nanoelements include, for example, nanocrystals, dendrimers, nanoparticles, nanowires, biological materials, proteins, molecules and organic nanotubes [66].

Although carbon nanotubes have been assembled into different nanostructures, convenient nanotools and fabrication methods to do it have not yet been developed. One obstacle has been the manipulation of individual nanoelements, which is often inefficient and tedious. This problem is particularly challenging when assembling complex nanostructures that require selecting and ordering millions of nanoelements across a large area [67].

To date, nanostructure assembly has focused on dispersing and manipulating nanoelements using atomic force or scanning tunneling microscopic methods. Although these methods are useful for fabricating simple nanodevices, neither is practical when selecting and patterning, for example, millions of nanoelements for more complex structures. The development of nanomachines or "nanoassemblers" which are programmed and used to order nanoelements for their assembly holds promise, although there have been few practical advancements with these machines.

The advancement of nanotechnology requires millions of nanoelements to be conveniently selected and simultaneously assembled. Three-dimensional nanostructure assembly also requires that nanoelements be ordered across a large area [68, 69].

Nanoelements have generated much interest due to their potential use in devices requiring nanoscale features such as new electronic devices, sensors, photonic crystals, advanced batteries, and many other applications. The realization of commercial applications, however, depends on developing high-rate and precise assembly techniques to place these elements onto desired locations and surfaces [70].

Different approaches have been used to carry out directed assembly of nanoelements in a desired pattern on a substrate, each approach having different advantages and disadvantages. In electrophoretic assembly, charged nanoelements are driven by an electric field onto a patterned conductor. This method is fast, with assembly typically taking less than a minute; however, it is limited to assembly on a conductive substrate [71]. Directed assembly can also be carried out onto a chemically functionalized surface. However, such assembly is a slow process, requiring up to several hours, because it is diffusion limited. Thus, there remains a need for a method of nanoelement assembly that is both rapid and not reliant on having either a conductive surface or a chemically functionalized surface [72].

In some cases, devices have a volume element having a larger diameter than the nanoelement arranged in epitaxial connection to the nanoelement. The volume element is being doped in order to provide a high charge carrier injection into the nanoelement and a low access resistance in an electrical connection. The nanoelement may be upstanding from a semiconductor substrate. A concentric layer of low resistivity material forms on the volume element forms a contact [73].

Semiconductor nanoelement devices show great promise, potentially outperforming standard electrical, opto-electrical, and sensor, etc., semiconductor devices. These devices can use certain nanoelement specific properties, 2-D, 1-D, or 0-D quantum confinement, flexibility in axial material variation due to less lattice match restrictions, antenna properties, ballistic transport, wave guiding properties, etc. Furthermore, in order to design first-rate semiconductor devices from nanoelements, transistors, light emitting diodes, semiconductor lasers, and sensors, and to fabricate efficient contacts, particularly with low access resistance, to such devices, the ability to dope and fabricate doped regions is crucial [74, 75].

As an example the limitations in the commonly used planar technology are related to difficulties in making field effect transistors (FET), with low access resistance, the difficulty to control the threshold voltage in the post-growth process, the presence of short-channel effects as the planar gate length is reduced, and the lack of suitable substrate and lattice-matched heterostructure material for the narrow band gap technologies [76].

One advantage of a nanoelement FET is the possibility to tailor the band structure along the transport channel using segments of different band gap and or doping levels. This allows for a reduction in both the source-to-gate and gate-to-drain access resistance. These segments may be incorporated directly during the growth, which is not possible in the planar technologies. The doping of nanoelements is challenged by several factors. Physical incorporation of dopants into the nanoelement crystal may be inhibited, but also the established carrier concentration from a certain dopant concentration may be lowered as compared to the corresponding doped bulk semiconductor. One factor that limits the physical incorporation and solubility of dopants in nanoelements is that the nanoelement growth temperatures very often are moderate [77].

For vapor-liquid-solid (VLS) grown nanoelements, the solubility and diffusion of dopant in the catalytic particle will influence the dopant incorporation. One related effect, with similar long-term consequences, is the out-diffusion of dopants in the nanoelement to surface sites. Though not limited to VLS grown nanoelements, it is enhanced by the high surface to volume ratio of the nanoelement. Also, the efficiency of the doping, the amount of majority charge carriers established by ionization of donors/acceptor atoms at a certain temperature may be lowered compared to the bulk semiconductor, caused by an increase in donor or acceptor effective activation energy, due to the small dimensions of the nanoelement. Surface depletion effects, decreasing the volume of the carrier reservoir, will also be increased due to the high surface to volume ratio of the nanoelement [78].

The above-described effects are not intended to establish a complete list, and the magnitudes of these effects vary with nanoelement material, dopant, and nanoelement dimensions. They may all be strong enough to severely decrease device performance.

2.6   CONCLUSION

The outcome of self-assembly is self-assembled aggregate. For nanoassembly, it goes one more step. The apparent initial outcome of nanoassembly is a nanoassembled system. But it is the nanoelements that make nanoassembled systems distinctive from self-assembled aggregates. Self-assembled aggregates have their own characteristic properties, which in many ways are effective, and many applications have been established using them over a wide range of scientific and technological fields. When a nanoassembly is involved with a directional interaction, the most likely nanoelement will be a film or surface-based nanoscale operation. The nanoassemblies that occur with external force, specific functional and asymmetric packing segments are critical for nanomechanical and nanocommunication elements. Electron tunneling and Coulomb blockade are good examples. Each nanofabricated system is a unique product of each fabrication system. Each nanoelement of the nanofabricated system is a unique expression of its building units.

KEYWORDS

- **asymmetric packing segments**
- **nanoassemblies**
- **nanoscale element**
- **self-assembly**

REFERENCES

1. Murphy, C. J., Coffer, J. L. (2002). *Quantum Dots: A Primer.* Applied Spectroscopy. 56(1), 16A–27A.
2. Carotenuto, G., Pepe, G. P., Nicolais, L. (2000). *Preparation and Characterization of Nano-sized Ag/PVP Composites for Optical Applications.* The European Physical Journal B-Condensed Matter and Complex Systems. 16(1), 11–17.
3. Winiarz, J. G., et al. (1999). *Observation of the Photorefractive Effect in a Hybrid Organic-inorganic Nanocomposite.* Journal of the American Chemical Society. 121(22), 5287–5295.
4. Riley, D. J. (2002). *Electrochemistry in Nanoparticle Science.* Current Opinion in Colloid & Interface Science. 7(3), 186–192.
5. Cleuziou, J. P., et al. (2011). *Electrical Detection of Individual Magnetic Nanoparticles Encapsulated in Carbon Nanotubes.* ACS nano. 5(3), 2348–2355.
6. Matsui, I. (2005). *Nanoparticles for Electronic Device Applications: A Brief Review.* Journal of Chemical Engineering of Japan. 38(8), 535–546.
7. Mozafari, M., Moztarzadeh, F. (2011). *Microstructural and Optical Properties of Spherical Lead Sulphide Quantum Dots-based Optical Sensors.* Micro & Nano Letters, IET. 6(3), 161–164.
8. Kim, S. (2007). *Directed Molecular Self-Assembly: Its Applications to Potential Electronic Materials.* Electronic Materials Letters. 3(3), 109–114.
9. Li, Y., Lu, D., Wong, C. P. (2009). *Electrical Conductive Adhesives with Nanotechnologies*: Springer. 433.
10. Brust, M., Kiely, C. J. (2002). *Some Recent Advances in Nanostructure Preparation from Gold and Silver Particles: A Short Topical Review.* Colloids and Surfaces A: Physicochemical and Engineering Aspects. 202(2), 175–186.
11. Whitesides, G. M., Grzybowski, B. (2002). *Self-assembly at All Scales.* Science. 295(5564), 2418–2421.
12. Claessens, C. G., Stoddart, J. F. (1997). *π–π Intractions in Self-Assembly.* Journal of Physical Organic Chemistry. 10(5), 254–272.
13. Zimmerman, S. C., et al. (1996). *Self-assembling Dendrimers.* Science. 271(5252). : 1095–1098.
14. Boncheva, M., Whitesides, G. M. (2005). *Making Things by Self-assembly.* MRS Bulletin-Materials Research Society. 30(10), 736–742.

15. Brinker, C. J., et al. (1999). *Evaporation-induced Self-assembly: Nanostructures Made Easy.* Advanced Materials. 11(7), 579–585.
16. Haynes, C. L., Van Duyne, R. P. (2001). *Nanosphere Lithography: A Versatile Nanofabrication Tool for Studies of Size-dependent Nanoparticle Optics.* The Journal of Physical Chemistry B. 105(24), 5599–5611.
17. Wang, Y., Herron, N. (1991). *Nanometer-sized Semiconductor Clusters: Materials Synthesis, Quantum Size Effects, and Photophysical Properties.* The Journal of Physical Chemistry. 95(2), 525–532.
18. New, R. M. H., Pease, R. F. W., White, R. L. (1995). *Physical and Magnetic Properties of Submicron Lithographically Patterned Magnetic Islands.* Journal of Vacuum Science & Technology B: Microelectronics and Nanometer Structures. 13(3), 1089–1094.
19. Yang, M. X., et al. (1998). *Lithographic Fabrication of Model Systems in Heterogeneous Catalysis and Surface Science Studies.* Langmuir. 14(6), 1458–1464.
20. Wang, Z. L., et al. (1998). *Shape Transformation and Surface Melting of Cubic and Tetrahedral Platinum Nanocrystals.* The Journal of Physical Chemistry B. 102(32), 6145–6151.
21. Gorer, S., Penner, R. M. (1999). *"Multipulse" Electrochemical/chemical Synthesis of CdS/S Core/shell Nanocrystals Exhibiting Ultranarrow Photoluminescence Emission Lines.* The Journal of Physical Chemistry B. 103(28), 5750–5753.
22. Bezryadin, A., Dekker, G., Schmid, C. (1997). *Electrostatic Trapping of Single Conducting Nanoparticles between Nanoelectrodes.* Applied Physics Letters. 71(9), 1273–1275.
23. Qi, M., et al. (2004). *A Three-dimensional Optical Photonic Crystal with Designed Point Defects.* Nature. 429(6991), 538–542.
24. Yablonovitch, E. (1987). *Inhibited Spontaneous Emission in Solid-state Physics and Electronics.* Physical Review Letters. 58(20), 2059–2061
25. John, S. (1987). *Strong Localization of Photons in Certain Disordered Dielectric Superlattices.* Physical Review Letters. 58(23), 2486–2489.
26. Xia, Y., Gates, B., Li, Z. Y. (2001). *Self-Assembly Approaches to Three-Dimensional Photonic Crystals.* Advanced Materials. 13(6), 409–413.
27. Seelig, E. W., et al. (2003). *Self-assembled 3D Photonic Crystals from ZnO Colloidal Spheres.* Materials Chemistry and Physics. 80(1), 257–263.
28. Shklover, V., Hofmann, H. (2005). *Methods of Self-assembling in Fabrication of Nanodevices,* in *Handbook of Semiconductor Nanostructures and Nanodevices,* American Scientific Publishers, North Lewis Way, California. 181–213.
29. Vincent, B. (2009). *Self-assembly, Self-organization: Nanotechnology and Vitalism.* Nano Ethics. 3(1), 31–42.
30. Vincent, B. (2006). *Two Cultures of Nanotechnology,* in *Nanotechnology Challenges: Implications for Philosophy, Ethics and Society,* World Scientific. 7–128.
31. Boeckl, M., Graham, D. (2006). *Self-Assembled Monolayers: Advantages of Pure Alkanethiols.* Material Matters. 1, 1–5.
32. Strong, L., Whitesides, G. M. (1988). *Structures of Self-assembled Monolayer Films of Organosulfur Compounds Adsorbed on Gold Single Crystals: Electron Diffraction Studies.* Langmuir. 4(3), 546–558.
33. Winfree, E., et al. (1998). *Design and Self-assembly of Two-dimensional DNA Crystals.* Nature. 394(6693), 539–544.
34. Yan, H., et al. (2003). *DNA-templated Self-assembly of Protein Arrays and Highly Conductive Nanowires.* Science. 301(5641), 1882–1884.
35. Love, J. C., et al. (2005). *Self-assembled Monolayers of Thiolates on Metals as a Form of Nanotechnology.* Chemical Reviews. 105(4), 1103–1170.

36. Wysocki, V. H., Jones, J. L., Ding, J. M. (1991). *Polyatomic Ion/surface Collisions at Self-assembled Monolayers Films.* Journal of the American Chemical Society. 113(23), 8969–8970.
37. Ulman, A., Eilers., J. E., Tillman, N. (1989). *Packing and Molecular Orientation of Alkanethiol Monolayers on Gold Surfaces.* Langmuir. 5(5), 1147–1152.
38. Smith, D. J., Miggio, E. T., Kenyon, G. L. (1975). *Simple Alkanethiol Groups for Temporary Blocking of Sulfhydryl Groups of Enzymes.* Biochemistry. 14(4), 766–771.
39. Schröter, A., Kalus, M., Hartmann, N. (2012). *Substrate-mediated Effects in Photothermal Patterning of Alkanethiol Self-assembled Monolayers with Microfocused Continuous-wave Lasers.* Beilstein Journal of Nanotechnology. 3(1), 65–74.
40. Klocek, J., et al. (2012). *Annealing Influence on Siloxane-Based Materials Incorporated with Fullerenes, Phthalocyanines, and Silsesquioxanes.* BioNanoScience. 2(1), 52–58.
41. Ulman, A. (1996). *Formation and Structure of Self-assembled Monolayers.* Chemical Reviews. 96(4), 1533–1554.
42. Ferreira, M., Cheung, J. H., Rubner, M. F. (1994). *Molecular Self-assembly of Conjugated Polyions: A New Process for Fabricating Multilayer Thin Film Heterostructures.* Thin Solid Films. 244(1), 806–809.
43. Cheng, J. H., Fou, A. F., Rubner, M. F. (1994). *Molecular Self-assembly of Conducting Polymers.* Thin Solid Films. 244(1), 985–989.
44. Antonietti, M., Förster, S. (2003). *Vesicles and Liposomes: A Self-Assembly Principle Beyond Lipids.* Advanced Materials. 15(16), 1323–1333.
45. Bianchi, E., et al. (2007). *Fully Solvable Equilibrium Self-assembly Process: Fine-tuning the Clusters Size and the Connectivity in Patchy Particle Systems.* The Journal of Physical Chemistry B. 111(40), 11765–11769.
46. Zhang, S. (2002). *Emerging Biological Materials through Molecular Self-assembly.* Biotechnology Advances. 20(5), 321–339.
47. Whitesides, G. M., Boncheva, M. (2002). *Beyond Molecules: Self-assembly of Mesoscopic and Macroscopic Components.* Proceedings of the National Academy of Sciences. 99(8), 4769–4774.
48. Huc, I., Lehn, J. M. (1997). *Virtual Combinatorial Libraries: Dynamic Generation of Molecular and Supramolecular Diversity by Self-assembly.* Proceedings of the National Academy of Sciences. 94(6), 2106–2110.
49. Sukhishvili, S. A. (2005). *Responsive Polymer Films and Capsules via Layer-by-layer Assembly.* Current Opinion in Colloid & Interface Science. 10(1), 37–44.
50. Kumar, A., Biebuyck, H. A., Whitesides, G. M. *Patterning Self-assembled Monolayers: Applications in Materials Science.* Langmuir. 10(5), 1498–1511.
51. Decher, G. (1997). *Fuzzy Nanoassemblies: Toward Layered Polymeric Multicomposites.* Science. 277(5330), 1232–1237.
52. Ai, H., et al. (2002). *Electrostatic Layer-by-layer Nanoassembly on Biological Microtemplates: Platelets.* Biomacromolecules. 3(3), 560–564.
53. Lee, Y. S. (2012). *Nanotechnology Systems,* in *Self-Assembly and Nanotechnology Systems: Design, Characterization, and Applications,* Wiley Online Library. 33–60.
54. Christensen, J. B., Tomalia, D. A. (2011). *Dendrimers as Quantized Nano-Modules in the Nanotechnology Field,* in *Designing Dendrimers,* John Wiley & Sons. 1–32.
55. Schwartz, M. (2010). *New Materials, Processes, and Methods Technology:* CRC Press. 712.
56. Mirkin, C. A., Tuominen, M. (2011). *Synthesis, Processing, and Manufacturing of Components, Devices, and Systems,* in *Nanotechnology Research Directions for Societal Needs in 2020,* Springer. 109–158.
57. Öztürk, S., Akata, B. (2009). *Oriented Assembly and Nanofabrication of Zeolite A Monolayers.* Microporous and Mesoporous Materials. 126(3), 228–233.

58. Takeuchi, K., Tajima, Y. (2001). *Nano-integration: An Ingenuity Driven Approach in Nanotechnology.* Riken Review. 38, 3–6.

59. Ekinci, K. L., Roukes, M. L. v*Nanoelectromechanical Systems.* Review of Scientific Instruments. 76(6), 061101–061101–12.

60. Badzey, R. L., et al. (2004). *A Controllable Nanomechanical Memory Element.* Applied Physics Letters. 85(16), 3587–3589.

61. Xu, S., et al. (2006). *Integrated Plasma-aided Nanofabrication Facility: Operation, Parameters, and Assembly of Quantum Structures and Functional Nanomaterials.* Vacuum. 80(6), 621–630.

62. Ho, D., Garcia, D., Ho, C. M. (2006). *Nanomanufacturing and Characterization Modalities for Bio-nano-informatics Systems.* Journal of Nanoscience and Nanotechnology. 6(4), 875–891.

63. Innocenzi, P., Malfatti, L., Falcaro, P. (2012). *Hard X-rays Meet Soft Matter: When Bottom-up and Top-down Get along Well.* Soft Matter. 8(14), 3722–3729.

64. Lee, Y. S. (2012). *Nanofabricated Systems: Combined to Function,* in *Self-Assembly and Nanotechnology Systems: Design, Characterization, and Applications,* Wiley Online Library. 333–357.

65. Busnaina, A., Miller, G. P. (2009). *Functionalized Nanosubstrates and Methods for Three-dimensional Nanoelement Selection and Assembly,* Google Patents.

66. Teredesai, P. V., et al. (2000). *Pressure-induced Reversible Transformation in Single-wall Carbon Nanotube Bundles Studied by Raman Spectroscopy.* Chemical Physics Letters. 319(3), 296–302.

67. Liu, Z., et al. (2000). *Organizing Single-walled Carbon Nanotubes on Gold Using a Wet Chemical Self-assembling Technique.* Langmuir. 16(8), 3569–3573.

68. Shin, S. R., et al. (2009). *Fullerene Attachment Enhances Performance of a DNA Nanomachine.* Advanced Materials. 21(19), 1907–1910.

69. Suo, Z., Lu, W. (2000). *Forces that Drive Nanoscale Self-assembly on Solid Surfaces.* Journal of Nanoparticle Research. 2(4), 333–344.

70. Abramson, A. R. et al., (2004). *Fabrication and Characterization of a Nanowire/polymer-based Nanocomposite for a Prototype Thermoelectric Device.* Journal of Microelectromechanical Systems. 13(3), 505–513.

71. Akabori, M., et al. (2003). *In GaAs Nano-pillar Array Formation on Partially Masked InP (111) B by Selective Area Metal-Organic Vapour Phase Epitaxial Growth for Two-dimensional Photonic Crystal Application.* Nanotechnology. 14(10), 1071–1074.

72. Huang, Y., et al. (2001). *Directed Assembly of One-dimensional Nanostructures into Functional Networks.* Science. 291(5504), 630–633.

73. Bockstaller, M. R., Mickiewicz, R. A., Thomas, E. L. (2005). *Block Copolymer Nanocomposites: Perspectives for Tailored Functional Materials.* Advanced Materials. 17(11), 1331–1349.

74. Voldman, S. (2009). *Nano Electrostatic Discharge.* Nanotechnology Magazine. 3(3), 12–15.

75. Vaia, R. A., Maguire, J. F. (2007). *Polymer Nanocomposites with Prescribed Morphology: Going Beyond Nanoparticle-filled Polymers.* Chemistry of Materials. 19(11), 2736–2751.

76. Thelander, C., et al. (2006). *Nanowire-based One-dimensional Electronics.* Materials Today. 9(10), 28–35.

77. Simpkins, B. S. et al. (2008). *Surface Depletion Effects in Semiconducting Nanowires.* Journal of Applied Physics. 103(10), 104313–104316.

78. Samuelson, L. I., Ohlsson, B. J. (2010). *Nanostructures and Methods for Manufacturing the Same,* Google Patents.

CHAPTER 3

NANOMATERIALS: PROPERTIES AND APPLICATION

CONTENTS

Abstract 56
3.1 Nanomaterials: Properties and Application Investigation 56
3.2 Thermal Properties of Nanomaterials 68
3.3 Electrical Properties 71
3.4 Magnetic Properties 75
3.5 Optical Properties 81
3.6 Acoustic Properties 84
3.7 Nanomaterial Applications 85
3.8 Conclusion 96
Keywords 96
References 97

ABSTRACT

In this chapter, the important properties of nanoelements, mechanical, thermal, electrical, magnetic, optical and acoustic properties are discussed. For this purpose, the nanomaterial properties are studied in detail. After properties investigation, different applications of nanomaterials in various industries are reviewed.

3.1 NANOMATERIALS: PROPERTIES AND APPLICATION INVESTIGATION

3.1.1 THE MECHANICAL PROPERTIES OF NANOSTRUCTURED MATERIALS

The bulk properties of materials such as density, modulus, field strength, thermal and electrical conductivity are innate and a small piece of the material has the same values for these properties as a large one. It is a basic assumption of continuum mechanics so it could be said that the material mechanical properties are scale independent. This assumption is applicable and causes to simplify the structure analysis with the most adequately accurate. The micromechanical description of materials which uses of classical mechanics to model the way the internal structure of a material influences its properties has followed the same path, assuming that the properties of the individual grains or crystals that make up the material could be averaged to get the overall properties without taking account of their scale. The classical property bounds (upper and lower estimates) of solid mechanics rest entirely on this assumption.

Mechanical properties such as the strength of metals can also be greatly improved by making them with nanoscale grains. Several basic attributes of materials are involved in defining their mechanical properties. One is strength, which includes characteristics with more precise definitions but basically determines how much a material deforms in response to a force. Others are (a) hardness, which is given by the amount another body such as a ball bearing or diamond is able to penetrate a material, and (b) wear resistance, which is determined by the rate at which a material erodes when in contact with another. These properties are dominated by the grain structure found in metals produced by normal processing.

In the nano-grained equivalent, about a quarter of the atoms are at a grain boundary. Clearly this change is going to have a marked effect on the mechanical properties of the material. Changes in mechanical properties with grain size were quantified over 50 years ago, but the modern ability to vary the grain size right down to the nanometer scale is likely to field large increases in performance (Fig. 3.1).

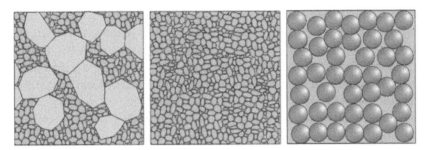

FIGURE 3.1   Material structure from macro to nanoscale: from left to right the figures, respectively, show: normal micro-grained bulk, mixture of coarse and fine grains and nano-grained bulk.

The nano structured materials field strength changes have a value that is up to the conventional material. This increase in strength is a notable improvement in field strength [1].

3.1.1.1   MECHANICAL BEHAVIOR STRESS, STRAIN, STIFFNESS, AND STRENGTH

Stress is something that is applied to a material by loading it. Strain, a change of shape, is the material's response. It depends on the magnitude of the stress and the way it is applied (the mode of loading). Stiffness is the resistance to change of shape that is elastic, meaning that the material returns to its original shape when the stress is removed. Strength is its resistance to permanent distortion or total failure. Stress and strain are not material properties. They describe a stimulus and a response. Stiffness (measured by the elastic modulus E, defined in a moment) and strength (measured by the field strength σ_y or tensile strength σ_{ts}) *are* material properties. Stiffness and strength are central to mechanical design.

The elastic module reflects the stiffness of the bonds that hold atoms together. There is not much you can do to change any of this, so the module of pure materials cannot be manipulated at all. If you want to control them you must either mix materials together, making composites, or disperse space within them, making foams.

3.1.1.2 MODES OF LOADING

Most engineering components carry loads. Their elastic response depends on the way the loads are applied. Usually one mode dominates, and the component can be idealized as one of the simply loaded cases in tie, column, beam, shaft, or shell (see Fig. 3.2). Ties carry simple axial tension, shown in (a) in the figure, columns do the same in simple compression, as in (b). Bending of a beam (c) creates simple axial tension in elements above the neutral axis (the center line, for a beam with a symmetric cross-section) and simple compression in those below. Shafts carry twisting or torsion (d), which generates shear rather than axial load. Pressure difference applied to a shell, such as the cylindrical tube shown at (e), generates biaxial tension or compression [2, 3].

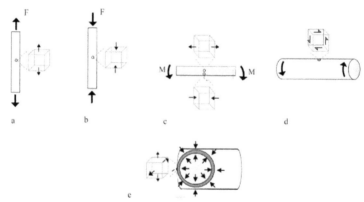

FIGURE 3.2   Modes of loading and states of stress: (a) tie, (b) column, (c) beam, (d) shaft, and (e) shell.

3.1.1.3 STRESS

Consider a force **F** applied as normal to the face of an element of material, as in Fig. 3.3 on the left of row (a) in the figure. The force is transmitted through the element and balanced by an equal but opposite force on the other side so that it is in equilibrium (it does not move). Every plane normal to **F** carries the force the tensile stress in the element (neglecting its self-weight) is:

$$\sigma = F/A \qquad (3.1)$$

If the sign of is reversed, the stress is compressive and given a negative sign. If, instead, the force lies parallel to the face of the element, three other forces are needed to maintain equilibrium (Fig. 3.3b). They create a state of shear in the element. The shaded plane, for instance, carries the shear stress of:

$$\tau = \frac{F_s}{A} \tag{3.2}$$

One further state of multi-axial stress is useful in defining the elastic response of materials that produced by applying equal tensile or compressive forces to all six faces of a cubic element, as in Fig. 3.3c. Any plane in the cube now carries the same state of stress. It is equal to the force on a cube face divided by its area. There is an unfortunate convention here. Pressures are positive when they push-the reverse of the convention for simple tension and compression.

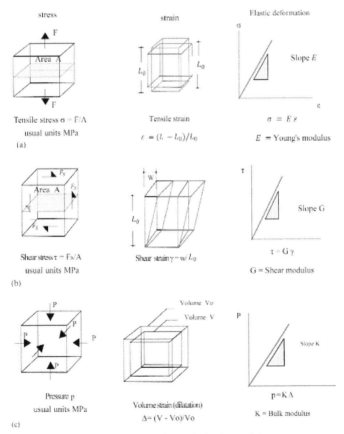

FIGURE 3.3   The definitions of stress, strain, and elastic module.

Engineering components can have complex shapes and can be loaded in many ways, creating complex distributions of stress. But no matter how complex, the stresses in any small element within the compression, and shear.

3.1.1.4 STRAIN

Strain is the response of materials to stress (second column of Fig. 3.3). A tensile stress σ applied to an element causes the element to stretch. If the element in Fig. 3.3a, originally of side L_0, stretches by $\Delta L = L - L_0$, the nominal tensile strain is:

$$\varepsilon = \frac{\delta L}{L_0} \tag{3.3}$$

A compressive stress shortens the element. The nominal compressive strain (negative) is defined in the same way. Since strain is the ratio of two lengths, it is dimensionless.

A shear stress causes a shear strain (Fig. 3.3b). If the element shears by a distance w, the shear strains.

$$\tan(\gamma) = \frac{w}{L_0} = \gamma \tag{3.4}$$

In practice $\tan\gamma \approx \gamma$ because strains are almost always small. Finally, a hydrostatic pressure causes an element of volume V to change in volume by ΔV. The volumetric strain, or dilatation (Fig. 3.3c), is:

$$\Delta = \frac{\delta V}{V} \tag{3.5}$$

The tensile strain is proportional to the tensile stress:

$$\sigma = E\varepsilon \tag{3.6}$$

Similarly, the shear strain is proportional to the shear stress:

$$\tau = G\gamma \tag{3.7}$$

and the dilatation Δ is proportional to the pressure:

$$P = K\Delta \tag{3.8}$$

3.1.1.5 STRENGTH AND DUCTILITY

If a material is loaded above its field strength, it deforms plastically or it fractures. Ductility is a measure of how much plastic strain a material can tolerate. It is measured in standard tensile tests by the elongation (the tensile strain at break) expressed as a percent. This is not a material property, because it depends on the sample dimensions [4].

3.1.1.6 HARDNESS

Tensile and compression tests are not always convenient; you need a large sample and the test destroys it. The hardness test (Fig. 3.4) avoids these problems, although it has problems of its own. In it, a pyramidal diamond or a hardened steel ball is pressed into the surface of the material, leaving a tiny permanent indent, the size of which is measured with a microscope. The indent means that plasticity has occurred, and the resistance to it, a measure of strength, is the load F divided by the area of the indent projected onto a plane perpendicular to the load. The indented region is surrounded by material that has not deformed, and this constrains it so that H is larger than the field strength.

$$H = \frac{F}{A} \qquad\qquad (3.9)$$

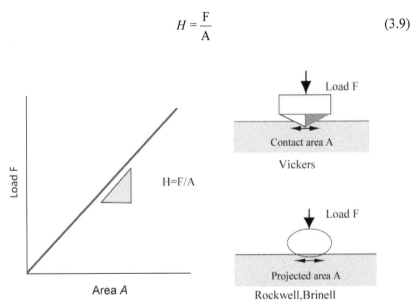

FIGURE 3.4   The hardness test (the Vickers test uses a diamond pyramid; the Rockwell and Brinell tests use a steel sphere).

The hardness test has the advantage of being nondestructive, so strength can be measured without destroying the component, and it requires only a tiny volume of material. But the information it provides is less accurate and less complete than the tensile test, so it is not used to provide critical design data [2].

3.1.1.7 MECHANICAL FREQUENCIES INCREASE IN SMALL SYSTEMS

Mechanical resonance frequencies depend on the dimensions of the system at hand. For the simple pendulum

$$\omega = \left(\frac{g}{l}\right)^{1/2} \tag{3.10}$$

$$T = \frac{2\pi}{\omega} \tag{3.11}$$

There is the miniature version, which has a much faster response time than the original one:

$$a = \frac{d^2 y}{dt^2} \tag{3.12}$$

$$\omega = \left[(g+a)/l\right]^{1/2} \tag{3.13}$$

A mass m attached to a rigid support by a spring of constant k has a resonance frequency

$$\omega = \left(\frac{k}{m}\right)^{1/2} \tag{3.14}$$

A. Simple harmonic oscillation occurs when a displacement of a mass in a given direction, x, produces a (oppositely directed) force:

$$F = -kx \tag{3.15}$$

According to Newton's Second Law:

$$F = ma = m\frac{d^2 x}{dt^2} \tag{3.15}$$

Applied to the mass on the spring, this gives the differential equation:

$$\frac{d^2 x}{dt^2} + \left(\frac{k}{m}\right)x = 0 \tag{3.17}$$

$$x = x_{max}\cos(\omega t + \delta) \tag{3.18}$$

is a solution of the equation for arbitrary amplitude x_{max} and arbitrary initial phase angle δ, but only when:

$$\omega = \left(\frac{k}{m}\right)^{1/2} \qquad (3.19)$$

The period of the motion is therefore:

$$T = 2\pi \left(m/k\right)^{1/2} \qquad (3.20)$$

The maximum values of the speed $v = dx/dt$ and the acceleration $a = d^2x/dt^2$ are seen to be:

$$v_{max} = x_{max}\,\omega \qquad (3.21)$$

$$a_{max} = x_{max}\,\omega^2 \qquad (3.22)$$

The total energy $E = U + K$ in the motion is constant and equal to:

$$E = \frac{1}{2}kx_{max}^2 \qquad (3.23)$$

In nanophysics, which is needed when the mass m is on an atomic scale, the same frequency $\omega = (k/m)^{1/2}$ is found, but the energies are restricted to:

$$E_n = \left(n + \frac{1}{2}\right)\hbar\omega \qquad (3.24)$$

where the quantum number n can take zero or positive integer values, $\hbar = h/2\pi$ and Planck's constant is $h = 6.67 \times 10^{-34}$ Joule·s. Simple harmonic oscillation is a more widely useful concept than one might think at first, because it is applicable to any system near a minimum in the system potential energy $U(x)$. Near x_0 the potential energy $U(x)$ can be closely approximated as a constant plus $k(x-x_0)^2/2$, leading to the same resonance frequency for oscillations of amplitude A in $(x-x_0)$. An important example in molecular bonding is where x_0 is the interatomic spacing.

More generally, the behavior applies whenever the differential equation appears, and the resonant frequency will be the square root of the coefficient of x in the equation. In the case of the pendulum, if $x \cong L$ is the horizontal displacement of the mass:

$$F \cong -g\frac{x}{L} \tag{3.25}$$

$$\omega = \left(\frac{g}{L}\right)^{1/2} \tag{3.26}$$

Considering the mass and spring to be three-dimensional, mass will vary as L_3 and k will vary as L, leading to $\omega \propto aL^{-1}$. Frequency inversely proportional to length scale is typical of mechanical oscillators such as a violin or piano string and the frequency generated by a solid rod of length L struck on the end. In these cases the period of the oscillation T is the time for the wave to travel $2L$, hence $T = 2L/v$. This is the same as $L = \lambda/2$, where $\lambda = vT$ is the wavelength. If the boundary conditions are different at the two ends, as in a clarinet, then the condition will be $L = \lambda/4$ with half the frequency. Hence for the stretched string:

$$\omega = 2\pi\left(\frac{v}{L}\right) \tag{3.27}$$

$$v = \left(\frac{F}{\rho}\right)^{1/2} \tag{3.28}$$

The speed of sound in a solid material is:

$$v = \left(\frac{E}{\rho}\right)^{1/2} \tag{3.29}$$

Young's modulus represents force per unit area (pressure stress) per fractional deformation (strain). Young's modulus is therefore a fundamental rigidity parameter of a solid, related to the bonding of its atoms.

This means, for example, that a pressure F/A of 101 kPa applied to one end of a brass bar of length $L = 0.1$ m would compress its length by

$$\Delta l = \frac{LF}{EA} = 11\mu m \tag{3.30}$$

Note that $\Delta l = 11$ μm, $E = 90$ GPa and $\rho = 10^4$ kg/m³, values similar to brass, correspond to a speed of sound $v = 3000$ m/s. On this basis the longitudinal resonant frequency of a 0.1 m brass rod is $f = v/2L = 15$ kHz. This frequency is in the ultrasonic range.

If one could shorten a brass rod to 0.1 micron in length, the corresponding frequency would be 15 *Ghz*, which corresponds to an electromagnetic wave with 2 cm wavelength. This huge change in frequency will allow completely different applications to be addressed, achieved simply by changing the size of the device.

A connection between macroscopic and nanometer scale descriptions can be made by considering a linear chain of N masses spaced by springs. The total length of the linear chain is:

$$L = N.a \qquad (3.31)$$

B. Vibrations on a Linear Atomic Chain of length $L = Na$

On a chain of N masses of length, and connected by springs of constant, denote the longitudinal displacement of the nth mass from its equilibrium position by u_n. The differential equation (Newton's Second Law) for the nth mass is:

$$m\,d^2 u_n / dt^2 + k\left(u_n + 1 - 2u_n - u_{n-1}\right) = 0 \qquad (3.32)$$

A traveling wave solution to this equation is:

$$u_n = u_0 \cos\left(\omega t + k'na\right) \qquad (3.33)$$

Here $k'na$ denotes $k'x = 2Px/\lambda$. Substitution of this solution into the difference equation reveals the auxiliary condition:

$$m\omega^2 = 4k \sin^2\left(k'a/2\right) \qquad (3.34)$$

This "dispersion relation" is the central result for this problem. One sees that the allowed frequencies depend upon the wavenumber $k' = 2\pi/\lambda$, as:

$$\omega = 2\left(k/m\right)^{1/2}\left|\sin\left(k'a/2\right)\right| \qquad (3.35)$$

The highest frequency, $2(k/m)^{1/2}$ occurs for $k'a/2 = \pi/2$ or $k' = \pi/a$; where the wavelength $\lambda = 2a$, and nearest neighbors move in opposite directions. The smallest frequency is at $\lambda = \pi/Na = \pi/L$ which corresponds to $L = \lambda/2$. Here one can use the expansion of $sin(x) \cong x$ for the small x. This gives:

$$\omega = 2\left(k/m\right)^{1/2} k'a/2 = a\left(k/m\right)^{1/2} k' \qquad (3.36)$$

Representing a wave velocity:

$$\omega/k' = v = a\left(k/m\right)^{1/2} \qquad (3.37)$$

Comparing this speed with $v = (E/\rho)^{1/2}$ for a thin rod, then $E/\rho = ka^2/m$. Young's modulus can thus be expressed in microscopic quantities as:

$$E = \rho ka^2 / m \qquad (3.38)$$

A cantilever clamped at one end and free at the other, such as a diving board, resists transverse displacement y (at its free end, $x = L$) with a force $-ky$. The effective spring constant for the cantilever is of interest to designers of scanning tunneling microscopes and atomic force microscopes, as well as to divers. The resonant frequency of the cantilever varies as L^{-2} according to the relation $\omega = 2\pi(0.56/L^2)(EI/\rho A)^{1/2}$. If t is the thickness of the cantilever in the y direction, then $I_A = \int A(\gamma)\gamma^2 d\gamma = wt^3/12$ where w is the width of the Cantilever. It can be shown that $K = 3YI/L^3$. It is possible to detect forces of a small fraction of a nano-Newton (nN).

The resonant frequency for a doubly clamped beam differs from that of the cantilever, but has the same characteristic L^{-2} dependence upon length:

$$\omega = \left(4.73/L\right)^2 \left(EI/\rho A\right)^{1/2} \qquad (3.39)$$

The measured resonant frequencies in the nano structure range from 15 MHz to 380 MHz.

The largest possible vibration frequencies are those of molecules, for example, the fundamental vibration frequency of the CO molecule is 6.42×10^{13} Hz (64.2 THz). Analyzing this vibration as two masses connected by a spring, the effective spring constant is 1860N/m [5–7].

3.1.1.8 MECHANICAL PROPERTIES FOR NANOELEMENT LAYERS

A nanoelement layer, which comprises of a great number of nanoscale helical springs, bars, or zigzags, can be fabricated by the Dynamic Oblique Deposition (DOD) method that exploits the atomic self-shadowing effect during physical vapor deposition with a highly oblique angle [8]. The size and geometric shape of the nanoelements can be controlled for engineering applications by the number of turns of the substrate and the incident angle of the vapor flux to the substrate during deposition [9]. A helical nanoelement layer with various heights and radii can be freely fabricated by the DOD method and the lateral and vertical stiffness of the nanoelement can be controlled by its geometric shape. This nanoelement layer has high potential for use as a low dielectric layer in microelectronic devices, microelectronic packaging, and nano- or micro-electromechanical (NEMS/MEMS) systems [10, 11]. A unique role of the nanoelement interfacial layer is the elimination of the stress concentration around an interfacial crack tip or a right-angled interfacial notch root. In fact, such stress concentration may be a cause of failure of electronic devices and systems due to these are fabricated by the deposition of many thin film layers of different materials and include unavoidably many generic wedges such as interfacial cracks and notches. Therefore, a clear understanding of the nanoelement interfacial layer as a discrete material for eliminating the conditions for stress concentration is very useful for fabricating highly reliable electronic devices and systems using the layer. Moreover, such inquiry helps to obtain the background of fracture mechanics for bridging continuous and discrete materials and extends our knowledge about fracture mechanics for continuous materials to that for discrete materials [12, 13].

For the numerical analysis, each helical nanoelement can be modeled by using a 2D beam element that has a vertical stiffness and lateral stiffness and the beam diameter and the distance between beams. All the other parts of the specimen can be modeled as 2D plane strain solid elements. A 2D beam Fig. 3.5. element with two nodes is used for the nanoelement and plane strain elements with eight nodes are used for all the other parts [14].

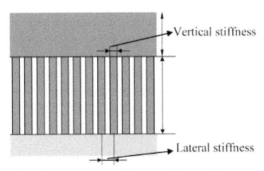

FIGURE 3.5   The helical nanoelement layer, its 2D modeling.

Stress analysis is conducted under plane strain condition. Each nanospring is replaced by the equivalent cylindrical beam with a diameter and a Young's modulus, which can be, derived from the vertical and lateral stiffness measurements of a single spring. The stress concentration at the free edge of an interface is induced by the mismatch of deformation on the interface. The mismatch is mainly due to the difference in Poisson's contractions between the different components. Since nanosprings are oriented vertically on the substrate and are separated from each other, the in-plane deformation of the thin film follows that of the adjoining component on the interface. In short, the difference of Poisson's contractions for the vertical deformation is zero. The stress concentration at the interface crack tip is induced by not only the mismatch of deformation on the interface but also the traction force from above and below the crack. However, no traction force is exerted on the crack tip in the spring model because the under part of the crack is constitutively separated from the crack tip. The apparent normal stress in the longitudinal direction of the nanospring is represented as follow [14]:

$$\sigma = \left(\frac{E\lambda}{h} \right)\left(\frac{D^2}{d^2} \right) \qquad (3.40)$$

3.2   THERMAL PROPERTIES OF NANOMATERIALS

3.2.1   MELTING POINT

The melting point of a material is a fundamental point of reference because it directly correlates with the bond strength. In bulk systems the surface-to-volume ratio is small and the curvature of the surface is negligible. Therefore, for a solid, in bulk form, surface effects can be disregarded. On the other hand, for the case of nanoscale solids, for which the ratio of surface to mass is large, the system may be regarded as containing surface phases in addition to the typical volume phases. In addition, for 0-D and 1-D nanomaterials, the curvature of the surface is usually very pronounced. Consequently, for nanomaterials the melting temperature is size dependent. In general, surface effects can be expressed mathematically by introducing an additional term $\Delta G_{Surface}$ to the total energy change $\Delta G_{Surface}$ resulting from the solid-liquid transformation. This is given by

$$\Delta G_{Total} = \Delta G_{Bulk} + \Delta G_{Surface} \qquad (3.41)$$

$$\Delta G_{Bulk} = \frac{L_0 (T_0 - T)}{T} V_L \qquad (3.42)$$

T is the melting point of the extended system where surface effects are included. When the surface of a body is increased, the change in surface energy is given by:

$$\Delta G_{Surface} = \gamma \Delta A \qquad (3.43)$$

Evidently, at the melting temperature, a layer of liquid with thickness t is formed on the surface and moves at a certain rate into the solid. During the change, (Fig. 3.6) a new liquid surface and liquid/solid interface are created, whereas the solid surface is destroyed. In other words, $\Delta G_{Surface}$ can be written as:

$$\Delta G_{Surface} = A_L \gamma_L + A_{SL} \gamma_{SL} - A_S \gamma_S \qquad (3.44)$$

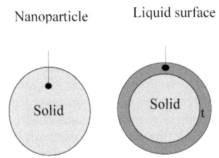

FIGURE 3.6   Upon formation of a liquid layer on the nanoparticle's surface, an interface between the liquid layer and the solid core develops.

At equilibrium, the solid core of radius r has the same chemical potential as the surrounding liquid layer of thickness t, which occurs when the differential $\partial \Delta G_{Total}/\partial t = 0$. For a sphere, this happens when:

$$\frac{L_0 (T_0 - T)}{T_0} = \frac{2\gamma_{SL}}{r - t} \qquad (3.45)$$

Assuming $t \to 0$ which represents the appearance of the first melting, the upper melting temperature for a sphere can be found from the expression.

$$T_M^{upper} = T_0 \left(1 - \frac{2\gamma_{SL}}{L_0 r} \right) \qquad (3.46)$$

in which the term $\dfrac{2\gamma_{SL}}{r}$ is associated with the increase in internal pressure resulting from an increase in the curvature of the particle with decreasing particle size. Due to, the variables are all positive quantities, this means that the upper melting temperature of a spherical nanoparticle decreases with decreasing particle size [2, 15].

3.2.2   THERMAL TRANSPORT

In addition to the melting temperature, many of the current applications of nanomaterials require knowledge about thermal transport. In some cases, such as microprocessors and semiconductor lasers, the goal is to transport heat away as quickly as possible, whereas for applications such as thermal barriers, the objective is to reduce thermal conduction. Heat is transported in materials by two different mechanisms: lattice vibration waves (phonons) and free electrons. In metals, the electron mechanism of heat transport is significantly more efficient than phonon processes due to the fact that metals possess a high number of free electrons and because electrons are not as easily scattered. In the case of nonmetals, phonons are the main mechanism of thermal transport due to the lack of available free electrons and because phonon scattering is much more efficient. In both metals and nonmetals, as the system length scale is reduced to the nanoscale, there are quantum confinement and classical scattering effects. In the case of bulk homogeneous solid nonmetal materials, the wavelengths of phonons are much smaller than the length scale of the microstructure [16]. However, in nanomaterials the length scale of the microstructure is similar to the wavelength of phonons. Therefore, quantum confinement occurs. In nanomaterials, quantum confinement comes in several flavors. In 0-D nanomaterials such as nanoparticles, quantum confinement occurs in three dimensions. In 1-D nanomaterials such as nanowires and nanotubes, confinement is restricted to two dimensions. In 2-D nanomaterials such as nanofilms and nanocoatings, quantum confinement takes place in one dimension. These confinement effects are similar for electron transport in nanomaterials. A good way to understand quantum confinement is to consider that the presence of nearby surfaces in 0-D, 1-D, and 2-D nanostructures causes a change in the distribution of the phonon frequencies as a function of phonon wavelength as well as the appearance of surface phonon modes [17, 18].

These processes lead to changes in the velocity with which the variations in the shape of the wave's amplitude propagate, the so-called group velocity. This is similar to a group or ring of waves forming when a stone hits the surface of water.

For nanoporous materials, the nanosize effect is determined by the number and size of the pores. Due to the porosity, these materials have low permittivity and thermal conductivity, which, in the case of microelectronic components, leads to an increase in the operation temperature and earlier circuit failure. The current problem is that it is still not theoretically understood how to treat nanoscale pores for thermal transport. One possibility is the similarity between the size of the pores and relevant phonon wavelengths, which suggests that phonons would not see a continuum field. However, experiments showed that the porosity did not play a role in heat transport except to reduce average density. This still remains to be seen. One final theory that has been gaining some respect is to consider the porous solid as a composite material comprising a matrix filled by voids [2].

3.3 ELECTRICAL PROPERTIES

The electronic conduction of electrons in systems considered large in size compared with the nanoscale. In this case, the conduction of electrons is delocalized, that is, electrons can move freely in all dimensions. As they travel their paths, the electrons are primarily scattered by various mechanisms, such as phonons, impurities, and interfaces, resembling a random walk process. However, as the system length scale is reduced to the nanoscale, two effects are of importance: (1) the quantum effect, where due to electron confinement the energy bands are replaced by discreet energy states, leading to cases where conducting materials can behave as semiconductors or insulators, and (2) the classical effect, where the mean-free path for inelastic scattering becomes comparable with the size of the system, leading to a reduction in scattering events [19, 20].

In 3-D nanomaterials, the three spatial dimensions are all above the nanoscale. Therefore, the two aforementioned effects can be neglected. However, bulk nanocrystalline materials exhibit a high grain boundary area-to-volume ratio, leading to an increase in electron scattering. As a consequence, nanosize grains tend to reduce the electrical conductivity.

In the case of 2-D nanomaterials with thickness at the nanoscale, quantum confinement will occur along the thickness dimension. Simultaneously, carrier motion is uninterrupted along the plane of the sheet. In fact, as the thickness is reduced to the nanoscale, the wave functions of electrons are limited to very specific values along the cross-section. This is because only electron wavelengths that are multiple integers of the thickness will be allowed. All other electron wavelengths will be absent. In other words, there is a reduction in the number of energy states available for electron conduction along the thickness direction. The electrons become trapped in what is called a potential well of width equal to the thickness [3, 21]. In general, the effects of confinement on the energy state for a 2-D nanomaterial with thickness at the nanoscale can be written as

$$E_n = \left[\frac{\pi^2 \hbar^2}{2mL^2} \right] n^2 \tag{3.47}$$

where $\hbar \equiv h/2$

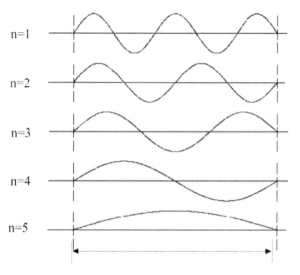

n=1

n=2

n=3

n=4

n=5

FIGURE 3.7 The energies and wave functions of the first five confined states for the case of an infinite-depth quantum well.

Equation 3.46 assumes infinite-depth potential well model (Fig. 3.7). As mentioned, the carriers are free to move along the plane of the sheet. Therefore, the total energy of a carrier has two components, namely a term related to the confinement dimension and a term associated with the unrestricted motion along the two other in-plane dimensions. To understand the energy associated with unrestricted motion, let's assume the z-direction to be the thickness direction and x and y the in-plane directions in which the electrons are delocalized. Under these conditions, the unrestricted motion can be characterized by two wave vectors k_x and k_y which are related to the electron's momentum along the x and y directions, respectively, in the form $\hbar x = \hbar k_x$ and $\hbar y = \hbar k_y$. The energy corresponding to these delocalized electrons is given by the so-called Fermi energy, which can be expressed as:

$$E_F = \left[\frac{\hbar^2 k_F^2}{2m} \right] \tag{3.48}$$

Where $k_F = \sqrt{k_x^2 + k_y^2}$.

At the temperature of absolute zero, all the conduction electrons are contained within a circle of radius k_F. As a result, the total energy of an electron (due to confinement and unrestricted motion) in a 2-D nanomaterial with thickness at the nanoscale can be given by:

$$E_n = \left[\frac{\pi^2 \hbar^2}{2mL^2} \right] n^2 + \left[\frac{\hbar^2 k_F^2}{2m} \right] \tag{3.49}$$

Since the electronic states are confined along the nanoscale thickness, the electron momentum is only relevant along the in-plane directions. As a result, scattering by phonons and impurities occurs mainly in-plane, leading to a 2-D electron conduction. However, for 2-D nanomaterials with nanocrystalline structure, the large amount of grain boundary area provides an additional source for in-plane scattering. So, the smaller the grain size, the lower the electrical conductivity of 2-D nanocrystalline materials.

In the case of 1-D nanomaterials, quantum confinement occurs in two dimensions, whereas unrestricted motion occurs only along the long axis of the nanotube/rod/wire. Contrary to a 2-D nanomaterial, which allows only one value of the principal quantum number n for each energy state, for a 1-D nanomaterial, the energy of a 2-D confinement depends on two quantum numbers, n_y and n_z, in the form:

$$E_{ny,nz} = \left[\frac{\pi^2 \hbar^2 n_y^2}{2mL_y^2} \right] + \left[\frac{\pi^2 \hbar^2 n_z^2}{2mL_z^2} \right] \tag{3.50}$$

Considering now the electron free motion along the x-direction (long axis), above equation can be modified to:

$$E_{ny,nz} = \left[\frac{\pi^2 \hbar^2 n_y^2}{2mL_y^2} \right] + \left[\frac{\pi^2 \hbar^2 n_z^2}{2mL_z^2} \right] + \left[\frac{\hbar^2 k_x^2}{2m} \right] \tag{3.51}$$

For 0-D nanomaterials, the motion of electrons is now totally confined along the three directions L_x, L_y and L_z. Therefore, the total energy can be given by:

$$E_{nx,ny,nz} = \left[\frac{\pi^2 \hbar^2 n_x^2}{2mL_x^2} \right] + \left[\frac{\pi^2 \hbar^2 n_y^2}{2mL_y^2} \right] + \left[\frac{\pi^2 \hbar^2 n_z^2}{2mL_z^2} \right] \tag{3.52}$$

In this fashion, all the energy states are discreet and no electron delocalization occurs. Under these conditions, metallic systems can behave as insulators due to the formation of an energy band gap, which is not allowed in the bulk form.

The electrical properties of 0-D, 1-D, 2-D, and 3-D nanomaterials as isolated entities are discussed. However, from a practical point of view, these materials need to be coupled to external circuits by electrodes. For 2-D and

3-D nanomaterials, ohmic contacts are possible. However, for 0-D and 1-D nanomaterials, the contact resistances between nanomaterials and the connecting leads are usually high.

In the electric circuit theory, this process is expressed by a definite relationship of physical parameters of an element, say, voltage versus current strength or electric charge versus voltage, and the number of appropriate units of measurement (ohm, farad, hertz, etc.) Electric circuit elements form branches, which are joined by means of ideal electric connections (i.e., connections free of resistance, inductance, capacitance) to form nodes and loops [22].

Conductors, semiconductors and insulators are desired as components for most of the applications. The reliability of these materials depends on the complete understanding of their electrical properties. The influence of solid-state structure, band structure, electron and hole behavior, and doping concentration on conductor, semiconductor and insulator behavior has been understood and can be found in the solid-state physics literature. Insulators are desired to act not just as electrical barriers but also as dopant diffusion barriers. Diffusion layers are also critical during the integration of interconnects into microelectronic devices.

The microscopic theory of electrical conduction was first proposed by Drude in 1900 and later developed by Lorentz in 1909. Here, the material is considered to be composed of a crystal lattice (ions) surrounded by an electron cloud or gas that is free to move (Fermi gas). This successfully predicts the electrical behavior in conductors (Ohm's law):

$$\rho = \frac{m\upsilon}{ne^2 \lambda'} \qquad (3.53)$$

The electrons do not behave as particles but follow a wave-particle behavior. The wave nature of the electrons allows for the correct prediction of the temperature and resistivity dependence on these parameters. It also predicts the conductive nature of materials describing them either as conductors, semiconductors or insulators. Classical theory does not permit this.

The Fermi energy is dependent on the carrier density. A higher carrier density implies a higher Fermi energy and therefore a smaller Fermi wavelength. Because of this dependence on carrier density the observation of the quantum phenomena for semiconductors occurs at dimensions in the order of a couple hundred nanometers while for metals this occurs in the order of few nanometers. Nano interconnects have two of the dimensions (thickness and width) small enough to show quantized behavior while the conduction in the other dimension (length) is considered to be continuous. In this type of interconnects the electrons move in a ballistic rather than diffusive manner. This means that the electrons are only scattered by the boundaries of interconnect [23].

Conventional Faraday capacitors store electric charge between parallel charged plates that are separated by an insulating dielectric material. Instead of flat parallel plates, capacitors that come in tubes use two metallic foils separated by an electrolyte-impregnated paper in a "sandwich" that is rolled up into the tube. For these devices, nanotube thin films can increase the surface area of the conducting foil due to the nanotubes' very small size, orderly alignment and high conductivity. "Nanotubes provide a huge surface area on which to store and release energy-that is what makes the difference" [22].

3.4   MAGNETIC PROPERTIES

This section is concerned with the magnetic properties of nanomaterials. In general, for any ferromagnetic material, the total energy can be written as the sum of various terms, in the form:

$$E_{Total} = E_{Exc} + E_{Ani} + E_{Dem} + E_{App} \qquad (3.54)$$

For macroscopic magnetic materials, a magnetostrictive energy must also be included in Eq. (3.52), but for nanoscale materials this energy can be neglected. This magnetization energy E_{total} can then be related to a magnetic field according to the expression:

$$E_{Total} = M.H \qquad (3.55)$$

The first term in Eq. (3.52) is due to the quantum mechanical interaction between atomic magnetic moments and represents the tendency for the magnetization vectors to align in one direction. In other words, if the magnetic moment is sufficiently large, the resulting magnetic field can drive a nearest neighbor to align in the same direction, provided the exchange energy is greater than the thermal energy. The second term in it represents the anisotropy energy that results from the spin's tendency to align parallel to specific crystallographic axes, called easy axes. Thus, a "soft" magnetic material will exhibit low anisotropy energy, whereas a "hard" magnetic material shows high anisotropy energy [24].

Though both the exchange energy and the anisotropy energy try to order the spins in a parallel configuration, the third term in the equation namely the demagnetization energy, which is related to the magnetic dipole character of spins, leads to the formation of magnetic domains. Thus, for macroscopic fer-romagnetic materials, all the magnetic moments are aligned in magnetic do-mains, although the magnetization vectors of different domains are not paral-lel to each other. Each domain is magnetized to saturation, with the moments typically aligned in an easy direction. Depending on the ratio of anisotropy to demagnetization energy, we can expect open (ratio < 1) or closure domain (ratio > 1) structures [25].

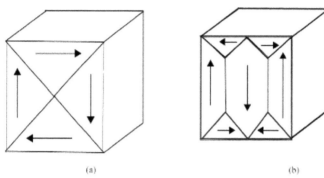

(a)   (b)

FIGURE 3.8   (a) Open domain structure in a macroscopic ferromagnetic material (b) Closure domain structure in a macroscopic ferromagnetic material.

Finally, the energy associated with an applied magnetic field, called Zeeman energy and represented by the last term in the equation, results from the tendency of spins to align with a magnetic field. Initially, as the magnetic field increases, the magnetization of the material increases. However, at some point, a satura-tion point, called saturation magnetization, is reached, above which an increase in magnetic field does not produce an increase in magnetization. The saturation magnetization is material and temperature dependent.

For nanocrystalline ferromagnetic materials (Fig. 3.8), an important con-sideration is the interaction among exchange energy, anisotropic energy, and demagnetization energy. For very small particles or grain sizes, the exchange forces are dominant due to strong coupling, causing all the spins in neighbor-ing grains to align, superseding in this way the anisotropic and demagnetiz-ing forces. Therefore, there is a critical grain size, below which the material will be single domain. Other magnetic properties are also strongly affected by scale. To address this point, first recall the magnetic response of a bulk fer-romagnetic material to an applied magnetic field. The hysteresis is associated with the fact that, on removal of the magnetic field, the magnetic domains do not revert to their original configuration. In other words, there is a remnant magnetization. On the other hand, the coercive field is the applied magnetic field that needs to be applied in the direction opposite the initial magnetic field, to bring the magnetization back to zero [2, 24–25]. By reducing the particle size or grain size to the nanoscale, the magnetization curve shown in Fig. 3.9 can be altered.

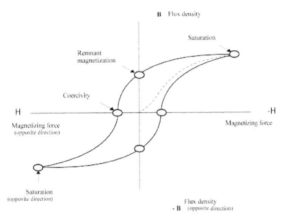

FIGURE 3.9   Magnetization versus applied magnetic field showing the hysteresis loop with (a) saturation magnetization, (b) remnant magnetization, and (c) coercive field.

3.4.1. NANOELEMENT MAGNETIZATION

One of the most exciting recent developments in magnetism has been the use of nanometer fabrication techniques to form nanometer scale magnets. These so-called nanomagnets, or nanoelements, possess by virtue of their extremely small size very different magnetic properties from their parent bulk material [26]. The magneto crystalline anisotropy of a bulk magnetic material governs its magnetic behavior and therefore is a key to its technological applicability. However, it is an intrinsic property of the material and cannot readily be tailored. In contrast, the magnetic behavior of a nanomagnet is also largely influenced by the interaction of the magnetization with its shape. This dependency provides the possibility of fine-tuning magnetic properties through shape-manipulation, which in turn requires very precise growth techniques [26, 27]. Fast magnetization reversal processes play an important role in magnetic recording applications. They inflict physical limits on data rates, areal storage densities, and signal to noise ratio, which makes their concepts understanding for the design of novel ultrahigh density magnetic storage applications. Recent advances in the fabrication and experimental investigation of thin films and patterned media have shown the influence of edge effects and inhomogeneous internal fields on the excitation of spin wave modes [28]. Magnetization reversal in quasi-single-domain magnetic particles by nonlinear spin-wave-like excitations has been explored using numerical micromagnetic simulations with limited discretization. These nonlinear spin-wave excitations have been shown to transfer excess Zeeman energy to uniaxial anisotropy and exchange energies, allowing the average magnetization to reverse. Micromagnetic simulations revealed processional oscillation effects of the magnetization in nanoelements during and after fast switching processes [29, 30]. There are various nanomagnets such as triangular, square and pentagonal nanomagnets, isolated circular nanomagnets and finally interacting circular nanomagnets arranged on a rectangular lattice [26].

3.4.2 MAGNETIC NANOELEMENT FABRICATION

Lithographic methods have been widely used to produce ordered arrays of nanoelements. The basic idea is to deposit a thin resist layer onto a substrate, parts of which are then chemically altered by exposing them to radiation. Finally, different techniques are used in order to transfer the generated pattern into an array of nanoelements. However, these nanoelements are not very well defined along the direction perpendicular to the original resist layer. In contrast, chemical methods are based on what is often referred to as the bottom up approach as the nanoparticles develop from smaller units. The challenge of fabricating nanoparticles of nonspherical geometry is to obtain a suitably anisotropic growth. Corresponding research on magnetic nanoparticles has led to the growth of a wide variety of shapes for hard magnetic iron compounds [27].

3.4.3 NANOMAGNETOMETRY

The high definition of the lithography means, however, that all of the particles in the array are virtually identical to each other and so the measured average properties for the array can also be interpreted as the individual properties of a single nanomagnet.

A large and static magnetic field is applied in the sample plane perpendicularly to the direction of magneto-optical sensitivity. A small transverse oscillating field is then applied in the direction of the magneto-optical sensitivity, in order to cause the magnetization to oscillate about H. The measured response can be written as:

$$\frac{\partial \varphi}{\partial H_t} \equiv \chi_t = \left(\frac{E''(\varphi)}{M_s} + H \right)^{-1} \tag{3.56}$$

The reciprocal of the measured magneto-optical response, $1/\chi_t$, is the total internal field in the direction of the magnetization, the externally applied field H plus the effective anisotropy field coming from the internal energy surface. A powerful probe of the energy surface of the nanomagnet can thus be obtained simply by measuring $1/\chi_t$ as a function of direction and H. The anisotropy field and the anisotropy energy of any system are related by [26] (Fig. 3.10):

$$U = \frac{2M_s V H_a}{n^2} \tag{3.57}$$

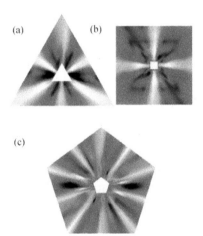

FIGURE 3.10   Anisotropy field inside different nanomagnets (a) triangular, (b) square and (c) pentagonal symmetry.

3.4.4   MICROMAGNETIC CONCEPT

The micromagnetic model, as introduced by Brown, approaches ferromagnetism on a mesoscopic scale, it only indirectly accounts for the underlying atomic structure of the material and assumes a continuous magnetization, which determines the state of the ferromagnetic structure. These torques are due to local magneto crystalline anisotropy, short-range exchange interaction, long-range magneto static interaction and an externally applied magnetic field. The former three contributions are material dependent, so that the model requires the input of corresponding parameters. These parameters are the exchange constant, anisotropy constants of different order and the saturation magnetization. The total energy of the system can be written as [27]:

$$E_{Total} = E_{Exch} + E_{demag} + E_{ext} \qquad (3.58)$$

The detailed distribution of the magnetization inside the nanoelements (a square nanoelement) is obtained through numerical integration of the Landau–Lifshitz–Gilbert equation of motion [29]:

$$\frac{dJ}{dt} = -\gamma J \times H + \frac{\alpha}{J_s} J \times \frac{dJ}{dt} \qquad (3.59)$$

$$H = -\frac{\delta E}{\delta J} \qquad (3.60)$$

where the total Gibbs' free energy:

$$E = \frac{A}{J_s^2}\left(\left(\nabla J_x\right)^2 + \left(\nabla J_y\right)^2 + \left(\nabla J_z\right)^2\right) - J.H_{ext} - \frac{1}{2}J.H_{demag} \qquad (3.61)$$

3.4.5 INTERACTING NANOMAGNETS

In order to demonstrate and investigate the phenomenon of magneto static interactions, a number of arrays of circular nanomagnets arranged on a rectangular lattice have to make. If the lattice spacing is sufficiently small, it must be considered the geometry of the lattice as well as the shape of the motif, due to the magnetic field emanating from one nanomagnet can influence its neighbors [37].

The measured average property of the lattice approximately is the same as the individual property of an isolated nanomagnet. In this case, the weak intrinsic uniaxial anisotropy is unable to stabilize the zero-field magnetization against thermal fluctuations and lead to a time-averaged remanence of zero, and hence the closed, super-paramagnetic hysteresis loops. Magneto static coupling between nanomagnets, especially between the nearest neighbors, becomes stronger. Finally it can overcome the thermal fluctuations. When this occurs, the spins essentially remain parallel and locked together in the x-direction even under zero applied field, leading to increased remanence in the x-direction loops.

Magneto static coupling is an anisotropic coupling, with an energy minimum (easy axis) when the dipoles are aligned with the line joining their centers (the x-direction) and an energy maximum (hard axis) when aligned perpendicular to this line (the y-direction). Consequently, whereas the remanence rises in the loops measured in the x-direction, the loops measured in the y-direction become increasingly sheared as a uniaxial hard direction appears [26].

3.5   OPTICAL PROPERTIES

In the case of semiconductor bulk materials, if an incident photon has energy greater than the band gap of the material, an electron may be excited from the valence band to the unfilled conduction band. Under these conditions, the photon is absorbed while a hole is left in the valence band when the electron jumps to the conduction band. Inversely, if an electron in the conduction band returns to the valence band and recombines with a hole, a photon is released with energy equal to the band gap of the semiconductor. However, at low temperatures, bulk semiconductors often show optical absorption just below the energy gap (Fig. 3.11). This process is associated with the formation of an electron and hole bound to each other, which is called an exciton. As for any other particle, the exciton has mobility and thus can move freely through the material. The binding between the electron and the hole arises from the difference in electrical charge between the electron (negative) and the hole (positive), leading to a Coulombic attractive force across the two particles, which can be written as [21, 31]:

$$F_C = \frac{-e^2}{4\pi\varepsilon_0 r^2} \tag{3.62}$$

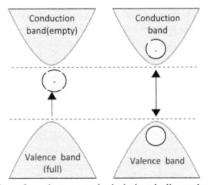

FIGURE 3.11   Creation of an electron and a hole in a bulk semiconductor material.

In nanomaterials, as the confinement is enhanced, the exciton binding energy increases, which reduces the possibility for exciton ionization at higher temperatures. As a result, strong excitonic states appear in the absorption spectra of nanomaterials at room temperature. As shown in Fig. 3.12, optical emission may also occur if the electron and the hole recombine, leading to the generation of a photon. If the photon energy is within 1.8 eV to 3.1 eV, the emitted light is in the visible range, a phenomenon called luminescence. Because of the quantum confinement in nanomaterials, the emission of visible light that's can be tuned by varying the nanoscale dimensions. An important effect of shape is related to whether the nanomaterial is 2-D, 1-D, or 0-D. In the case of a 2-D nanomaterial there are longitudinal plasmons along the plane of the sheet and a transverse plasmon across the thickness of the sheet. For a 1-D nanomaterial, there is a longitudinal plasmon along the long axis, whereas transverse plasmons exist across the diameter of the material. In the case of 0-D nanomaterials, the production of well-defined nonspherical shapes is usually quite challenging but of great interest. However, another alternative is to create core-shell nanoparticles. Finally, it is important to consider the environment surrounding the nanoparticles. When the refractive index or dielectric constant of the material embedding the nanoparticles increases, the Plasmon frequency decreases, generating a red shift. Using the aforementioned surface plasmon properties associated with metallic nanoparticles, light absorption and light emission can be significantly enhanced. As a result, metallic nanoparticles can be used as structural and chemical labels, photothermal therapy, and colorimetric chemical sensors [3, 32, 33].

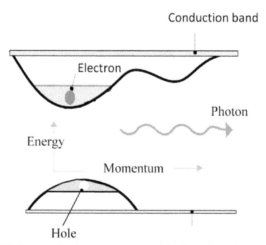

FIGURE 3.12 Emission of a photon upon recombination of an electron-hole pair.

3.5.1 OPTICAL WAVE INTERACTION OF NANOELEMENTS

These days one of the most important problems in nanotechnology and nanophotonics is optical wave interaction with metallic and nonmetallic nanoparticles. Certain metals such as Ag, Au behave as plasmonic materials in the optical frequencies, thus their plasma frequency is in the visible or ultraviolet (UV) regimes, and their permittivity has a negative real part. As a result, the interaction of optical signals with plasmonic nanoparticles involves surface Plasmon resonances [34]. These particles may be much smaller than the optical wavelength, so: "can such metallic (and nonmetallic) nanoparticles be treated as circuit "lumped nanoelements," such as nanoinductors, nanocapacitors, and nanoresistors in the optical regimes?" The conventional circuits in the lower frequency domains (such as in the rf and lower frequency range) indeed involve elements that are much smaller than the wavelength of operation, and the circuit theory may be regarded as the "approximation" to the Maxwell equations in the limit of such small sizes [35].

The circuit concepts and elements may be quantitatively extended to the optical frequencies when dealing with nanoparticles. Conducting materials (such as metals) behave differently at these higher frequencies so a mere scaling of the circuit component concepts conventionally used at lower frequencies may not work at frequencies beyond the far infrared. Future applications of such optical nanoelements could be in areas such as biological circuits, nanooptics, optical information storage, biophotonics, and molecular signaling [36, 37].

3.5.2 NANOCIRCUIT ANALOGY

To begin, we consider a nanosphere homogeneous material with dielectric function, which is in general a complex quantity. The sphere is assumed to be much smaller than the wavelength of operation in vacuum and in the material.

Because of the small size of the particle (with respect to wavelength), the scattered electromagnetic fields in the vicinity of the sphere and the total fields inside it may be obtained with very good approximation using the well-known time-harmonic, quasistatic approach. This leads to the following approximate expressions for the fields inside and outside the sphere [35, 37].

$$H_{int} = \frac{3\varepsilon_0 H_0}{\varepsilon + 2\varepsilon_0} \tag{3.63}$$

$$H_{ext} = H_0 + H_{dip} = H_0 + \frac{\left[3u\left(c.u\right) - c\right]}{4\pi\varepsilon_0 r^3} \tag{3.64}$$

with:

$$c = \frac{4\pi\varepsilon_0 R^3\left(\varepsilon - \varepsilon_0\right)H_0}{\varepsilon + 2\varepsilon_0} \qquad u = \frac{\bar{r}}{r}$$

At every point on the surface of the sphere the normal component of the displacement current is continuous:

$$-i\omega\left(\varepsilon - \varepsilon_0\right)H_0 . \hat{n} = -i\omega\varepsilon_0 H_{dip} . \hat{n} + i\omega\varepsilon H_{res} . \hat{n} \tag{3.65}$$

The residual field internal to the nanosphere when the incident field is subtracted from the total internal field can be written as:

$$H_{res} \equiv H_{int} - H_0 \tag{3.66}$$

The conventional circuits in the lower frequencies, relying on the conduction current circulating in metallic wires along the lumped elements, cannot be straightforwardly scaled down to the infrared and optical frequencies, at which conducting metallic materials behave quite differently. However, introducing plasmonic and nonplasmonic nanoparticles as basic elements of optical nanocircuits, in which effectively the "displacement" current can similarly "circulate," may provide analogous functionalities at the optical frequencies [37].

3.6 ACOUSTIC PROPERTIES

Materials interact strongly with radiation that has a wavelength comparable with their internal structure and/or dimensions. Therefore, in the case of nanomaterials for which the characteristic structure and dimensions are below 100 nm, there is a wide range of electromagnetic radiation within the visible, ultraviolet, and X-rays regimes that are affected by the nanoscale. On the other hand, acoustic waves, which exhibit wavelengths that range from microns to kilometers, have little or no direct effect on the properties of nanomaterials. As a result, the discussion of acoustic properties is limited for the case of nanomaterials. However, it should be pointed out that sound waves are used to produce nanomaterials. In addition, since the properties of nanomaterials are in many cases very different from traditional materials, acoustic waves can have a distinct indirect effect, such as in the case of seismic waves [38].

3.7 NANOMATERIAL APPLICATIONS

Nanotechnology is entering many industry sectors including medicine, plastics, energy, electronics, and aerospace. New applications for nanomaterials are being worked on in hundreds of laboratories around the globe (Table 3.1). Some will find useful applications that will revolutionize much of how we will live in the years ahead.

TABLE 3.1 Application of Nanomaterial in Different Industry

	Fillers for Point Systems
Chemical industry	Coating Systems based on Nanocomposites
	Magnetic fluids
Automotive Industry	Light weight construction
	Painting
	Catalysts
	Sensors
Medicine	Drug delivery systems
	Active agents
	Medical rapid tests
	Antimicrobial agents and coatings
	Agents in cancer therapy
Electronic Industry	Data memory
	Displays
	Laser diodes
	Glass fibers
	Filters
	Conductive, antistatic coatings
Energy Sources	Fuel cells
	Solar cells
	Batteries
	Capacitors
Cosmetics	Sun protection creams
	Tooth paste

3.7.1 COMPOSITES

An important use of nanoparticles and nanotubes is in composites, materials that combine one or more separate components and which are designed to exhibit overall the best properties of each component. This multifunctionality applies not only to mechanical properties, but extends to optical, electrical and magnetic ones. Currently, carbon fibers and bundles of multiwalled CNTs are used in polymers to control or enhance conductivity, with applications such as anti static packaging. The use of individual CNTs in composites is a potential long-term application. CNTs have exceptional mechanical properties, particularly high tensile strength and light weight. An obvious area of application would be in nanotube-reinforced composites, with performance beyond current carbon-fiber composites. One current limit to the introduction of CNTs in composites is the problem of structuring the tangle of nanotubes in a well-ordered manner so that use can be made of their strength. Another challenge is generating strong bonding between CNTs and the matrix, to give good overall composite performance and retention during wear or erosion of composites. The surfaces of CNTs are smooth and relatively unreactive, and so tend to slip through the matrix when it is stressed. One approach that is being explored to prevent this slippage is the attachment of chemical side-groups to CNTs, effectively to form 'anchors.' Another limiting factor is the cost of production of CNTs. However, the potential benefits of such light, high strength material in numerous applications for transportation are such that significant further research is likely [39, 40].

A particular type of nanocomposite is where nanoparticles act as fillers in a matrix. The need for plastic products to improve their basic viscoplastic characteristics, in order to allow them to compete with traditional materials, is being fulfilled by the development of composite materials. This is being done by incorporating such inorganic fillers as glass fiber, talcum or kaolin. Mineral-based reinforcement additives in dosage quantities of between about 20 and 60% are mixed into these polymers to give them their required characteristics. A disadvantage is the associated increase in the density of these composite materials. At the end of the 1980s, Toyota developed a totally new concept, with the help of nanoclays, for optimizing the characteristics of plastics destined for use in the car industry. This development relies on bentonite, a layered clay material that has been in industrial use for over a century. Functional polymers are far more versatile than classic construction materials. Besides improving the performance of systems in medical technology, telecommunication or optoelectronic, functional polymers are also found in every day products like paint and concrete. Traditionally functional polymers are produced from special monomers or by carefully controlling the polymer architecture. Even tiny amounts of functional polymers can have a dramatic impact. At Degussa, for instance, such systems are used to disperse pigments or to agglomerate extremely fine particles. The crucial factor in these systems is the specific interaction between the macromolecules and the particles at the interface. Consequently, properties like flow behavior of concrete can be significantly improved by adding small amounts of special comb like polymers. These polymers are highly effective because of the well-dosed coordination of the functional groups of the lateral chains responsible for regulating viscosity. By covering the surface of the cement grains the macromolecules prevent premature fusion of the particles [41, 42].

3.7.2 CLAYS

Clays containing naturally occurring nanoparticles have long been important as construction materials and are undergoing continuous improvement.

In the late 1980s it was discovered that adding 5% in weight of nano-sized clays to Nylon 6, a synthetic polymer, greatly increased its mechanical and thermal properties. Since then, polymer-clay nanocomposites have been widely studied and many commercial products are available. These hybrid materials are made of organic polymer matrices and clay fillers. Clays are a type of layered silicates that are characterized by a fine 2D crystal structure. Among these, mica has been the most studied. Mica is made up of large sheets of silicate held together by relatively strong bonds. Semectic clays, such as montmorillonite, have relatively weak bonds between layers [43–45].

For these layered silicates to be useful as fillers in nanocomposites, the layers must be separated, and the clay mixed thoroughly in the polymer matrix. This is not trivial as silicate clays are inherently hydrophilic, whereas polymers tend to be hydrophobic. The solution is to exchange the cations that keep the layers in the silicate together with larger inorganic ions that can thus open the galleries between the layers (intercalation). When the silicate layers are completely separated, the material is called exfoliated. In the case of intercalation, extended polymer chains are present between clay layers, resulting in a multilayer structure with alternating polymer/clay phases at repeated distances of few nanometers. In the exfoliated state the silicate layers are totally separated and dispersed in a continuous polymer matrix [45, 46]. The presence of the cations is necessary to compensate for the overall negative charge of the single layers. The layers are 20–200 nm in diameter laterally and come into aggregates called tactoids, which can be about 1 nm or more thick. Naturally occurring clays include montmorillonite (MMT) and hecrite, and their synthetic equivalents are saponite and laponite, respectively. Clay particle based composites – containing plastics and nano-sized flakes of clay – are also finding applications such as use in car bumpers [47, 48].

3.7.3   COATINGS AND SURFACES

Nano-powders and nanoparticle dispersions have seen increasing applications in coatings. Due to their small size, very even coating can be achieved by painting nanoparticle dispersions onto a surface and baking off the residual solvent. Coatings with thickness controlled at the nano- or atomic scale have been in routine production for some time, for example in materials for optoelectronic devices, or in catalytically active and chemically functionalized surfaces [49–50].

Long-term antimicrobial activity can be imparted in many coating formulations through the incorporation of nanomaterials [51]. The desire for permanent coatings to impart long-term antimicrobial or bacteria-stat properties to coated products has been expressed in a variety of industries, including healthcare, industrial and institutional cleaning, food processing, food service, and general paints and coatings [52]. Nano zincoxide, copper oxide or doped zinc oxides can be fully dispersed into a wide variety of coating formulations, including urethanes, acrylics and vinyl acetates and have shown utility in UV curable and thermosetting coatings as well as in water-based coating systems [53].

Recently developed applications include the self-cleaning window, which is coated in highly activated titanium dioxide, engineered to be highly hydrophobic (water repellent) and anti bacterial, and coatings based on nanoparticulate oxides that catalytically destroy chemical agents. Wear and scratch-resistant hard coatings are significantly improved by nanoscale intermediate layers (or multilayers) between the hard outer layer and the substrate material. The intermediate layers give good bonding and graded matching of elastic and thermal properties, thus improving adhesion. A range of enhanced textiles, such as breathable, waterproof and stain- resistant fabrics, have been enabled by the improved control of porosity at the nanoscale and surface roughness in a variety of polymers and inorganics [54].

3.7.4   PAINTS

Incorporating nanoparticles in paints could improve their performance, for example by making them lighter and giving them different properties. Thinner paint coatings ('light weighting'), used for example on aircraft, would reduce their weight, which could be beneficial to the environment. However, the whole life cycle of the aircraft needs to be considered before overall benefits can be claimed. It may also be possible to substantially reduce solvent content of paints, with resulting environmental benefits. New types of fouling- resistant marine paint could be developed and are urgently needed as alternatives to tributyltin (TBT), now that the ecological impacts of TBT have been recognized. Anti-fouling surface treatment is also valuable in process applications such as heat exchange, where it could lead to energy savings. If they can be produced at sufficiently low cost, fouling-resistant coatings could be used in routine duties such as piping for domestic and industrial water systems. It remains speculation whether very effective antifouling coatings could reduce the use of biocides, including chlorine. Other novel, and more long-term, applications for nanoparticles might lie in paints that change color in response to change in temperature or chemical environment, or paints that have reduced infrared absorptivity and so reduce heat loss [55, 56].

Concerns about the health and environmental impacts of nanoparticles may require the need for the durability and abrasion behavior of nano-engineered paints and coatings to be addressed, so that abrasion products take the form of coarse or microscopic agglomerates rather than individual nanoparticles.

The potential of nanoparticles to react with pollutants in soil and groundwater and transform them into harmless compounds is being researched. In one pilot study the large surface area and high surface reactivity of iron nanoparticles were exploited to transform chlorinated hydrocarbons (some of which are believed to be carcinogens) into less harmful end products in groundwater. It is also hoped that they could be used to transform heavy metals such as lead and mercury from bio-available forms into insoluble forms [57, 58].

3.7.5   FUEL CELLS

Engineered surfaces are essential in fuel cells, where the external surface properties and the pore structure affect performance. The hydrogen used as the immediate fuel in fuel cells may be generated from hydrocarbons by catalytic reforming, usually in a reactor module associated directly with the fuel cell. This is likely to be increasingly important as traditional energy sources continue to escalate in cost. Many challenges that we are now facing in the energy area (improvements needed in solar panels, hydrogen fuel cells, rechargeable batteries, etc.) can be solved through the use of nano-engineered materials. Some of these materials are developed through direct inspiration from Nature, such as the new types of solar photovoltaic cells, which try to imitate the natural nanomachinery of photosynthesis. Another interesting example is that of using battery electrodes with self-assembling nanostructures grown by genetic engineered viruses. It has been reported that the use of rare earth metal oxides has utility in various aspects of fuel cell design, but especially as components in electrodes and as low-temperature electrolytes in solid oxide fuel cells (SOFC). Nano-ceria and mixed rare earth metal oxides have shown promise in testing for such applications [59, 60].

The potential use of nano-engineered membranes to intensify catalytic processes could enable higher-efficiency, small-scale fuel cells. These could act as distributed sources of electrical power. It may eventually be possible to produce hydrogen locally from sources other than hydrocarbons, which are the feed stocks of current attention.

3.7.6 CATALYSTS

In general, nanoparticles have a high surface area, and hence provide higher catalytic activity. Nanotechnologies are enabling changes in the degree of control in the production of nanoparticles, and the support structure on which they reside. It is possible to synthesis metal nanoparticles in solution in the presence of a surfactant to form highly ordered monodisperse films of the catalyst nanoparticles on a surface. This allows more uniformity in the size and chemical structure of the catalyst, which in turn leads to greater catalytic activity and the production of fewer by-products. It may also be possible to engineer specific or selective activity. These more active and durable catalysts could find early application in cleaning up waste streams. This will be particularly beneficial if it reduces the demand for platinum-group metals, whose use in standard catalytic units is starting to emerge as a problem, given the limited availability of these metals [61].

The use of nanomaterials based on rare earth metal oxides allows for the preparation of thinner active layers, which can mean less precious metal usage. These nanomaterials also allow for the preparation of higher solids dispersions that are very stable, minimizing the number of coating steps and losses due to flocculated dispersions. Automotive catalytic converters are a key focus area for catalyst performance. One way to achieve lower emissions in a cost effective manner is to use cocatalysts that provide good oxygen storage capability and thermal stability in thinner layers. Nano-ceria and mixed rare earth metal oxides meet the criteria necessary to enhance catalytic converter performance when properly incorporated into a catalyst system, and because they are dense, single phase individual crystals, there is nothing to collapse during thermal cycling [61, 62].

3.7.7 LUBRICANTS

Nanospheres of inorganic materials could be used as lubricants, in essence by acting as nanosized 'ball bearings.' The controlled shape is claimed to make them more durable than conventional solid lubricants and wear additives. Whether the increased financial and resource cost of producing them is offset by the longer service life of lubricants and parts remains to be investigated (along the lines of the methodology). It is also claimed that these nanoparticles reduce friction between metal surfaces, particularly at high normal loads. If so, they should find their first applications in high-performance engines and drivers; this could include the energy sector as well as transport. There is a further claim that this type of lubricant is effective even if the metal surfaces are not highly smooth. Again, the benefits of reduced cost and resource input for machining must be compared against production of nanolubricants. In all these applications, the particles would be dispersed in a conventional liquid lubricant; design of the lubricant system must therefore include measures to contain and manage waste [63].

3.7.8 MAGNETIC MATERIALS

Nanotechnology as a term applied in the field of magnetism may be regarded as either relatively mature, or as a subject at a nascent stage in its development. Many of the permanent magnets in common use today, in devices ranging from high-efficiency motors to fridge magnets, have properties dictated by the physical nanostructure of the material, and the subtle and complex magnetic interactions that this produces. The use of nanostructured magnets has already enabled very significant savings in energy consumption and weight for motors, which contributes strongly to the green economy. Data storage density has increased with a compound growth rate of 60%, giving disk drives for the mass PC market in excess of 400 Gbyte. The Internet, image handling in cameras and data storage drive the demand for increases in data storage capacity.

Fujitsu recently demonstrated 100 Gbit/in^2 data storage capacity in a form of conventional longitudinal recording media [21].

It has been shown that magnets made of nanocrystalline grains possess unusual magnetic properties due to their extremely large grain interface area (high coercively can be obtained because magnetization flips cannot easily propagate past the grain boundaries). This could lead to applications in motors, analytical instruments like magnetic resonance imaging (MRI), used widely in hospitals, and microsensors. Overall magnetization, however, is currently limited by the ability to align the grains' direction of magnetization.

Nanoscale-fabricated magnetic materials also have applications in data storage. Devices such as computer hard disks depend on the ability to magnetize small areas of a spinning disk to record information. If the area required to record one piece of information can be shrunk in the nanoscale (and can be written and read reliably), the storage capacity of the disk can be improved dramatically. In the future, the devices on computer chips, which currently operate using flows of electrons, could use the magnetic properties of these electrons, called spin, with numerous advantages. Recent advances in novel magnetic materials and their nanofabrication are encouraging in this respect [21].

3.7.9 MEDICAL IMPLANTS

The ability of block copolymers to form nanoparticles and nanostructures in aqueous solutions makes them particularly useful for biomedical applications, such as therapeutics delivery, tissue engineering and medical imaging. In the field of therapeutic delivery, materials that can encapsulate and release drugs are needed. Hydrogels are very useful for the controlled release of drugs and block copolymer hydrogels are particularly advantageous for the possibility of conferring some stimuli-activated properties, such as temperature-sensitivity. Block copolymers form nanostructures with both hydrophilic and hydrophobic areas, so they can form vesicles that can encapsulate and carry both hydrophobic and hydrophilic therapeutic agents. Micelles formed using block copolymers have a hydrophilic corona that makes them more resistant to the interaction of proteins, in particular plasma proteins. Therefore, these types of micelles exhibit long circulation times in vivo. Insoluble domains can also be engineered to exploit the sensitivity of specific hydrophobic polymers to external stimuli such as pH, oxidative species, temperature and hydrolytic degradation. Block copolymers are also of interest for preparing scaffolds for tissue engineering. For instance very long micelles that mimic the natural extracellular have recently been prepared exploiting the self-assembly properties of a peptide copolymer [64].

Current medical implants, such as orthopedic implants and heart valves, are made of titanium and stainless steel alloys, primarily because they are biocompatible. Unfortunately, in some cases these metal alloys may wear out within the lifetime of the patient. Nanocrystalline zirconium oxide (zirconia) is hard, wear-resistant, bio-corrosion resistant and bio-compatible. It therefore presents an attractive alternative material for implants. It and other nanoceramics can also be made as strong, light aerogels by sol–gel techniques. Nanocrystalline silicon carbide is a candidate material for artificial heart valves primarily because of its low weight, high strength and inertness [65, 66].

3.7.10   CERAMICS

Ceramics are hard, brittle and difficult to machine. However, with a reduction in grain size to the nanoscale, ceramic ductility can be increased. Zirconia, normally a hard, brittle ceramic, has even been rendered superplastic (for example, able to be deformed up to 300% of its original length). Nanocrystalline ceramics, such as silicon nitride and silicon carbide, have been used in such automotive applications as high-strength springs, ball bearings and valve lifters, because they can be easily formed and machined, as well as exhibiting excellent chemical and high-temperature properties. They are also used as components in high-temperature furnaces. Nanocrystalline ceramics can be pressed into complex net shapes and sintered at significantly lower temperatures than conventional ceramics [67–69].

High-performance ceramics are sought in many applications, like highly efficient gas turbines, aerospace materials, automobiles etc. The field of ceramics that focuses on improving their mechanical properties is referred to as structural ceramics. Nanocomposite technology is also applicable to functional ceramics such as ferroelectric, piezoelectric, varistor and ion-conducting materials. In this case the properties of these nanocomposites relate to the dynamic behavior of ionic and electronic species in electro-ceramic materials.

Among these materials here we limit the review to nanocomposite with enhanced magnetic properties [69].

3.7.11 WATER PURIFICATION

Nano-engineered membranes could potentially lead to more energy-efficient water purification processes, notably in desalination by reverse osmosis. Nano-engineered membranes filters are characterized by thin membrane layers with uniformly sized pores and for most applications the membrane layer is sustained by a support. Inorganic membrane and in particular ceramic membranes have a number of advantages above polymeric membranes like high temperature stability, relative inert to chemicals, applicable at high pressures, easy to sterilize and recyclable. However, they have not been used extensively because of their high costs and relatively poor control in pore size distribution.

Compared to other microfiltration membranes, nano-engineered membranes have an extremely small flow resistance. An accumulation of retained particles in front of the membrane (cake layer formation) will – more than for other membranes – greatly increase the flow resistance. It is therefore important to keep the surface free of particles during filtration. This is usually done by applying a cross-flow in which larger particles will be removed from the membrane surface. Permeate flow reversal (back pulsing) is a more advanced method to remove smaller particles from the surface of the membrane filter. The required cross-flow velocity to remove larger particles is dependent on several variables such as the ratio between the particle and the pore size, the transmembrane pressure and cross-flow channel dimensions. An additional advantage of a clean membrane surface is that the retention characteristics of the filtration process is only determined by the membrane layer itself and not by the additional permeation characteristics of the cake layer.

3.7.12 SUNSCREENS AND COSMETICS

Nanomaterials can act as sun blockers to protect human skin in formulations that go on smooth, silky and clear. Eliminating unnecessary exposure to the harmful UV rays of the sun has increasingly become a key health concern. Nanosized titanium dioxide and zinc oxide are currently used in some sunscreens, as they absorb and reflect ultraviolet (UV) rays and yet are transparent to visible light and so are more appealing to the consumer [70].

3.7.13 MILITARY BATTLE SUITS

Enhanced nanomaterials form the basis of a state-of-the-art 'battle suit' that is being developed by the Institute of Soldier Nanotechnologies at Massachusetts Institute of Technology, USA (MIT, 2004). A short-term development is likely to be energy-absorbing materials that will withstand blast waves; longer-term are those that incorporate sensors to detect or respond to chemical and biological weapons (for example, responsive nanopores that 'close' upon detection of a biological agent). There is speculation that developments could include materials, which monitor physiology while a soldier is still on the battlefield, and uniforms with potential medical applications, such as splints for broken bones. In the field of surface engineering there is constant and widespread research seeking to develop new, advantageous processes and materials. Among these research initiatives, thermal-sprayed nanostructured oxide coatings emerge as one of the very few that have matured and proven themselves to be practically advantageous for both military and industrial applications. Although many commercial, non nanostructured thermal spray coatings provide adequate performance for certain applications, nanostructured coatings will provide solutions to improve on existing applications or to allow for new applications that require the unique and added benefits that can only be attained through this approach [71, 72].

Now, the percent of researches about some examples of nanomaterial application is presented in the Table 3.2.

TABLE 3.2 The Amount of Research about Some Examples of Nanomaterial Applications.

Nanomaterial application	Percent of researches
Biomedical Devices	0.653019
Drug-delivery	1.100262
Flat-panel displays	0.122127
Medical Imaging	0.561563
High-energy batteries	0.540372
Sensors	2.028764
Paints and pigments	0.085322
Automobile components	0.083091
Ceramics	0.907311
High-power magnets	0.117108
Smart structures	0.457838
CMP slurries	0.012269
Penetrators/warheads	0.001673

TABLE 3.2 *(Continued)*

Nanomaterial application	Percent of researches
Catalysts	26.53398
Phosphors	2.743125
Dielectrics	15.77674
Fuel cells	11.63277
Optical devices	26.81616
Cutting tools	3.908075
Cosmetics	2.974554
Aerospace components	2.943883

3.8   CONCLUSION

In nanomaterials, quantum confinement comes in several flavors. In 0-D nano-materials such as nanoparticles, quantum confinement occurs in three dimensions. In 1-D nanomaterials such as nanowires and nanotubes, confinement is restricted to two dimensions. Mechanical resonance frequencies depend on the dimensions of the system. For 2-D and 3-D nanomaterials, ohmic contacts are possible. However, for 0-D and 1-D nanomaterials, the contact resistances between nanomaterials and the connecting leads are usually high. There are various nanomagnets such as triangular, square and pentagonal nanomagnets, isolated circular nanomagnets and finally interacting circular nanomagnets arranged on a rectangular lattice. New applications for nanomaterials are being worked on in hundreds of laboratories around the globe. Chemical, Automotive, Medicine, Electronic industries as well as energy sources are discussed as special nanomaterial applications.

KEYWORDS

- **mechanical resonance frequencies**
- **nanomaterial applications**
- **nanomaterial properties**
- **quantum confinement**

REFERENCES

1. Binns, C. (2010). *Introduction to Nanoscience and Nanotechnology*. Vol. 14, Wiley.
2. Ashby, M. F., Shercliff, H., Cebon, D. (2013). *Materials: Engineering, Science, Processing and Design; North American Edition*. 3 ed.: Butterworth-Heinemann.
3. Schodek, D. L., Ferreira, P., Ashby, M. F. (2009). *Nanomaterials, Nanotechnologies and Design: An Introduction for Engineers and Architects*: Butterworth-Heinemann.
4. Ashby, M. F., Jones, D. R. H. (2012). *Engineering Materials 2, An Introduction to Microstructures and Processing*. 4 ed.: Butterworth-Heinemann.
5. Askeland, D. R., Fulay., P. P., Wright, W. J (2011). *The Science and Engineering of Materials*: Cengage Learning.
6. Wolf, E. L. (2008). *Nanophysics and Nanotechnology: An Introduction to Modern Concepts in Nanoscience*: John Wiley & Sons.
7. Mitchell, B. S. (2004). *An Introduction to Materials Engineering and Science for Chemical and Materials Engineers*. illustrated ed: John Wiley & Sons.
8. Suzuki, M., Taga, Y. (2001). *Integrated Sculptured Thin Films*. Japanese Journal of Applied Physics. 40, 358–367.
9. Lakhtakia, A. (2002). *Sculptured Thin Films: Accomplishments and Emerging Uses*. Materials Science and Engineering: C. 19(1), 427–434.
10. Hirakata, H., et al. (2007). *Anisotropic Deformation of Thin Films Comprised of Helical Nanosprings*. International Journal of Solids and Structures. 44(11), 4030–4038.
11. Plawsky, J. L., et al. (2007). *Mechanical and Transport Properties of Low-k Dielectrics*, in *Dielectric Films for Advanced Microelectronics*, John Wiley & Sons, Ltd. 137–197.
12. Sumigawa, T., et al. (2008). *Disappearance of Stress Singularity at Interface Edge Due to Nanostructured Thin Film*. Engineering Fracture Mechanics. 75(10), 3073–3083.
13. Sumigawa, T., et al. (2009). *Effect of Interface Layer Consisting of Nanosprings on Stress Field Near Interface Edge*. Engineering Fracture Mechanics. 76(9), 1336–1344.
14. Sueda, T., et al. (2011). *Stress Singularity Transition of Generic Wedges Due to Nanoelement Layers*. Engineering Fracture Mechanics. 78(16), 2789–2799.
15. Kondepudi, D. (2008). *Introduction to Modern Thermodynamics*: Wiley.
16. Cahill, D. G., et al. (2003). *NanoscaleThermal Transport*. Journal of Applied Physics. 93(2), 793–818.
17. Tiwari, J. N., Tiwari, R. N., Kim, K. S. (2012). *Zero-Dimensional, One-Dimensional, Two-Dimensional and Three-Dimensional Nanostructured Materials for Advanced Electrochemical Energy Devices*. Progress in materials science. 57(4), 724–803.
18. Perea, D. E., et al. (2006). *Three-Dimensional Nanoscale Composition Mapping of Semiconductor Nanowires*. Nano letters. 6(2), 181–185.
19. Edelstein, A. S., Cammaratra, R. C. (1998). *Nanomaterials: Synthesis, Properties and Applications, Second Edition*. 2 ed: Taylor & Francis. 616.
20. Darrin, A. G., Barth, J. L. (2011). *Systems Engineering for Microscale and Nanoscale Technologies*: CRC Press. 592.
21. Kelsall, R., Hamley I. W., Geoghegan, M. (2013). *Nanoscale Science and Technology* 2005. John Wiley & Sons.
22. Gorokhov, V., *The Roots of the Theoretical Models of the Nanotechnoscience in the Electric Circuit Theory*. Advances in Historical Studies. 2, 19–31.
23. Dupas, C., Lahmani, M. (2007). *Nanoscience: Nanotechnologies and Nanophysics*: Springer.
24. Jiles, D. C. (1998). *Introduction to Magnetism and Magnetic Materials, Second Edition*: Taylor & Francis.

25. Liu, J. P., et al. (2010). *Nanoscale Magnetic Materials and Applications*: Springer Science+Business Media, LLC.
26. Cowburn, R. P. (2000). *Property Variation with Shape in Magnetic Nanoelements.* Journal of Physics D: Applied Physics. 33(1), R1–R16.
27. Knittel, A., et al. (2010). *Micromagnetic Studies of Three-dimensional Pyramidal Shell Structures.* New Journal of Physics. 12(11), 113048–113074.
28. Jorzick, J., et al. (2002). *Spin Wave Wells in Nonellipsoidal Micrometer Size Magnetic Elements.* Physical Review Letters. 88(047204–047204–4), 13.
29. Fidler, J., et al. (2004). *Magnetostatic Spin Waves in Nanoelements.* Physica B: Condensed Matter. 343(1), 200–205.
30. Safonov, V. L., Bertram, H. N. (2000). *Intrinsic Mechanism of Nonlinear Damping in Magnetization Reversal.* Journal of Applied Physics. 87(9), 5508–5510.
31. Seal, S. (2010). *Functional Nanostructures: Processing, Characterization, and Applications*: Springer.
32. Rodriguez, J. A., Fernández-García, M. (2007). *Synthesis, Properties, and Applications of Oxide Nanomaterials*: John Wiley & Sons.
33. Poole, C. P., Owens, F. J. (2003). *Introduction to Nanotechnology*. illustrated ed: John Wiley & Sons.
34. Lamprecht, B., et al. (2001). *Surface Plasmon Propagation in Microscale Metal Stripes.* Applied Physics Letters. 79(1), 51–53.
35. Silveirinha, M. G. et al., (2008). *Nanoinsulators and Nanoconnectors for Optical Nanocircuits.* Journal of Applied Physics. 103(6), 064305-064305-24.
36. Engheta, N. (2007). *Circuits with Light at Nanoscales: Optical Nanocircuits Inspired by Metamaterials.* Science. 317(5845), 1698–1702.
37. Engheta, N., Salandrino, A., Alù, A. (2005). *Circuit Elements at Optical Frequencies: Nanoinductors, Nanocapacitors, and Nanoresistors.* Physical Review Letters. 95(9), 095504–095504–4.
38. Fletcher, N. H., Rossing, T. D. (1998). *The Physics of Musical Instruments.* illustrated, reprint ed: Springer.
39. Singh, (2007). *Science and Technology For Upsc*: McGraw-Hill Education (India) Pvt Limited.
40. Grady, B. P. (2011). *Carbon Nanotube-Polymer Composites: Manufacture, Properties, and Applications*: Wiley.
41. Lines, M. G. (2008). *Nanomaterials for Practical Functional Uses.* Journal of Alloys and Compounds. 449(1), 242–245.
42. Paipetis, A., Kostopoulos, V. (2012). *Carbon Nanotube Enhanced Aerospace Composite Materials: A New Generation of Multifunctional Hybrid Structural Composites*: Springer. 496.
43. Kato, A., Usik, A. (2005). *Polymer-Clay Nanocomposites.* Inorganic Polymeric Nanocomposites and Membranes. 179, 135–195.
44. Fornes, T. D., Paul, D. R. (2003). *Modeling Properties of Nylon 6/Clay Nanocomposites Using Composite Theories.* Polymer. 44(17), 4993–5013.
45. Alexandre, M., Dubois, P. (2000). *Polymer-Layered Silicate Nanocomposites: Preparation, Properties and Uses of a New Class of Materials.* Materials Science and Engineering: R: Reports. 28(1), 1–63.
46. LeBaron, P. C., Wang, Z., Pinnavaia, T. J. (1999). *Polymer-Layered Silicate Nanocomposites: An Overview.* Applied clay science. 15(1), 11–29.
47. Utracki, L. A., Sepehr, M., Boccaleri, E. (2007). *Synthetic,Layered Nano-Particles for Polymeric Nanocomposites (PNC's).* Polymers for advanced technologies. 18(1), 1–37.

48. Forestier, L., et al. (2010). *Textural and Hydration Properties of a Synthetic Montmorillonite Compared with a Natural Na-Exchanged Clay Analogue.* Applied clay science. 48(1), 18–25.

49. Choy, K. L. (2003). *Chemical Vapour Deposition of Coatings.* Progress in materials science. 48(2), 57–170.

50. Baer, D. R., Burrows, P. E., El-Azab, A. A. (2003). *Enhancing Coating Functionality Using Nanoscience and Nanotechnology.* Progress in Organic coatings. 47(3), 342–356.

51. Saxl, O. (2013). *Nanotechnology: Applications and Markets, Present and Future,* in *Ellipsometry at the Nanoscale,* Springer 705–730.

52. Dumiterscu, L., et al., *Nanocomposite Ecomaterials Based on Polymers and Alumina Nanoparticles for Wood Preservation.* RECENT ADVANCES in ENVIRONMENT, ECOSYSTEMS and DEVELOPMENT: 180–185.

53. Joshi, M., Bhattacharyya, A. (2011). *Nanotechnology–A New Route to High-Performance Functional Textiles.* Textile Progress. 43(3), 155–233.

54. Davis, S. S. (1997). *Biomedical Applications of Nanotechnology—Implications for Drug Targeting and Gene Therapy.* Trends in biotechnology. 15(6), 217–224.

55. Bryan, G. W., et al. (1986). *The Decline of the Gastropod Nucella Lapillus Around South-West England: Evidence for the Effect of Tributyltin From Antifouling Paints.* Journal of the Marine Biological Association of the United Kingdom. 66(3), 611–640.

56. Gibbs, P. E., Bryan, G. W. (1986). *Reproductive Failure in Populations of the Dog-Whelk, Nucella Lapillus, Caused by Imposex Induced byTributyltin from Antifouling Paints.* Journal of the Marine Biological Association of the United Kingdom. 66(4), 767–777.

57. Evans, S. M., Leksono, T., McKinnell, P. D. (1995). *Tributyltin Pollution: A Diminishing Problem Following Legislation Limiting the Use of TBT-Based Anti-Fouling Paints.* Marine Pollution Bulletin. 30(1), 14–21.

58. Zoeteman, B. C. J., Greef., E. De., Brinkmann, F. J. J. (1981). *Persistency of Organic Contaminants in Groundwater, Lessons from Soil Pollution Incidents in the Netherlands.* Science of the Total Environment. 21, 187–202.

59. Carmo, M., et al. (2005). *Alternative Supports for the Preparation ofCatalysts for Low-Temperature Fuel Cells: the Use of CarbonNanotubes.* Journal of Power Sources. 142(1), 169–176.

60. Park, D. E., et al. (2007). *Micromachined Methanol SteamReforming System as a Hydrogen Supplier for Portable Proton Exchange Membrane Fuel Cells.* Sensors and Actuators A: Physical. 135(1), 58–66.

61. Somorjai, G. A., Philippot, K., Chaudret, B. (2012). *Nanomaterials in Catalysis*: John Wiley & Sons.

62. Bond, G. C., Webb, G. (1989). *Catalysis.* Vol. 8, Royal Society of Chemistry. 203.

63. Rudnick, L. R., V*Lubricant Additives: Chemistry and Applications, Second Edition.* 2 ed: CRC Press. 790.

64. León, B., Jansen, J. (2009). *Thin Calcium Phosphate Coatings for Medical Implants*: Springer.

65. Mändl, S., Rauschenbach, B. (2002). *Improving the Biocompatibility of Medical Implants with Plasma Immersion Ion Implantation.* Surface and Coatings Technology. 156(1–3), 276–283.

66. Shellock, F. G. (2002). *Biomedical Implants and Devices: Assessment of Magnetic Field Interactions with a 3.0-Tesla MR System.* Journal of Magnetic Resonance Imaging. 16(6), 721–732.

67. Roy, R. V., *Aids in Hydrothermal Experimentation: II, Methods of Making Mixtures for Both "Dry" and "Wet" Phase Equilibrium Studies.* Journal of the American Ceramic Society. 39(4), 145–146.
68. DeVries, R. C., Roy, R., Osborn, E. F. (1955). *Phase Equilibria in the System CaO-TiO2–SiO2.* Journal of the American Ceramic Society. 38(5), 158–171.
69. Koch, C. C. (2006). *Nanostructured Materials: Processing, Properties and Applications*: William Andrew.
70. Nordlund, J. J., Halder, R. M., (2006). *Sunscreens and Cosmetics.* The Pigmentary System: Physiology and Pathophysiology, Second Edition: 1188–1190.
71. Jordan, E. H., et al. (2001). *Fabrication and Evaluation of Plasma Sprayed Nanostructured Alumina–Titania Coatings with Superior Properties.* Materials Science and Engineering: A. 301(1), 80–89.
72. Gell, M., et al., *Development and Implementation of Plasma Sprayed Nanostructured Ceramic Coatings.* Surface and Coatings Technology. 146, 48–54.

CHAPTER 4

MODELING AND SIMULATION

CONTENTS

Abstract 102
4.1   Modeling and Simulation Principles 102
4.2   Models 105
4.3   Simulation 108
4.4   Conclusion 128
Keywords 129
References 129

ABSTRACT

A group of interacting, interrelated, or interdependent elements form a system. The first step in studying a system is building a model. A model is a representation of an object, a system, or an idea in some form other than that of the entity itself. To build a model, important factors that act on the system must be included and unimportant factors that only make the model harder to build, understand, and solve should be omitted. Proper balance between accuracy and complexity must be achieved and a model must be both. Simulation is a numerical technique for conducting experiments on a digital computer, which involves certain types of mathematical and logical models that describes the behavior of a business or economic system (or some component thereof) over extended periods of real time.

4.1 MODELING AND SIMULATION PRINCIPLES

4.1.1 SYSTEMS

4.1.1.1 WHAT IS A SYSTEM?

A system is a set of components, which are related by some form of interaction, which act together to achieve some objective or purpose. In a system, components are the individual parts or elements that collectively make up the system and relationships are the cause-effect dependencies between components. An objective is the desired state or outcome, which the system is attempting to achieve.

4.1.1.2 STUDY A SYSTEM

A detailed study to determine whether, to what extent, and how automatic data-processing equipment should be used. It usually includes an analysis of the existing system and the design of the new system, including the development of system specification, which provides a basis for the selection of the equipment. There are some common steps for studying the behavior of a system (Fig. 4.1):

1. observe the behavior of a system;
2. formulate a hypothesis about system behavior;
3. design and carry out experiments to prove or disprove the validity of the hypothesis;
4. often a model of the system is used;
5. measure/estimate performance;

6. improve operation;
7. prepare for failures.

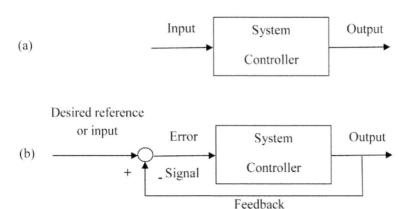

FIGURE 4.1   A schematic of a system.

4.1.1.3   AN EXAMPLE OF THE SYSTEM COMPONENTS DETERMINATION

A set of interacting components or entities operating is collected together to achieve a common goal or objective. For example in a manufacturing system its components are the machine centers, inventories, conveyor belts, production schedule and items produced and a telecommunication system is made of the messages, communication network servers.

4.1.1.4   VARIOUS TYPES OF SYSTEMS

Systems can be classified in a variety of ways. There are natural and artificial systems, adaptive and nonadaptive systems. An adaptive system reacts to changes in its environment, whereas a nonadaptive system does not. Analysis of an adaptive system requires a description of how the environment induces a change of state. Now, some various types of system are reviewed, below:

4.1.1.4.1   NATURAL VS. ARTIFICIAL SYSTEMS

A natural system exists as a result of processes occurring in the natural world (e.g., river, universe) and an artificial system owes its origin to human activity (e.g., space shuttle, automobile).

4.1.1.4.2 STATIC VS. DYNAMIC SYSTEMS

A static system has structure but no associated activity (e.g., bridge, building) and a dynamic system involves time-varying behavior for complex systems (e.g., machine, U.S. economy). It deals with internal feedback loops and time delays that affect the behavior of the entire system [1].

4.1.1.4.3 OPEN-LOOP VS. CLOSED-LOOP SYSTEMS

In all systems there will be an input and an output. Inputs are variables that influence the behavior of the system and outputs are variables, which determined by the system and may influence the surrounding environment. Signals flow from the input through the system and product an output.

An open-loop system cannot control or adjust its own performance but a closed-loop system controls and adjusts its own performance in response to outputs generated by the system through feedback (Fig. 4.2). Feedback is the system function that obtains data on system performance (outputs), compares the actual performance to the desired performance (a standard or criterion), and determines the corrective action necessary. The controller acts on the error signal and uses the information to product the signal that actually affects the system, which we are trying to control.

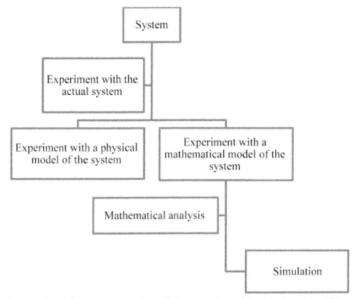

FIGURE 4.2   Graphical representation of the open-loop (a) and close-loop (b) systems.

We consider both internal and external relationships. The internal relationships connect the elements within the system, while the external relationships connect the elements with the environment, that is, with the world outside the system [2, 3].

The system is influenced by the environment through the input it receives from the environment. When a system has the capability of reacting to changes in its own state, we say that the system contains feedback. A nonfeedback, or open-loop, system lacks this characteristic.

The attributes of the system elements define its state. If the behavior of the elements cannot be predicted exactly, it is useful to take random observations from the probability distributions and to average the performance of the objective. We say that a system is in equilibrium or in the steady state if the probability of being in some state does not vary in time. There are still actions in the system, that is, the system can still move from one state to another, but the probabilities of its moving from one state to another are fixed. These fixed probabilities are limiting probabilities that are realized after a long period of time, and they are independent of the state in which the system started. A system is called stable if it returns to the steady state after an external shock in the system. If the system is not in the steady state, it is in a transient state [2].

4.2 MODELS

The first step in studying a system is building a model. A model is a representation of an object, a system, or an idea in some form other than that of the entity itself. A crucial step in building the model is constructing the objective function, which is a mathematical function of the decision variables. For the complex systems a model describes the behavior of systems by using the construct theories or hypotheses, which could be accounted for the observed behavior. So the model can predict future behavior and the effects that will be produced by changes in the system due to the analysis of the proposed systems.

4.2.1 MODELING APPROACH

Computational models, which also called simulation models, used to design new systems study and improve the behavior of existing systems. They allow the use of an interactive design methodology (sometimes called computational steering) so used in most branches of science and engineering (Fig. 4.3).

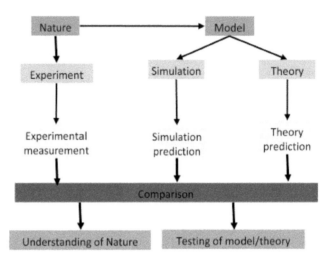

FIGURE 4.3   Modeling and experimental.

There are many types of models such as:

1. A scale model of the real system, for example, a model aircraft in a wind tunnel or a model railway.

2. A physical model in different physical system to the real one, for example, colored water in tubes has been used to simulate the flow of coal in a mine. More common in the use of electrical circuits – analog computers are based on this idea.

3. Mathematical Model: A description of a system where the relationship between variables of the system are expressed in a mathematical form

4. Deterministic vs. stochastic models: In deterministic models, the input and output variables are not subject to random fluctuations, so that the system is at any time entirely defined by the initial conditions and in stochastic models, at least one of the input or output variables is probabilistic or involves randomness

Among the models mathematical models are the most applicable models, which have many advantages such as:

• Enable the investigators to organize their theoretical beliefs and empirical observations about a system and to deduce the logical implications of this organization.

• Lead to improved system understanding.

• Bring into perspective the need for detail and relevance.

• Expedite the analysis.

- Provide a framework for testing the desirability of system modifications
- Allow for easier manipulation than the system itself permits.
- Permit control over more sources of variation than direct study of a
- An additional advantage is that a mathematical model describes a problem more concisely than, for instance, a verbal description does.

To build a model, important factors that act on the system must be included and unimportant factors that only make the model harder to build, understand, and solve should be omitted. For a continuous model, a set of equations that describe the behavior of a system as a continuous function of time t are written. Models use statistical approximations for systems that cannot be modeled using precise mathematical equations. While building a model it must be taken to ensure that it remains a valid representation of the problem. In order to get this purpose, a scientific model necessarily embodies elements of two conflicting attributes-realism and simplicity.

4.2.2 COMPUTATIONAL MODELS, ACCURACY, AND ERRORS

Proper balance between accuracy and complexity must be achieved and a model must be both. An accurate representation of the physical system must be Simple enough to implement as a program and solve on a computer in a reasonable amount of time. On the one hand, the model should serve as a reasonably close approximation to the real system and incorporate most of the important aspects of the system. On the other hand, the model must not be so complex that it is impossible to understand and manipulate. Adding details to the model makes the solution more difficult and converts the method for solving a problem from an analytical to an approximate numerical one [4].

In addition, it is not even necessary for the model to approximate the system to indicate the measure of effectiveness for all various alternatives. It needs to be a high correlation between the prediction by the model and what would actually happen with the real system. To ascertain whether this requirement is satisfied or not, it is important to test and establish control over the solution.

Usually, we begin testing the model by reexamining the formulation of the problem and revealing possible flaws. Another criterion for judging the validity of the model is determining whether all mathematical expressions are dimensionally consistent. A third useful test consists of varying input parameters and checking that the output from the model behaves in a plausible manner. The fourth test is the so-called retrospective test. It involves using historical data to reconstruct the past and then determining how well the resulting solution would have performed if it had been used. Comparing the effectiveness of this hypothetical performance with what actually happened then indicates how well the model predicts the reality. However, a disadvantage of retrospective testing is that it uses the same data that guided formulation of the model. Unless the past is a true replica of the future, it is better not to resort to this test at all.

Suppose that the conditions under which the model was built change. In this case the model must be modified and control over the solution must be established. Often, it is desirable to identify the critical input parameters of the model, that is, those parameters subject to changes that would affect the solution, and to establish systematic procedures to control them. This can be done by sensitivity analysis, in which the respective parameters are varied over their ranges to determine the degree of variation in the solution of the model [4].

After constructing a mathematical model for the problem under consideration, the next step is to derive a solution from this model. There are analytic and numerical solution methods. An analytic solution is usually obtained directly from its mathematical representation in the form of formula.

A numerical solution is generally an approximate solution obtained as a result of substitution of numerical values for the variables and parameters of the model. Many numerical methods are iterative, that is, each successive step in the solution uses the results from the previous step [5].

4.3   SIMULATION

The process of conducting experiments on a model of a system in lieu of either (i) direct experimentation with the system itself, or (ii) direct analytical solution of some problem associated with the system. A simulation of a system is the operation of a model of the system, as an imitation of the real system. A tool to evaluate the performance of a system, existing or proposed, under different configurations of interest and over a long period of time so a simulation of an industrial process is to learn about its behavior under different operating conditions in order to improve the process. Simulation is indeed an invaluable and very versatile tool in those problems where analytic techniques are inadequate. Simulation is a numerical technique for conducting experiments on a digital computer, which involves certain types of mathematical and logical models that describes the behavior of a business or economic system (or some component thereof) over extended periods of real time. Simulation does not require that a model be presented in a particular format. It permits a considerable degree of freedom so that a model can bear a close correspondence to the system being studied. The results obtained from simulation are much the same as observations or measurements that might have been made on the system itself.

4.3.1 REASONS FOR SIMULATION

Simulation allows experimentation, although computer simulation requires long programs of some complexity and is time consuming. Yet what are the other options? The answer is direct experimentation or a mathematical model. Direct experimentation is costly and time consuming, yet computer simulation can be replicated taking into account the safety and legality issues. On the other hand, one can use mathematical models yet mathematical models cannot cope with dynamic effects. Also, computer simulation can sample from nonstandard probability distribution. One can summarize the advantages of computer simulation [6, 7]:

Simulation, first, allows the user to experiment with different scenarios and, therefore, helps the modeler to build and make the correct choices without worrying about the cost of experimentations. The second reason for using simulation is the time control. The modeler or researcher can expand and compress time, just like pressing a fast forward button on life. The third reason is like the rewind button: seeing a scene over and over will definitely shed light on the answer of the question, "why did this happen?"

The fourth reason is "exploring the possibilities." Considering that the package user would be able to witness the consequences of his/her actions on a computer monitor and, as such, avoid jeopardizing the cost of using the real system; therefore, the user will be able to take risks when trying new things and diving in the decision pool with no fears hanging over her/his neck.

As in chess, the winner is the one who can visualize more moves and scenarios before the opponent does. In business the same idea holds. Making decisions on impulse can be very dangerous, yet if the idea is envisaged on a computer monitor then no harm is really done, and the problem is diagnosed before it even happens. Diagnosing problems is the fifth reason why people need to simulate.

Likewise, the sixth reason tackles the same aspect of identifying constraints and predicting obstacles that may arise, and is considered as one major factor why businesses buy simulation software. The seventh reason addresses the fact that many times decisions are made based on "someone's thought" rather than what is really happening.

When studying some simulation packages, the model can be viewed in 3-D. This animation allows the user "to detect design flaws within systems that appear credible when seen on paper or in a 2-D CAD drawing." The ninth incentive for simulation is to "visualize the plan."

It is much easier and more cost effective to make a decision based on predictable and distinguished facts. Yet, it is a known fact that such luxury is scarce in the business world [6].

Nevertheless, before trying out the "what if" scenario many would rather have the safety net beneath them. Therefore, simulation is used for "preparing for change" [7].

In addition, the 13th reason is evidently trying different scenarios on a simulated environment; proving to be less expensive, as well as less disturbing, than trying the idea in real life. Therefore, simulation software does save money and effort, which denotes a wise investment. In any field listing the requirements can be of tremendous effort, for the simple reason that there are so many of them. As such, the 14th reason crystallizes in avoiding overlooked requirements and imagining the whole scene, or the trouble of having to carry a notepad to write on it when remembering a forgotten requirement. While these recited advantages are of great significance, yet many disadvantages still show their effect, which are also summarized. The first hardship faced in the simulation industry is [6–8]:

4.3.2   DANGERS OF SIMULATION

Becoming too enthusiastic about a model and forget about the experimental frame. Force reality into the constraints of a model and forget the model's level of accuracy. Also it should not be forgotten that all models have simplifying assumptions. If two modelers work together and cannot agree on a model, which can be due to the human nature, a consequence of it is the difficulty of interpreting the results of the simulation. It is the simple fact that simulation is not the solution for all problems. Hence, certain types of problems can be solved using mathematical models and equations.

Simulation may be used inappropriately - Simulation is used in some cases when an analytical solution is possible, or even preferable. This is particularly true in the case of small queuing systems and some probabilistic inventory systems, for which closed form models (equations) are available.

Although of all dangerous of simulation, recent advances in simulation methodologies, availability of software, and technical developments have made simulation one of the most widely used and accepted tools in system analysis and operations research.

4.3.3 MODEL TRAINING

Simulation models can provide excellent training when designed for that purpose. Used in this manner, the team provides decision inputs to the simulation model as it progresses. The team, and individual members of the team, can learn from their mistakes, and learn to operate better. Moreover, training any team using a simulated environment is less expensive than real life. Some of model building requires special training. Model building is an art that is learned over time and through experience. Furthermore, if two models of the same system are constructed by two competent individuals, they may have similarities, but it is highly unlikely that they will be identical.

4.3.4 SIMULATION APPROACHES

There are four significant simulation approaches or methods used by the simulation community:
   a) Process interaction approach.
   b) Event scheduling approach.
   c) Activity scanning approach.
   d) Three-phase approach.

There are other simulation methods, such as transactional-flow approach, that are known among the simulation packages and used by simulation packages like Pro Model, Arena, Extend, and Witness. Another, known method used specially with continuous models is stock and flow method. A brief description is given in following paragraphs, yet an explicit description is better read in detail:

4.3.4.1 PROCESS INTERACTION APPROACH

The simulation structure that has the greatest intuitive appeal is the process interaction method. In this method, the computer program emulates the flow of an object (for example, a load) through the system. The load moves as far as possible in the system until it is delayed, enters an activity, or exits from the system. When the load's movement is halted, the clock advances to the time of the next movement of any load.

This flow, or movement, describes in sequence all of the states that the object can attain in the system. In a model of a self-service laundry, for example, a customer may enter the system, wait for a washing machine to become available, wash his or her clothes in the washing machine, wait for a basket to become available, unload the washing machine, transport the clothes in the basket to a dryer, wait for a dryer to become available, unload the clothes into a dryer, dry the clothes, unload the dryer, and then leave the laundry. Each state and event is simulated. Process interaction approach is used by many commercial packages [9].

4.3.4.2   TRANSACTION FLOW APPROACH

Transaction flow approach was first introduced by GPSS in 1962. Transaction flow is a simpler version of process interaction approach, as the following clearly states: "Gordon's transaction flow world-view was a cleverly disguised form of process interaction that put the process interaction approach within the grasp of ordinary users." In transaction flow approach models consist of entities (units of traffic), resources (elements that service entities), and control elements (elements that determine the states of the entities and resources). Discrete simulators, which are generally, designed for simulating detailed processes, such as call centers, factory operations, and shipping facilities, rely on such approach [9].

4.3.4.3   EVENT SCHEDULING APPROACH

The basic concept of the event scheduling method is to advance time to the moment when something happens next (that is, when one event ends, time is advanced to the time of the next scheduled event). An event usually releases a resource. The event then reallocates available objects or entities by scheduling activities, in which they can now participate. For example, in the self-service laundry, if a customer's washing is finished and there is a basket available, the basket could be allocated immediately to the customer, who would then begin unloading the washing machine. Time is advanced to the next scheduled event (usually the end of an activity) and activities are examined to see whether any can now start as a consequence. Event scheduling approach has one advantage and one disadvantage as: "The advantage was that it required no specialized language or operating system support. Event-based simulations could be implemented in procedural languages of even modest capabilities." While the disadvantage "of the event-based approach was that describing a system as a collection of events obscured any sense of process flow." As such, "In complex systems, the number of events grew to a point that following the behavior of an element flowing through the system became very difficult" [9].

4.3.4.4 ACTIVITY SCANNING APPROACH

Another simulation modeling structure is activity scanning. Activity scanning is also known as the two-phase approach. Activity scanning produces a simulation program composed of independent modules waiting to be executed. In the first phase, a fixed amount of time is advanced, or scanned. In phase two, the system is updated (if an event occurs). Activity scanning is similar to rule-based programming (if the specified condition is met, then a rule is executed) [9].

4.3.5 SIMULATION ASPECTS

Three aspects need to be considered when planning a computer simulation project:
- a. time-flow handling;
- b. behavior of the system;
- c. change handling.

The flow of time in a simulation can be handled in two manners: the first is to move forward in equal time intervals. Such an approach is called time-slice. The second approach is next-event which increments time in variable amounts or moves the time from state to state. On one hand, there is less information to keep in the time-slice approach. On the other hand, the next-event approach avoids the extra checking and is more general [10].

The behavior of the system can be deterministic or stochastic: deterministic system, of which its behavior would be entirely predictable, whereas, stochastic system, of which its behavior cannot be predicted but some statement can be made about how likely certain events are to occur.

The change in the system can be discrete or continuous. Variables in the model can be thought of as changing values in four ways [10]:
1. Continuously at any point of time: thus, change smoothly and values of variables at any point of time.
2. Continuously changing but only at discrete time events: values change smoothly but values accessible at predetermined time.
3. Discretely changing at any point of time: state changes are easily identified but occur at any time.
4. Discretely changing at any point of time: state changes can only occur at specified point of time.

Others define 3 and 4 as discrete event simulation as follows: "a discrete-event simulation is one in which the state of a model changes at only a discrete, but possibly random, set of simulated time points." Mixed or hybrid systems with both discrete and continuous change do exist. Actually simulation packages try to include both.

In the natural sciences one makes to model the complex processes occurring in nature as accurately as possible. The first step in this direction is the description of nature. It serves to develop an appropriate system of concepts. However, in most cases, only observation is not enough to find the underlying principles. Most processes are too complex and cannot be clearly separated from other processes that interact with them. Instead, if it is possible, the scientist creates the conditions under the process is to be observed in an experiment. This method allows discovering how the observed event depends on the chosen conditions and allows inferences about the principles underlying the behavior of the observed system. The goal is the mathematical formulation of the underlying principles by using a theory of the phenomena under investigation and describes how certain variables behave independence of each other and how they change under certain conditions over time. This is mostly done by means of differential and integral equations. The resulting equations, which encode the description of the system or process, are referred to as a mathematical model.

A model that has been confirmed does not only permit the precise description of the observed processes, but also allows the prediction of the results of similar physical processes within certain bounds. Thereby, experimentation, the discovery of underlying principles from the results of measurements, and the translation of those principles into mathematical variables and equations go hand in hand. Theoretical and experimental approaches are therefore most intimately connected. The phenomena that can be investigated in this way in physics and chemistry extend over very different orders of magnitudes. They can be found from the smallest to the largest observable length scales, from the investigation of matter in quantum mechanics to the study of the shape of the universe. The occurring dimensions range from the nanometer range (10^{-9} meters) in the study of properties of matter on the molecular level to 10^{23} meters in the study of galaxy clusters. Similarly, the time scales that occur in these models (that is, the typical time intervals in which the observed phenomena take place) are vastly different. They range in the mentioned examples from 10^{-12} or even 10^{-15} seconds to 10^{17} seconds, thus from picoseconds or even femtoseconds up to time intervals of several billions of years. The masses occurring in the models are just as different, ranging between 10^{-27} kilograms for single atoms to 10^{40} kilograms for entire galaxies.

The wide range of the described phenomena shows that experiments cannot always be conducted in the desired manner. For example in astrophysics, there are only few possibilities to verify models by observations and experiments and to thereby confirm them, or in the opposite case to reject models, to falsify them. On the other hand, models that describe nature sufficiently well are often so complicated that no analytical solution can be found.

Take for example the case of the Vander Waals equation to describe dense gases or the Boltzmann equation to describe the transport of rarefied gases. Therefore, one usually develops a new and simplified model that is easier to solve. However, the validity of this simplified model is in general more restricted. To derive such models one often uses techniques such as averaging methods, successive approximation methods, matching methods, asymptotic analysis and homogenization. Unfortunately, many important phenomena can only be described with more complicated models. But then these theoretical models can often only be tested and verified in a few simple cases. As an example consider again planetary motion and the gravitational force acting between the planets according to Newton's law. As is known, the orbits following from Newton's law can be derived in closed form only for the two-body case. For three bodies, analytical solutions in closed form in general no longer exist. This is also true for our planetary system as well as the stars in our galaxy.

Many models, for example in materials science or in astrophysics, consist of a large number of interacting bodies (called particles), as for example atoms and molecules. In many cases the number of particles can reach several millions or more. But large numbers of particles do not only occur on a microscopic scale. These are some of the reasons why computer simulation has recently emerged as a third way in science besides the experimental and theoretical approach. Over the past years, computer simulation has become an indispensable tool for the investigation and prediction of physical and chemical processes. In this context, computer simulation means the mathematical prediction of technical or physical processes on modern computer systems [11].

The following procedure is typical in this regard: A mathematical-physical model is developed from observation. The derived equations in most cases valid for continuous time and space, are considered at selected discrete points in time and space. For instance, when discretizing in time, the solution of equations is no longer to be computed at all points in time, but is only considered at selected points along the time axis. Differential operators, such as derivatives with respect to time, can then be approximated by difference operators. The solution of the continuous equations is computed approximately at those selected points. The more densely those points are selected, the more accurately the solution can be approximated. Here, the rapid development of computer technology, which has led to an enormous increase in the computing speed and the memory size of computing systems, now allows simulations that are more and more realistic. The results can be interpreted with the help of appropriate visualization techniques. If corresponding results of physical experiments are available, then the results of the computer simulation can be directly compared. This leads to a verification of the results of the computer simulation or to an improvement in the applied methods or the model (for instance by appropriate changes of parameters of the model or by changing the used equations).

Altogether, for a computer experiment, one needs a mathematical model. But the solutions are now obtained approximately by computations, which are carried out by a program on a computer. This allows studying models that are significantly more complex and therefore more realistic than those accessible by analytical means. Furthermore, this allows avoiding costly experimental setups. In addition, situations can be considered that otherwise could not be realized because of technical shortcomings or because they are made impossible by their consequences. For instance, this is the case if it is hard or impossible to create the necessary conditions in the laboratory, if measurements can only be conducted under great difficulties or not at all, if experiments would take too long or would run too fast to be observable, or if the results would be difficult to interpret. In this way, computer simulation makes it possible to study phenomena not accessible before by experiment. If a reliable mathematical model is available that describes the situation at hand accurately enough, it does in general not make a difference for the computer experiment. Obviously this is different, if the experiment would actually have to be carried out in reality. Moreover, the parameters of the experiment can easily be changed. And the behavior of solutions of the mathematical model with respect to such parameters changes can be studied with relatively little effort [11].

In nanotechnology numerical simulation can help to predict properties of new materials that do not yet exist in reality (Fig. 4.4). And it can help to identify the most promising or suitable materials. The trend is towards virtual laboratories in which materials are designed and studied on a computer. Moreover, simulation offers the possibility to determine mean or average properties for the material macroscopic characterization. At whole it can be said, computer experiments act as a link between laboratory experiments and mathematical-physical theory.

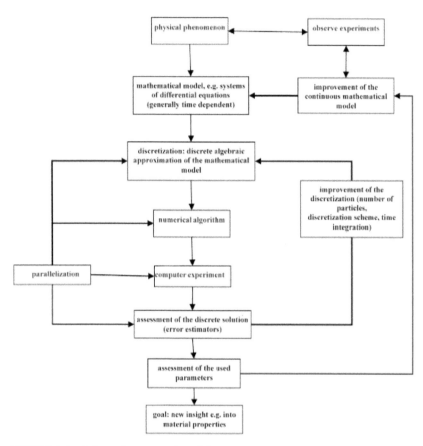

FIGURE 4.4 Schematic presentation of the typical approach for numerical simulation.

Each of the partial steps of a computer experiment must satisfy a number of requirements. First and foremost, the mathematical model should describe reality as accurately as possible. In general, certain compromises between accuracy in the numerical solution and complexity of the mathematical model have to be accepted. In most cases, the complexity of the models leads to enormous memory and computing time requirements, especially if time-dependent phenomena are studied. Depending on the formulation of the discrete problem, several nested loops have to be executed for the time dependency, for the application of operators, or also for the treatment of nonlinearities.

Current researches therefore have focus in particular on the development of methods and algorithms that allow to compute the solutions of the discrete problem as fast as possible (multilevel and multiscale methods, multiple methods, fast Fourier transforms) and that can approximate the solution of the continuous problem with as little memory as possible. More realistic and therefore in general more complex models require faster and more powerful algorithms. Another possibility to run larger problems is the use of vector computers and parallel computers. Vector computers increase their performance by processing similar arithmetical instructions on data stored in a vector in an assembly line-like fashion. In parallel computers, several dozen in many thousands of powerful processors (The processors in use today have mostly a RISC (reduced instruction set computer) processor architecture. They have fewer machine instructions compared to older processors, allowing a faster, assembly line-like execution of the instructions which are assembled into one computing system [12]. These processors can compute concurrently and independently and can communicate with each other to improve portability of programs among parallel computers from different manufacturers and to simplify the assembly of computers of various types to a parallel computer, which has uniform standards for data exchange between computers. A reduction of the required computing time for a simulation is achieved by distributing the necessary computations to several processors. Up to a certain degree, the computations can then be executed concurrently. In addition, parallel computer systems in general have a substantially larger main memory than sequential computers. Hence, larger problems can be treated [11].

Figure 4.5 shows the development of the processing speed of high performance. The performances in flop/s is plotted versus the year, for the fastest parallel computer in the world, and for the computers at position 100 and 500 in the list of the fastest parallel computers in the world. Personal computers and workstations have seen a similar development of their processing speed. Because of that, satisfactory simulations have become possible on these smaller computers [11].

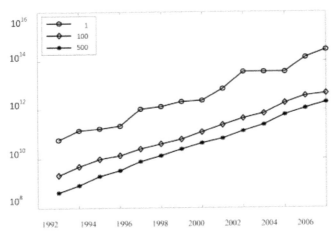

FIGURE 4.5 Development of processing speed over the last years (parallel Linpack benchmark); fastest (1), 100th fastest (100) and 500th fastest (500) computer in the world; up to now the processing speed increases tenfold about every four years.

4.3.6 PARTICLE MODELS

An important area of numerical simulation deals with so-called particle models. These are simulation models in which the representation of the physical system consists of discrete particles and their interactions. For instance, systems of classical mechanics can be described by the positions, velocities, and the forces acting between the particles. In this case the particles do not have to be very small either with respect to their dimensions or with respect to their mass, as possibly suggested by their name. Rather they are regarded as fundamental building blocks of an abstract model. For this reason the particles can represent atoms or molecules. The particles carry properties of physical objects, as for example mass, position, velocity or charge. The state and the evolution of the physical system are represented by these properties of the particles and by their interactions, respectively [13].

Figure 4.6 shows the simulation result of the large-scale structure formation of the universe. The model consists of 32,768 particles that each represents several hundred galaxies. Figure 4.7 shows a protein, which consists of 12,386 particles representing single atoms. The laws of classical mechanics are used in many particle models. The use of Newton's second law results in a system of ordinary differential equations of second order describing how the acceleration of any particle depends on the force acting on it. The force results from the interaction with the other particles and depends on their position. If the positions of the particles change relative to each other, then in general also the forces between the particles change. The solution of the system of ordinary differential equations for given initial values then leads to the trajectories of the particles. This is a deterministic procedure, meaning that the trajectories of the particles are in principle uniquely determined for all times by the given initial values.

But why is it reasonable at all to use the laws of classical mechanics when at least for atomic models the laws of quantum mechanics should be used? Should not Schrodinger's equation be employed as equation of motion instead of Newton's laws? And what does the expression "interaction between particles" actually mean, exactly?

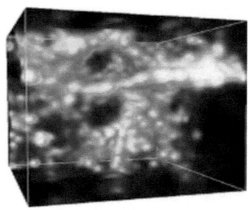

FIGURE 4.6   Result of a particle simulation of the large-scale structure of the universe.

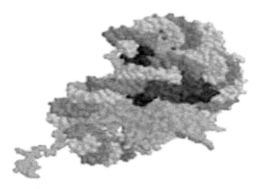

FIGURE 4.7   A view of a protein.

If one considers a system of interacting atoms, which consists of nuclei and electrons, one can in principle determine its behavior by solving the Schrodinger equation with the appropriate Hamilton operator. However, an analytic or even numerical solution of the Schrodinger equation is only possible in a few simple special cases. Therefore, approximations have to be made. The most prominent approach is the Born-Oppenheimer approximation. It allows a separation of the equations of motions of the nuclei and of the electrons. The intuition behind this approximation is that the significantly smaller mass of the electrons permits them to adapt to the new position of the nuclei almost instantaneously. The Schrodinger equation for the nuclei is therefore replaced by Newton's law. The nuclei are then moved according to classical mechanics, but using potentials that result from the solution of the Schrodinger equation for the electrons. For the solution of this electronic Schrodinger equation approximations have to be employed. Such approximations are for instance derived with the Hartree-Fock approach or with density functional theory. This approach is known as ab initio molecular dynamics. However, the complexity of the model and the resulting algorithms enforces a restriction of the system size to a few thousand atoms [13].

A further simplification is the use of parameterized analytical potentials that just depend on the position of the nuclei (classical molecular dynamics). The potential function itself is then determined by fitting it to the results of quantum mechanical electronic structure computations for a few representative model configurations and subsequent force matching or by fitting to experimentally measured data. The use of these very crude approximations to the electronic potential hyper-surface allows the treatment of systems with many millions of atoms. However, in this approach quantum mechanical effects are lost to a large extent.

4.3.7   PHYSICAL SYSTEMS FOR PARTICLE MODELS

The following list gives some examples of physical systems that can be represented by particle systems in a meaningful way. They are therefore amenable to simulation by particle methods:

Solid State Physics: The simulation of materials on an atomic scale is primarily used in the analysis of known materials and in the development of new materials. Examples for phenomena studied in solid state physics are the structure conversion in metals induced by temperature or shock, the formation of cracks initiated by pressure, shear stresses, etc. in fracture experiments, the propagation of sound waves in materials, the impact of defects in the structure of materials on their load-bearing capacity and the analysis of plastic and elastic deformations.

Fluid Dynamics: Particle simulation can serve as a new approach in the study of hydro-dynamical instabilities on the microscopic scale, as for instance, the Rayleigh-Taylor or Rayleigh-Benard instability. Furthermore, molecular dynamics simulations allow the investigation of complex fluids and fluid mixtures, as for example emulsions of oil and water, but also of crystallization and of phase transitions on the microscopic level.

Biochemistry: The dynamics of macromolecules on the atomic level is one of the most prominent applications of particle methods. With such methods it is possible to simulate molecular fluids, crystals, amorphous polymers, liquid crystals, zeolites, nuclear acids, proteins, membranes and many more biochemical materials.

Astrophysics: In this area, simulations mostly serve to test the soundness of theoretical models. In a simulation of the formation of the large-scale structure of the universe, particles correspond to entire galaxies. In a simulation of galaxies, particles represent several hundred to thousand stars. The force acting between these particles results from the gravitational potential [13].

4.3.8 COMPUTER SIMULATION OF PARTICLE MODELS

In the computer simulation of particle models, the time evolution of a system of interacting particles is determined by the integration of the equations of motion. Here, one can follow individual particles, see how they collide, repel each other, attract each other, how several particles are bound to each other, are binding to each other, or are separating from each other. Distances, angles and similar geometric quantities between several particles can also be computed and observed over time. Such measurements allow the computation of relevant macroscopic variables such as kinetic or potential energy, pressure, diffusion constants, transport coefficient, structure factors, spectral density functions, distribution functions, and many more. In most cases, variables of interest are not computed exactly in computer simulations, but only up to certain accuracy. Because of that, it is desirable to achieve:

- an accuracy as high as possible with a given number of operations;
- a given accuracy with as few operations as possible; or
- a ratio of effort (number of operations) to achieved accuracy, which is as small as possible.

Clearly the last alternative includes the first two as special cases. A good algorithm possesses a ratio of effort (costs, number of operations, necessary memory) to benefit (achieved accuracy) that is as favorable as possible. As a measure for the ranking of algorithms one can use the quotient:

$$\frac{effort}{benfit} \neq \frac{operations}{achieved\ accuracy}$$

This is a number that allows the comparison of different algorithms. If it is known how many operations are minimally needed to achieve certain accuracy, this number shows how far a given algorithm is from optimal. The minimal number of operations to achieve a given accuracy ε is called ε-complexity. The ε-complexity is thus a lower bound for the number of operations for any algorithm to achieve an accuracy of ε [14].

4.3.9 OPTIMIZATION

Its objective is to select the best possible decision for a given set of circumstances without having to enumerate all of the possibilities and involves maximization or minimization as desired. In optimization decision variables are variables in the model, which you have control over. Objective function is a function (mathematical model) that quantifies the quality of a solution in an optimization problem. Constraints must be considered, conditions that a solution to an optimization problem must satisfy and restrict decision variables are determined by defining relationships among them. It must be found the values of the decision variables that maximize (minimize) the objective function value, while staying within the constraints. The objective function and all constraints are linear functions (no squared terms, trigonometric functions, ratios of variables) of the decision variables [15, 16].

4.3.10 SIMULATION MODEL DEVELOPMENT

Step 1. Identify Problem:
- Enumerate problems with an existing system.
- Produce requirements for a proposed system.

Step 2. a) Formulate Problem:
- Define overall objectives of the study and specific issues to be addressed.
- Define performance measures.

b) Quantitative criteria on the basis of which different system configurations will be evaluated and compared:

- Develop a set of working assumptions that will form the basis for model development.
- Model boundary and scope (width of model).
- Determines what is in the model and what is out.
- Level of detail (depth of model).
- Specifies how in-depth one component or entity is modeled.
- Determined by the questions being asked and data availability.
- Decide the time frame of the study.
- Used for one-time or over a period of time on a regular basis.

Step 3. a) Collect and Process Real System Data:
- Collect data on system specifications, input variables, performance of the existing system, etc.
- Identify sources of randomness (stochastic input variables) in the system.
- Select an appropriate input probability distribution for each stochastic input variable and estimate corresponding parameters.

b) Standard distributions.

c) Empirical distributions.

d) Software packages for distribution fitting.

Step 4. a) Formulate and Develop a Model:
- Develop schematics and network diagrams of the system.

b) How do entities flow through the system
- Translate conceptual models to simulation software acceptable form.
- Verify that the simulation model executes as intended.

c) Build the model right (low-level checking)
- Traces.
- Vary input parameters over their acceptable ranges and check the output.

Step 5. a) Validate Model:
- Check whether the model satisfies or fits the intended usage of system (high-level checking).

b) Build the right model
- Compare the model's performance under known conditions with the performance of the real system.
- Perform statistical inference tests and get the model examined by system experts.

- Assess the confidence that the end user places on the model and address problems, if any.

Step 6. Document Model for Future Use:
- Objectives, assumptions, inputs, outputs, etc.

Step 7. a) Select Appropriate Experimental Design:
- Performance measures.
- Input parameters to be varied.
  b) Ranges and legitimate combinations
- Document experiment design.

Step 8. Establish Experimental Conditions for Runs:
- Whether the system is stationary (performance measure does not change over time) or nonstationary (performance measure changes over time).
- Whether a terminating or a nonterminating simulation run is appropriate.
- Starting condition.
- Length of warm-up period.
- Model run length.
- Number of statistical replications.

Step 9. Perform Simulation Runs

Step 10. a) Analyze Data and Present Results:
- Statistics of the performance measure for each configuration of the model
  b) Mean standard deviation, range, confidence intervals, etc.
- Graphical displays of output data
  c) Histograms scatter plot, etc.
- Document results and conclusions

Step 11. Recommend Further Courses of Actions:
- Other performance measures
- Further experiments to increase the precision and reduce the bias of estimators
- Sensitivity analysis
- How sensitive the behavior of the model is to changes of model parameters

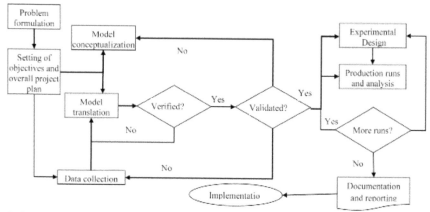

FIGURE 4.8   Steps in a simulation study.

4.3.11 SIMULATION LANGUAGES

SLAM introduced in 1979 by Pritsker and Pegden. SIMAN introduced in 1982 by Pegden, first language to run both a mainframe as well as a micro computer. In primary computers, accessibility and interactions were limited. GASP IV introduced by Pritsker, Triggered a wave of diverse applications, which is significant in the evaluation of simulation.

Many programming systems have been developed, incorporating simulation languages. Some of them are general-purpose in nature, while others are designed for specific types of systems. FORTRAN, ALGOL, and PL/1 are examples of general-purpose languages, while GPSS, SIMSCRIPT, and SIMULA are examples of special simulation languages. Programming can be done in general purpose languages such as Java, simulation languages like SIMAN or use simulation packages, Arena.

Four choices are existed: simulation language, general-purpose language, extension of general purpose, simulation package. Simulation language is built in facilities for time steps, event scheduling, data collection, reporting. General-purpose is known to developer, available on more systems, flexible. The major difference is the cost tradeoff. Simulation language requires startup time to learn, while general purpose may require more time to add simulation flexibility. Recommendation may be for all analysts to learn one simulation language so understand those "costs" and can compare. Extension of general-purpose is collection of routines and tasks commonly used. Often, base language with extra libraries that can be called and Simulation packages allow definition of model in interactive fashion. Get results in one day. Tradeoff is in flexibility, where packages can only do what developer envisioned, but if that is what is needed then is quicker to do so.

Now some of advantages and disadvantages of common languages are listed as:

a) General Simulation Languages: Arena, Extend, GPSS, SIMSCRIPT, SIMULINK (Matlab), etc.
   - Advantages: Standardized features in modeling, shorter development cycle for each model, Very readable code
   - Disadvantages: Higher software cost (up-front), Additional training required, limited portability

b) Special Purpose Simulation Packages: Manufacturing (AutoMod, FACTOR/AIM, etc.), Communications network (COMNET III, NETWORK II.5, etc.), Business (BP$IM, Process Model, etc.), Health care (MedModel).
   - Advantages: Very quick development of complex models, short learning cycle, little programming.
   - Disadvantages: High cost of software, limited scope of applicability, limited flexibility [17–18].

4.3.12 SIMULATION CHECKLIST

You should check the simulation process in each step as:

a) Checks before developing simulation;
   - Is the goal properly specified?
   - Is detail in model appropriate for goal?
   - Does the team include the right mix (leader, modeling, programming, background)?
   - Has sufficient time been planned?

b) Checks during simulation development;
   - Is random number random?
   - Is model reviewed regularly?
   - Is model documented?

c) Checks after simulation is running;
   - Is simulation length appropriate?
   - Are initial transients removed?
   - Has the model been verified?
   - Has the model been validated?
   - Are there any surprising results? If yes, have they been validated?

4.3.13 COMMON MISTAKES IN SIMULATION

In appropriate level of detail: Level of detail often potentially unlimited but more detail requires more time to develop and often to run. It can be introduced more bugs, making more inaccurate not less. Often, more detailed viewed as "better" but may not be the case. So more detail requires more knowledge of input parameters and getting input parameters wrong may lead to more inaccuracy. Therefore, start with less detail, study sensitivities and introduce detail in high impact areas.

Improper language: Choice of language can have significant impact on time to develop, special-purpose languages can make implementation, verification and analysis easier, C^{++}Sim, JavaSim, SimPy (thon).

Unverified models: Simulations generally need large computer programs unless special steps taken, bugs or errors.

Invalid models: Unless no errors occur, the model does not represent real system, you need to validate models by analytic, measurement or intuition.

Improperly handled initial conditions: Often, initial trajectory is not representative of steady state so can lead to inaccurate results. In this case, typically you want to discard, but need a method to do it so effectively.

Too short simulation runs: Attempting to save time makes even more dependent upon initial conditions. Therefore, correct length depends upon the accuracy desired (confidence intervals).

Poor random number generators and seeds: "Home grown" are often not random enough to make artifacts. So the best is to use well-known one and choose seeds that are different [19, 20].

4.4 CONCLUSION

Simulation is the imitation of the operation of a real-world process or system over time. The act of simulating something first requires that a model be developed, this model represents the key characteristics or behaviors/functions of the selected physical or abstract system or process. The model represents the system itself, whereas the simulation represents the operation of the system over time. The behavior of the system can be deterministic or stochastic: deterministic system, of which its behavior would be entirely predictable, whereas, stochastic system, of which its behavior cannot be predicted but some statement can be made about how likely certain events are to occur. The wide range of the described phenomena shows that experiments cannot always be conducted in the desired manner. A model that has been confirmed does not only permit the precise description of the observed processes, but also allows the prediction of the results of similar physical processes within certain bounds. The simulation structure that has the greatest intuitive appeal is the process interaction method. The simulation of materials on an atomic scale is primarily used in the analysis of known materials and in the development of new materials. Key issues in simulation include acquisition of valid source information about the relevant selection of key characteristics and behaviors, the use of simplifying approximations and assumptions within the simulation, and fidelity and validity of the simulation outcomes.

KEYWORDS

- **accuracy**
- **assumptions**
- **optimization**
- **simulation process**
- **system modeling**
- **validation**

REFERENCES

1. Kostami, V., Ward, A. R. (2010). *Analysis and Comparison of Inventory Systems: Dynamic vs Static Policies*: 1–40.
2. Ghosh, (2004). *Control Systems: Theory And Applications*: Pearson Education. 628.
3. Huang, B. K. (1994). *Computer Simulation Analysis of Biological and Agricultural Systems*: Taylor & Francis. 880.
4. Kroese, D. P., Taimre, Z., Botev, I. (2013). *Handbook of Monte Carlo Methods*: Wiley. 768.
5. Rubinstein, R. Y., Kroese, D. P. (2011). *Simulation and the Monte Carlo Method*: Wiley. 372.
6. Banks, J. (*1999). Introduction to Simulation*. in *Simulation Conference Proceedings Winter*(1999):IEEE.
7. Banks, J. (*2000). Introduction to Simulation*. in *Simulation Conference. Proceedings. Winter*(2000) : IEEE.
8. Drappa, A., Ludewig, J. *Simulation in Software Engineering Training*. in *Proceedings of the 22nd international conference on Software engineering* (2000), ACM.
9. Abu-Taieh, E. M. O. (2008). *Computer Simulation Using Excel without Programming*: Universal Publishers.
10. Pidd, M., Cassel, R. A. (1998). *Three Phase Simulation in Java*. in *Proceedings of the 30th conference on Winter simulation*: IEEE Computer Society Press.
11. Griebel, M., Knapek, S., Zumbusch, G. (2010). *Numerical Simulation in Molecular Dynamics: Numerics, Algorithms, Parallelization, Applications*: Springer.

12. Peter, C., Daura, X., van Gunsteren, W. F. (2000). *Peptides of Aminoxy Acids: A Molecular Dynamics Simulation Study of Conformational Equilibria under Various Conditions.* Journal of the American Chemical Society. **122**(31), 7461–7466.
13. Sheikh, A. A. R. E., Ajeeli, A. A., Abu-Taieh, E. M. O. (2008). *Simulation and Modeling: Current Technologies and Applications*: IGI Global Snippet.
14. Traub, J., Wasilkowski, G., Wozniakowski, H. (1988). *Information-Based Complexity,* Academic Press, New York.
15. Law, A. M., McComas, M. G. (2002). *Simulation Optimization: Simulation-Based Optimization.* in *Proceedings of the 34th conference on Winter simulation: exploring new frontiers*: Winter Simulation Conference.
16. Banks, J. (1998). *Handbook of Simulation: Principles, Methodology, Advances, Applications, and Practice*: John Wiley & Sons. 849.
17. Gray, M. A. (2007). *Discrete Event Simulation: A Review of SimEvents.* Computing in Science and Engineering. **9**(6), 62–66.
18. Huntsinger, R. C. (2003). *Simulation Languages and Applications*, in *Modeling and Simulation: Theory and Practice*, Springer. 145–154.
19. Jain, R. (1991). *The Art of Computer Systems Performance Analysis: Techniques for Experimental Design, Measurement, Simulation, and Modeling*: Wiley. 685.
20. Yilmaz, L., Ören, T. (2009). *Agent-Directed Simulation and Systems Engineering*: John Wiley & Sons. 600.

CHAPTER 5

MOLECULAR SIMULATION FOR NANOMATERIALS

CONTENTS

Abstract 132
5.1 Outline of Molecular Simulation and Microsimulation Methods 132
5.2 Molecular Dynamics Method 133
5.3 Monte Carlo Method 144
5.4 Brownian Dynamics Method 149
5.5 Dissipative Particle Dynamics Method 153
5.6 Lattice Boltzmann Method 159
5.7 Conclusion 163
Keywords 164
References 164

ABSTRACT

Molecular Simulation covers all aspects of research related to, or of importance to, molecular modeling and simulation (including informatics, theoretical and experimental work). Molecular Simulation exists to bring together the most significant papers concerned with applications of simulation methods, and original contributions to the development of simulation methodology from biology and biochemistry, chemistry, chemical engineering, materials and nanomaterials, medicine, physics and information science. Molecular Simulation is of interest to all researchers using or developing simulation methods (Monte Carlo and molecular dynamics methods, Brownian dynamics method, dissipative particle dynamics method and Lattice Boltzmann methods).

5.1   OUTLINE OF MOLECULAR SIMULATION AND MICROSIMULATION METHODS

The real progress in nanotechnology, has been due to a series of advances in a variety of supplementary areas, such as: the discoveries of atomically precise materials such as nanotubes and fullerenes, the ability of the scanning probe and the development of manipulation techniques to image and manipulate atomic and molecular configurations in real materials, the conceptualization and demonstration of individual electronic and logic devices with atomic or molecular level materials, the advances in the self-assembly of materials to be able to put together larger functional or integrated systems, and finally the advances in computational nanotechnology, physics and chemistry based modeling and simulation of possible nanomaterials, devices and applications. It turns out that at the nanoscale, devices and systems sizes have condensed sufficiently small, so that, it is possible to describe their behavior justly identically. The simulation technologies have become also predictive in nature, and many pioneer concepts and designs, which have been first proposed, based on modeling and simulations, and then are followed by their realization or verification through experiments. Microscopic analysis methods are needful in order to get new functional materials and study physical phenomena on a molecular level in the nanotechnology science. These methods treat the constituent species of a system, such as molecules and fine particles. Macroscopic and microscopic quantities of interest are derived from analyzing the behavior of these species [1–3]. These approaches, called "molecular simulation methods," are represented by the Monte Carlo (MC) and molecular dynamics (MD) methods. MC methods exhibit a powerful ability to analyze thermodynamic equilibrium, but are unsuitable for investigating dynamic phenomena. MD methods are useful for thermodynamic equilibrium but are more advantageous for investigating the dynamic properties of a system in a nonequilibrium situation [4, 5]. This chapter inspects MD and MC methods of nonspherical particle dispersion in a three-dimensional system and also pretexts Brownian dynamics (BD) methods that can simulate the Brownian motion of dispersed particles and dissipative particle dynamics (DPD) and at last Lattice Boltzmann methods, in which a liquid system is regarded as composed of virtual fluid particles.

5.2   MOLECULAR DYNAMICS METHOD

Spherical particle dispersion can be treated straight-forwardly in simulations because only the translational motion of particles is important, and the treatment of the rotational motion is basically unnecessary. MD simulations become much more complicated, if the translational and rotational motion has to be simulated for axisymmetric particle dispersion in comparison with the spherical particle system. Simulation techniques for a dispersion composed of nonspherical particles with a general shape may be obtained by generalizing the methods employed to axisymmetric particle dispersion. It is, therefore, very important to understand the MD method for the axisymmetric particle system [6–7].

5.2.1   SPHERICAL PARTICLE SYSTEMS

The concept of the MD method is rather straight forward and logical. Molecular dynamics simulation consists of the numerical, step-by-step, solution of the classical equations of motion which is generally governed by Newton's equations of motion.

In MD simulations, particle motion is simulated on a computer according to the equations of motion. If one molecule moves exclusively on a classical mechanics level, a computer is not essential because mathematical calculation by hand is enough to solve the motion of the molecule. However, since molecules in a real system are numerous and interact with each other, such mathematical analysis is infeasible. In this case, computer simulations become a towering tool for a microscopic analysis.

If the mass of molecule i and the force acting on molecule i by the ambient molecules and an external field are considered, then the motion of a particle is described by Newton's equation of motion:

$$m_i \frac{d^2 r_i}{dt^2} = f_i \qquad (5.1)$$

If a system is composed of N molecules, there are N sets of similar equations, and the motion of N molecules interacts through forces acting among the molecules.

For solving the set of N equations of motion on computer differential equations such as Eq. (5.1) are inappropriate. Computers can easily solve simple equations, such as algebraic ones, but are quite ill-conditioned at intuitive solving procedures such as a trial and error approach to find solutions. Accordingly, Eq. (5.1) will be transformed into an algebraic equation. To do this, the second-order differential term in Eq. (5.1) must be expressed as an algebraic expression, using the following Taylor series expansion:

$$x(t) + hx(t) + h\frac{dx(t)}{dt} + \frac{1}{2!}h^2 \frac{d^2 x(t)}{dt^2} + \frac{1}{3!}h^3 \frac{d^3 x(t)}{dt^3} + .. \qquad (5.2)$$

Eq. (5.2) insinuates that x at time (t+h can be expressed as the sum of x itself, the first-order differential, the second-order differential, and so on, multiplied by a constant for each term. If x does not significantly change with time, the higher-order differential terms can be neglected for a sufficiently small value of the time interval h. In order to approximate the second-order differential term in Eq. (5.1) as an algebraic expression, another form of the Taylor series expansion is needed:

$$x(t - h) = x(t) - h\frac{dx(t)}{dt} + \frac{1}{2!}h^2 \frac{d^2 x(t)}{dt^2} - \frac{1}{3!}h^3 \frac{d^3 x(t)}{dt^3} + .. \qquad (5.3)$$

If the first-order differential term is eliminated from Eqs. (5.2) and (5.3), so the second-order differential term can be solved as:

$$\frac{d^2 x(t)}{dt^2} = \frac{x(t + h) - 2x(t) + x(t - h)}{h^2} + 0(h^2) \qquad (5.4)$$

The last term on the right-hand side of this equation points the accuracy of the approximation, and, in this situation, terms higher than h² are neglected. If the second order differential is approximated as:

$$\frac{d^2 x(t)}{dt^2} = \frac{x(t+h) - 2x(t) + x(t-h)}{h^2} \tag{5.5}$$

This expression is called the "central difference approximation." With this approximation and the notation $r_i = (x_i, y_i, z_i)$ for the molecular position and $f_i = (f_{xi}, f_{yi}, f_{zi})$ for the force acting on particle i, the equation of the x-component of Newton's equation of motion can be written as:

$$x_i(t+h) = 2x_i(t) - x_i(t-h) + \frac{h^2}{2m_i} f_{xi}(t) \tag{5.6}$$

Similar equations are contented for the other components. Since Eq. (5.6) is a simple algebraic equation, the molecular position at the next time step can be evaluated using the present and previous positions and the present force. If a system is composed of N molecules, there are 3N algebraic equations for distinguishing the motion of molecules; these numerous equations are solved on a computer, where the motion of the molecules in a system can be followed up the time variable. Equation (5.6) does not require the velocity terms for determining the molecular position at the next time step. This scheme is called the "Verlet method" [8, 9]. If the velocity be required, then it can be evaluated from the central difference approximation as:

$$V_i(t) = \frac{r_i(t+h) - r_i(t-h)}{2h} \tag{5.7}$$

This approximation can be derived by eliminating the second-order differential terms in Eqs. (5.2) and (5.3). However, it has already been noted that the velocities are unnecessary for evaluating the position at the next time step, a scheme using the positions and velocities simultaneously may be more desirable to keep the system temperature constant. This method would be explained in the following paragraphs.

If it take into account that the first- and second-order differentials of the position are equal to the velocity and acceleration, respectively, with the neglect of differential terms equal to or higher than third-order in Eq. (5.2) directed to the following equation:

$$r_i(t+h) = r_i(t) + hv_i(t) + \frac{h^2}{2m_i} f_i(t) \tag{5.8}$$

This equation determines the position of the molecules, but the velocity term arises on the right-hand side, so that another equation is necessary for specifying the velocity. The first-order differential of the velocity is equal to the acceleration:

$$v_i(t+h) = v_i(t) + \frac{h^2}{m_i} f_i(t) \tag{5.9}$$

In order to improve accuracy, the force term in Eq. (5.9) is somewhat modified and the following equation obtained:

$$v_i(t+h) = v_i(t) + \frac{h}{m_i}(f_i(t) + f_i(t+h)) \tag{5.10}$$

Using of Eqs. (5.8) and (5.10) for determining the motion of molecules is called the "velocity Verlet method" [10, 11]. It is well known that the velocity Verlet method is significantly great in regard to the stability and accuracy of a simulation.

Consider another representative scheme. The application of the central difference approximation to the first-order differentials of the position, the velocity and the acceleration leads to the following equations:

$$r_i(t+h) = r_i(t) + h v_i\left(t + \frac{h}{2}\right) \tag{5.11}$$

$$v_i\left(t + \frac{h}{2}\right) = v_i\left(t - \frac{h}{2}\right) + \frac{h}{m_i}(f_i(t)) \tag{5.12}$$

The outline of following the positions and velocities of the molecules with Eqs. (5.11) and (5.12) is called the "leapfrog method." This name arises from the evaluation of the positions and forces, and then the velocities, by using time steps in a leapfrog manner. This method is also a significantly superior scheme in regard to stability and accuracy as well as the velocity Verlet method [12, 13].

The MD method is applicable to both equilibrium and nonequilibrium physical phenomena, which makes it a powerful computational tool that can be used to simulate many physical phenomena (if computing power is sufficient) [14, 15].

The main procedure for conducting the MD simulation could be presented using the velocity Verlet method in the following steps [16]:

1. Specify the initial position and velocity of all molecules.
2. Calculate the forces acting on molecules.

3. Evaluate the positions of all molecules at the next time step from Eq. (5.8).
4. Evaluate the velocities of all molecules at the next time step from Eq. (5.10).
5. Repeat the procedures from Step 2.

In the above procedure, the positions and velocities will be evaluated at every time interval in the MD simulation.

The method of evaluating the system averages will be shown, which are incumbent to make a comparison with experimental or theoretical values. Since microscopic quantities such as positions and velocities are evaluated at every time interval in MD simulations, a quantity evaluated from such microscopic values. For instance, the pressure will be different from that measured experimentally. In order to compare with experimental data, instant pressure is sampled at each time step, and these values are averaged during a short sampling time to outcome a macroscopic pressure. This average can be expressed as

$$\overline{A} = \sum_{n=1}^{N} A_n / N$$

(5.13)

In which A_n is the nth sampled value of an arbitrary physical quantity, and $\overline{A}$, called the "time average," is the mathematical average of N sampling data.

5.2.2 NONSPHERICAL PARTICLE SYSTEMS

5.2.2.1 CASE OF TAKING INTO ACCOUNT THE INERTIA TERMS

For the case of nonspherical particles, it needs to deliberate the translational motion of the center of mass of a particle and the rotational motion about an axis through the center of mass. Axisymmetric particles are very useful as a particle model for simulations [17], so we will focus on the axisymmetric particle model in this section. As shown in Fig. 5.1, the important rotational motion is to be treated about the short axis line. The equations of motion concerning the translational and rotational motion can be written as:

$$m \frac{d^2 r_i}{dt^2} = f_i$$

(5.14)

$$I \frac{dw_i}{dt} = T_i$$

(5.15)

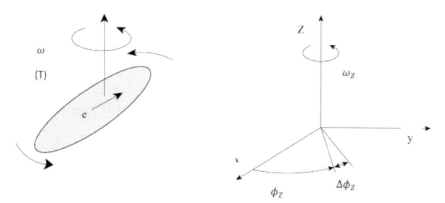

FIGURE 5.1 Linear particle and angular velocity: (A) the axisymmetric particle and (B) the coordinate system.

The translational velocity is related to the position vector as $v_i = dr_i / dt$, now consider the meaning of a quantity ϕ_i, which is related to the angular velocity as $w_i = d\phi_i / dt$. It is assumed that during a short time interval Δt, ϕ_i changes into $(\phi_i + \Delta \phi_i)$ where $\Delta \phi_i$ is expressed as $\Delta \phi_i = (\Delta \phi_{ix}, \Delta \phi_{iy}, \Delta \phi_{iz})$. As shown in Fig. 5.1B, w_z is related to the rotational angle in the xy-plane about the z-axis, $\Delta \phi_z$. The other components have the same meanings, so it can be related in the following expression:

$$\Delta \phi_i = \phi_i(t + \Delta t) - \phi_i(t) = \Delta t w_i(t) \tag{5.16}$$

Consider that it seems to be more direct and more intuitive to use the unit vector e_i denoting the particle direction rather than the quantity ϕ_i. The change in e_i during an infinitesimal time interval, Δe_i, can be written using the angular velocity w_i as:

$$\Delta e_i(t) = e_i(t + \Delta t) - e_i(t) = \Delta t w_i(t) \times e_i(t) \tag{5.17}$$

From Eqs. (5.16) and (5.17), e_i can be related to ϕ_i as:

$$\Delta e_i(t) = \Delta \phi_i(t) \times e_i(t) \tag{5.18}$$

Equation (5.17) leads to the governing equation specifying the change of the particle direction:

$$\frac{de_i(t)}{dt} = w_i(t) \times e_i(t) \tag{5.19}$$

Equation (5.15) for the angular velocity and Eq. (5.19) for the particle direction govern the rotational motion of an axisymmetric particle.

In order to solve Eqs. (5.15) and (5.19) for the rotational motion on a computer, these equations have to be translated into finite difference equations. To do it, as already explained, the first- and second-order differentials have to be expressed as algebraic expressions using the finite difference approximations based on Taylor series expansions. General finite difference expressions are as follows:

$$\frac{dx(t)}{dt} = \frac{x(t+\Delta t) - x(t)}{\Delta t} + 0(\Delta t), \frac{dx(t)}{dt} = \frac{x(t) - x(t-\Delta t)}{\Delta t} + 0(\Delta t) \quad (5.20)$$

$$\frac{dx(t)}{dt} = \frac{x(t+\Delta t) - x(t-\Delta t)}{2\Delta t} + 0((\Delta t)^2)$$

$$\frac{d^2 x(t)}{dt^2} = \frac{x(t+\Delta t) - 2x(t) + x(t-\Delta t)}{(\Delta t)^2} + 0((\Delta t)^2) \quad (5.21)$$

The simplest algorithm can be obtained using the forward finite difference approximation in Eq. (5.20) as:

$$e_i(t+\Delta t) = e_i(t) + \Delta t w_i(t) \times e_i(t) \quad (5.22)$$

This algorithm is quite attributive and understandable, but in practice does not have enough accuracy, since the error of the forward finite difference approximation is of the order of Δt. In order to improve the accuracy, the following algorithm has already been presented.

If the new vector function $u_i(t)$ such as $u_i(t) = w_i(t) \times e_i(t)$ is introduced, Eq. (5.19) can be written as:

$$\frac{de_i(t)}{dt} = u_i(t) \quad (5.23)$$

By conducting the operator $\times e$ from the right side on the both sides of Eq. (5.15), the following equation is obtained:

$$\frac{dw_i(t)}{dt} \times e_i(t) = \frac{1}{I} T_i(t) \times e_i(t) \quad (5.24)$$

The left-hand side of this equation leads to:

$$\frac{dw_i}{dt} \times e_i(t) = \frac{dw_i \times e_i(t)}{dt} - w_i \times \frac{de_i}{dt} = \frac{du_i}{dt} - w_i \qquad (5.25)$$

By substituting this equation into Eq. (5.24), the following equation can be obtained:

$$\frac{du_i(t)}{dt} = \frac{1}{I}T_i(t) \times e_i(t) + w_i(t) \times u_i(t) = \frac{1}{I}T_i(t) \times e_i(t) - |w_i(t)|^2 e_i(t) = \frac{1}{I}T_i(t) \times e_i(t) + \lambda_i(t)e_i(t) \quad (5.26)$$

In the transformation from the first to the second expressions on the right-hand side, we have used the identity $a \times (b \times c) = (a.c)b - (a.b)c$ in evaluating $w \times (w \times e)$. The quantity $\lambda_i(t)$ in the third expression has been introduced in order to satisfy the following relationship:

$$e_i.u_i = e_i.(w_i \times e_i) = 0 \qquad (5.27)$$

We have now completed the transformation of the variables from e_i and w_i to e_i and u_i for solving the rotational motion of particles.

According to the leap frog algorithm, Eqs. (5.23) and (5.26) reduce to the following algebraic equations:

$$e_i(t + \Delta t) = e_i(t) + \Delta t u_i(t + \Delta t / 2) \qquad (5.28)$$

$$u_i(t + \Delta t / 2) = u_i(t - \Delta t / 2) + \Delta t \frac{T_i(t) \times e_i(t)}{I} + \Delta t \lambda_i(t) e_i(t) \quad (5.29)$$

Another equation is necessary for determining the value of $\lambda_i(t)$. The velocity $u_i(t)$ can be evaluated from the arithmetic average of $u_i(t + \Delta t / 2)$ and $u_i(t - \Delta t/2)$, and the expression is finally written using Eq. (5.29) as:

$$u_i(t) = \frac{u_i(t + \Delta t / 2) + u_i(t - \Delta t / 2)}{2} = u_i(t - \Delta t / 2) + \frac{\Delta t}{2} \cdot \frac{T_i(t) \times e_i(t)}{I} + \frac{\Delta t}{2} \lambda_i(t) e_i(t) \quad (5.30)$$

Since $u_i(t)$ has to satisfy the orthogonality condition shown in Eq. (5.27), the substitution of Eq. (5.30) into Eq. (5.27) leads to the equation of $\lambda_i(t)$ as:

$$\lambda_i(t) = -\frac{2}{\Delta t} \cdot e_i(t).u_i(t - \Delta t / 2) \qquad (5.31)$$

In obtaining this expression, the identity $a.(b \times c) = 0$ has been used to evaluate $e.(T \times c)$

Now all the equations have been derived for determining the rotational motion of axisymmetric particles. With the value $\lambda_i(t)$ in Eq. (5.31), u_i at $(t + \Delta t / 2)$ is first evaluated from Eq. (5.29), and then e_i at $(t + \Delta t)$ is obtained from Eq. (5.28). This procedure shows that the solution of $u_i(t + \Delta t / 2)$ gives rise to the values of $e_i(t + \Delta t)$ and $T_i(t + \Delta t)$, and these solutions lead to $u_i(t + 3\Delta t / 2)$, and so forth. This algorithm is another example of a leapfrog algorithm.

For the translational motion, the velocity Verlet algorithm may be used, and the particle position $r_i(t + \Delta t)$ and velocity $v_i(t + \Delta t)$) can be obtained as:

$$r_i(t + \Delta t) = r_i(t) + \Delta t v_i(t) + \frac{(\Delta t)^2}{2m} fi(t) \qquad (5.32)$$

$$v_i(t + \Delta t) = v_i(t) + \frac{\Delta t}{2m} f_i(t) + \{ f_i(t + \Delta t) \}$$

These equations can be derived in a straightway manner from the finite difference approximations in Eqs. (5.20) and (5.21).

All the equations for specifying the translational and rotational motion of axisymmetric particles are shown for the case of taking into account the inertia terms. The main procedure for conducting the MD simulation is as follows:

1. Specify the initial configuration and velocity of the axisymmetric particles for the translational and rotational motion [18].
2. Calculate the forces and torques acting on particles.
3. Evaluate the positions and velocities of the translational motion at $(t + \Delta t)$ from Eq. (1.32).
4. Evaluate $\lambda_i(t)(i = 1, 2, ..., N)$ from Eq. (1.31).
5. Evaluate $u_i(i = 1, 2, ..., N)$ at $((t + \Delta t / 2)$ from Eq. (1.29).
6. Evaluate the unit vectors $e_i(i = 1, 2, ..., N)$ at $(t + \Delta t)$ from Eq. (1.28).
7. Advance one time step to repeat the procedures from step 2.

By following this procedure, the MD method for axisymmetric particles with the inertia terms can simulate the positions and velocities, and the directions and angular velocities, at every time interval Δt.

5.2.2.2   CASE OF NEGLECTED INERTIA TERMS

When treating a colloidal dispersion or a polymeric solution, the Stokesian dynamics and BD methods are usually employed as a microscopic or mesoscopic analysis tool. In these methods, dispersed particles or polymers are modeled as idealized spherical or dumbbell particles, but the base liquid is usually assumed to be a continuum medium and its effect is included in the equations of motion of the particles or the polymers only as friction terms. If particle size approximates to or is smaller than micron-order, the inertia terms may be considered as negligible. In this section, we treat this type of small particles and neglect the inertia terms. For the case of axisymmetric particles moving in a quiescent fluid, the translational and angular velocities of particle i, v_i and w_i, are written as [19, 20]:

$$v_i = \frac{1}{\eta}\left\{\frac{1}{X^A}e_i e_i + \frac{1}{Y^A}(I - e_i e_i)\right\}.F_i \tag{5.33}$$

$$w_i = \frac{1}{\eta}\left\{\frac{1}{X^C}e_i e_i + \frac{1}{Y^C}(I - e_i e_i)\right\}.T_i \tag{5.34}$$

In which X^A, Y^A, X^C, and Y^C are the resistance functions specifying the particle shape. If the long- and short-axis lengths are denoted by 2a and 2b, respectively, the eccentricity is denoted:

by $s(=(a^2-b^2)^{\frac{1}{2}}/a$, the resistance functions for the spheroid particle are written as:

$$X^A = 6\pi a.\frac{8}{3}.\frac{s^3}{-2s+(1+s^2)L}, Y^A = 6\pi a.\frac{16}{3}.\frac{s^3}{-2s+(3s^2-1)L} \tag{3.35}$$

$$X^C = 8?a^3.\frac{4}{3}.\frac{(s^3(1-s^2))}{(2s-(1-s^2)L)}, Y^C = 8\pi a^3.\frac{4}{3}.\frac{(s^3(1-s^2))}{(-2s-(1+s^2)L)} \tag{3.36}$$

The function of the eccentricity and is expressed as:

$$L = L(s) = \ln\frac{1+s}{1-s} \tag{5.37}$$

In the case of $s \ll 1$, Eqs. (5.35) and (5.36) are approximated using Taylor series expansions as:

$$X^A = 6\pi a\left(1 - \frac{2}{5}s^2 + \ldots\right), Y^A = 6\pi a\left(1 - \frac{3}{10}s^2 + \ldots\right) \quad (5.38)$$

$$X^C = 8\pi a^3\left(1 - \frac{6}{5}s^2 + \ldots\right), X^C = 8\pi a^3\left(1 - \frac{96}{10}s^2 + \ldots\right) \quad (5.39)$$

In the limit, the well-known Stokes drag formula for a spherical particle in a quiescent fluid can be obtained from Eqs. (5.33), (5.34), (5.38) and (5.39):

$$v_i = \frac{1}{6\pi\eta a}F_i, w_i = \frac{1}{8\pi a^3 \eta}T_i \quad (5.40)$$

It is possible to pursue the motion of an axisymmetric particle using Eqs. (5.33) and (5.34), but further simplified equations can be used for the present axisymmetric particle. For an axisymmetric particle, the translational motion can be decomposed into the motion in the long axis direction and that in a direction normal to the particle axis. Similarly, the rotational motion can be decomposed into the rotation about the particle axis and that about a line normal to the particle axis through the mass center. If the force acting on the particle is expressed as the sum of the force $F_i^{\parallel}$ parallel to the particle axis and the force $F_i^{\perp}$ normal to that axis, then these forces can be expressed using the particle direction vector as:

$$F_i^{\parallel} = e_i(e_i.F_i) = e_i e_i.F_i, F_i^{\perp} = F_i - F_i^{\parallel} = (I - e_i e_i).F_i \quad (5.41)$$

With these expressions, the velocities $v_i^{\parallel}$ and $v_i^{\perp}$ parallel and normal to the particle axis, respectively, can be written from Eq. (5.33) as:

$$v_i^{\parallel} = \frac{1}{\eta X^A}F_i^{\parallel}, v_i^{\perp} = \frac{1}{\eta Y^A}F_i^{\perp} \quad (5.42)$$

Similarly, the angular velocities $w_i^{\parallel}$ and $w_i^{\perp}$ about the long and short axes, respectively, are written from Eq. (5.34) as:

$$w_i^{\parallel} = \frac{1}{\eta X^C}T_i^{\parallel}, w_i^{\perp} = \frac{1}{\eta Y^C}T_i^{\perp} \quad (5.43)$$

According to the Eqs. (5.42) and (5.43), $v_i^{\perp}, v_i^{\parallel}, w_i^{\parallel}$ and $w_i^{\perp}$ can be evaluated from values of $F_i^{\perp}, F_i^{\parallel}, T_i^{\parallel}$ and $T_i^{\perp}$. The translational velocity and angular velocity are then obtained as:

$$v_i = v_i^{\perp} + v_i^{\parallel}, \ w_i = w_i^{\perp} + w_i^{\parallel} \tag{5.44}$$

With the solutions of the translational and angular velocities at the time step t shown in Eq. (5.44), the position vector and the particle direction vector at the next time step $(t + \Delta t)$ can finally be obtained as:

$$r_1(t + \Delta t) = r_i(t) + \Delta t \, v_i(t) \tag{5.45}$$

$$e_i(t + \Delta t) = e_i(t) + \Delta t w_i(t) \times e_i(t) \tag{5.46}$$

Lastly, the main procedure for the simulation can be shown in the following steps:

1. Specify the initial configuration and velocity of all axisymmetric particles for the translational and rotational motion.
2. Calculate all the forces and torques acting on particles.
3. Evaluate $F_i^{\perp}, F_i^{\parallel}, T_i^{\parallel}$; and $T_i^{\perp}(i = 1, 2, ..., N)$ from Eq. (5.41) and similar equations for the torques.
4. Calculate $v_i^{\perp}, v_i^{\parallel}, w_i^{\parallel}$; and $w_i^{\perp}(i = 1, 2, ..., N)$ from Eqs. (5.42) and (5.43).
5. Calculate v_i and $w_i(i = 1, 2, ..., N)$ from Eq. (5.44).
6. Calculate r_i and $e_i(i = 1, 2, ..., N)$ at the next time step $(t + \Delta t)$ from Eqs. (5.45) and (5.46).
7. Advance one time step and repeat the procedures from Step 2.

5.3 MONTE CARLO METHOD

In the MD method, the motion of molecules (particles) is simulated according to the equations of motion and therefore it is applicable to both thermodynamic equilibrium and nonequilibrium phenomena. In contrast, the MC method generates a series of microscopic states under a certain stochastic law, irrespective of the equations of motion of particles. Since the MC method does not use the equations of motion, it cannot include the concept of explicit time, and it is only a simulation technique for phenomena in thermodynamic equilibrium. Accordingly, it is unsuitable for the MC method to deal with the dynamic properties of a system, which are dependent on time. In the following paragraphs, we explain important points of the concept of the MC method [21, 22].

The microscopic states for thermodynamic equilibrium in a practical situation are discussed by considering a two-particle attractive system using Fig. 5.2. As shown in Fig. 5.2A, if the two particles overlap, then a repulsive force or significant interaction energy arises. As shown in Fig. 5.2B, for the case of close proximity, the interaction energy becomes low and an attractive force acts on the particles. If the two particles are sufficiently distant, as shown in Fig. 5.2C, the interactive force is negligible and the interaction energy can be regarded as zero. In actual phenomena, microscopic states which induce a significantly high energy, as shown in Fig. 5.2A, seldom appear, but microscopic states which give rise to a low-energy system, as shown in Fig. 5.2B, frequently arise. However, this does not mean that only microscopic states that induce a minimum energy system appear. Consider the fact that oxygen and nitrogen molecules do not gather in a limited area, but distribute uniformly in a room. It is seen from this discussion that, for thermodynamic equilibrium, microscopic states do not give rise to a minimum of the total system energy, but to a minimum free energy of a system. For example, in the case of a system specified by certain properties microscopic states arise such that the following Helmholtz free energy F becomes a minimum:

$$F = E - TS \qquad (5.47)$$

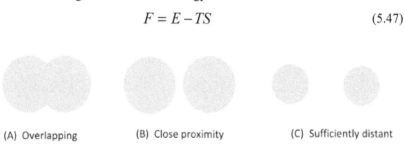

(A) Overlapping   (B) Close proximity   (C) Sufficiently distant

FIGURE 5.2 Typical energy situations for a two particle system.

In the preceding example, the reason why oxygen or nitrogen molecules do not gather in a limited area can be explained by taking into account the entropy term on the right-hand side in **Eq. (5.47)**. That is, the situation in which molecules do not gather together and form flocks but expand to fill a room gives rise to a large value of the entropy. Hence, according to the counterbalance relationship of the energy and the entropy, real microscopic states arise such that the free energy of a system is at minimum.

Next, consider how microscopic states arise stochastically. We here treat a system composed of N interacting spherical particles, the quantities of temperature and volume are given values and assumed to be constant.

The total interaction energy of the system can be expressed as a function of the particle positions. It can be expressed as $U = U(r_1, r_2, \ldots, r_N)$. For the present system specified by the appearance of a microscopic state that the particle $i(i = 1, 2, \ldots, N)$ exits within the small range of $r_i \sim (r_i + \Delta r_i)$ is governed by the probability density function $\rho(r_1, r_2, \ldots, r_N)$.

This can be expressed from statistical mechanics as:

$$\rho(r_1, r_2, \ldots, r_N) = \frac{exp\{-U(r_1, r_2, \ldots, r_N)/kT\}}{\int v \ldots \int v^{exp\{-U(r_1, r_2, \ldots, r_N)/kT\}dr_1\,dr_2\ldots dr_N}} \qquad (5.48)$$

If a series of microscopic states is generated with an occurrence according to this probability, a simulation may have physical meaning. However, this approach is impracticable, as it is extraordinarily difficult and almost impossible to evaluate analytically the definite integral of the denominator in Eq. (1.48). In fact, if we were able to evaluate this integral term analytically, we would not need a computer simulation because it would be possible to evaluate almost all physical quantities analytically.

The "Metropolis method" [21] overcomes this difficulty for MC simulations. In the Metropolis method, the transition probability from microscopic states i to j, P_{ij}, is expressed as:

$$p_{ij} = 1\left(\text{for } p_j / p_i \geq 1\right) \text{ and} \qquad (5.49)$$

$$\frac{\rho_j}{\rho_i}\left(\text{for } \rho_j / \rho_j < 1\right)$$

The ratio of ρ_j / ρ_i is obtained from Eq. (5.48) as:

$$\frac{\rho_j}{\rho_i} = exp\left\{-\frac{1}{kT}(U_j - U_i)\right\} = exp\left[-\frac{1}{kT}\left\{U\left(r_1^i, r_2^i, \ldots, r_N^i\right) - U\left(r_1^i, r_2^i, \ldots, r_N^i\right)\right\}\right] \qquad (5.50)$$

In the above equations, U_i and U_j are the interaction energies of microscopic states i and j, respectively. The superscripts attached to the position vectors denote the same meanings concerning microscopic states. Eq. (5.49) connotes that, in the transition from microscopic states i to j, new microscopic state j is adopted to the system energy decreases or increases, with the probability $\rho_j / \rho_i < 1$. As clearly demonstrated by Eq. (5.50), for ρ_j / ρ_i the denominator in Eq. (5.48) is not required in Eq. (5.50), because ρ_j is divided by ρ_i and the term is canceled through this operation. This is the main reason for the great success of the Metropolis method for MC simulations. That a new microscopic state is adopted with the probability ρ_j/ρ_i, even in the case of the increase in the interaction energy, verifies the accomplishment of the minimum free energy condition for the system. In other words, the adoption of microscopic states, making an increase in the system energy, corresponds to an increase in the entropy.

The above discussion is directly applicable to a system composed of nonspherical particles. The transition probability from microscopic states i to j, can P_{ij} be written in similar form to Eq. (5.49). The exact expression of ρ_j/ρ_i becomes:

$$
\begin{aligned}
\frac{\rho_j}{\rho_i} &= exp\left\{-\frac{1}{kT}(U_j - U_i)\right\} \\
&= exp\left[-\frac{1}{kT}\left\{U\left(r_1^j, r_2^j, r_N^j, e_1^j, e_2^j, \ldots, e_N^j\right) - U\left(r_1^i, r_2^i, r_N^i, e_1^i, e_2^i, \ldots, e_N^i\right)\right\}\right] \quad (5.51)
\end{aligned}
$$

The main procedure for the MC simulation of a nonspherical particle system is as follows [18, 23]:

1. Specify the initial position and direction of all particles.
2. Regard this state as microscopic state i, and calculate the interaction energy U_i.
3. Choose an arbitrary particle in order or randomly and call this particle "particle α."
4. Make particle α move translationally using random numbers and calculate the interaction energy U_j for this new configuration.
5. Adopt this new microscopic state for the case of $U_j \le U_i$ and go to Step 7.
6. Calculate ρ_j/ρ_i in Eq. (1.51) for the case of $U_j > U_i$ and take a random number R_1 from a uniform random number sequence distributed from zero to unity.
6.1. If $R_1 \le \rho_j/\rho_i$, adopt this microscopic state j and go to Step 7.
6.2. If. $R_1 \le \rho_j/\rho_i$, reject this microscopic state, regard previous state i as new microscopic state j, and go to Step 7.
7. Change the direction of particle α using random numbers and calculate the interaction energy U_k for this new state.
8. If $U_k \le U_j$, adopt this new microscopic state and repeat from Step 2.
9. If $U_k \le U_j$, calculate ρ_k/ρ_j in Eq. (1.51) and take a random number R_2 from the uniform random number sequence.
9.1. If $R_2 \le \rho_k/\rho_j$, adopt this new microscopic state k and repeat from Step 2.

9.2. If $R_2 > \rho_k/\rho_j$, reject this new state, regard previous state j as new microscopic state k, and repeat from Step 2.

Although the treatment of the translational and rotational changes is carried out separately in the above algorithm, a simultaneous procedure is also possible in such a way that the position and direction of an arbitrary particle are simultaneously changed, and the new microscopic state is adopted according to the condition in Eq. (5.49). However, for a strongly interacting system, the separate treatment may be found to be more executable in many cases.

It is now briefly explained how the translational move is made using random numbers during a simulation. If the position vector of an arbitrary particle α in microscopic state i is denoted by $r_\alpha = (x_\alpha, y_\alpha, z_\alpha)$, this particle is moved to a new position $r'_\alpha = (x'_\alpha, y'_\alpha, z'_\alpha)$, by the following equations using random.

Numbers R_1, R_2 and R_3 taken from a random number sequence ranged from zero to unity:

$$x'_\alpha = x_\alpha + R_1 \delta r_{max}, \, y'_\alpha = y_\alpha + R_2 \delta r_{max}, \, z'_\alpha = z_\alpha + R_3 \delta r_{max} \quad (5.52)$$

These equations imply that the particle is moved to an arbitrary position, determined by random numbers, within a cube centered at the particle center with side length of $2\delta r_{max}$. A series of microscopic states is generated by moving the particles according to the above-mentioned procedure.

Finally, the method of evaluating the average of a physical quantity in MC simulations will be shown. These averages, called "ensemble averages," are different from the time averages that are obtained from MD simulations [24, 25]. If a physical quantity A is a function of the microscopic states of a system, and A_n is the nth sampled value of this quantity in an MC simulation, then the ensemble average $\langle A \rangle$ can be evaluated from the equation:

$$\langle A \rangle = \sum_{n=1}^{N} A_n / N \quad (5.53)$$

In actual simulations, the sampling procedure is not carried out at each time step but at regular intervals. This may be more efficient because if the data have considerable correlations they are less likely to be sampled by taking a longer interval for the sampling time. The ensemble averages obtained in this way may be compared with experimental data.

5.4 BROWNIAN DYNAMICS METHOD

A dispersion or suspension composed of fine particles dispersed in a base liquid is a difficult case to be treated by simulations in terms of the MD method, because the characteristic time of the motion of the solvent molecules is considerably different from the dispersed particles. Simply speaking, if we observe such dispersion based on the characteristic time of the solvent molecules, we can see only the active motion of solvent molecules around the quiescent dispersed particles.

Clearly the MD method is quite insubstantial as a simulation technique for particle dispersions. One approach to dominate this difficulty is to not focus on the motion of each solvent molecule, and attend to the solvent molecules as a continuum medium and consider the motion of dispersed particles in such a medium. In this approach, the influence of the solvent molecules is included into the equations of motion of the particles as random forces. We can observe such random motion when pollen moves at a liquid surface or when dispersed particles move in a functional fluid. The BD method simulates the random motion of dispersed particles that is induced by the solvent molecules; thus, such particles are called "Brownian particles" [26–29].

If particle dispersion is so significantly dilute that each particle can be regarded as moving independently, the motion of this Brownian particle is governed by the following Langevin equation:

$$m\frac{dv}{dt} = f - \xi v - f^B \tag{5.54}$$

This equation is valid for spherical particle dispersion. In Eq. (5.54), the friction coefficient is expressed as $\xi = 3\pi\eta d$ and the force exerted by an external field, $f^B \left(= \left(f_x^B, f_y^B, f_z^B\right)\right)$ is the random force due to the motion of solvent molecules. This random force has the following stochastic properties:

$$f_x^B(t) = f_y^B(t) = f_z^B(t) = 0 \tag{5.55}$$

$$\left\{f_x^B(t)\right\}^2 = \left\{f_y^B(t)\right\}^2 = \left\{f_z^B(t)\right\}^2 = 2\xi kT\delta(t-t') \tag{5.56}$$

In which $\delta(t-t')$ is the Dirac delta function. In Eq. (5.56) larger random forces act on Brownian particles at a higher temperature because the mean square average of each component of the random force is in proportion to the system temperature. At a higher temperature the solvent molecules move more actively and induce larger random forces.

In order to simulate the Brownian motion of particles, the basic equation in Eq. (5.54) has to be transformed into an algebraic equation, as in the MD method. If the time interval is sufficiently short, the change in the forces is negligible and Eq. (5.54) can be regarded as a simple first-order differential equation. Hence, Eq. (1.54) can be solved by standard textbook methods of differential equations. So algebraic equations can finally be obtained as:

$$r(t+h) = r(t) + \frac{m}{\xi} v(t) \left\{ 1 - exp\left(-\frac{\xi}{m} h \right) \right\} \qquad (5.57)$$

$$+ \frac{1}{\xi} f(t) \left\{ h - \frac{m}{\xi} \left(1 - exp\left(-\frac{\xi}{m} h \right) \right) \right\} + \Delta r^{B}$$

Where Δr^{B} and Δv^{B} are a random displacement and velocity due to the motion of solvent molecules. The relationship of the x-components of Δr^{B} and Δv^{B} can be expressed as a two-dimensional normal distribution (similarly for the other components).

Now this expression is not investigated but instead of it a method is considered that is superior in regard to the extension of the BD method to the case with multibody hydrodynamic interactions. The BD method based on Eqs. (5.57) and (5.58) is applicable for physical phenomena in which the inertia term is a governing factor. Since the BD method with multibody hydrodynamic interactions among the particles is very complicated, an alternative method is attended that treats the friction forces between the particles and a base liquid, and the nonhydrodynamic interactions between the particles. This simpler type of simulation method is sometimes used as a first-order approximation because of the complexity of treating hydrodynamic interactions. A representative nonhydrodynamic force is the magnetic force influencing the magnetic particles in a ferrofluid Although the BD method based on the Ermak-McCammon analysis takes into account multibody hydrodynamic interactions among particles, we apply this analysis method to the present dilute dispersion without hydrodynamic interactions, and can derive the basic equation of the position vector $r_i (i = 1, 2, ..., N)$ of Brownian particle i as [30, 31]:

$$r_i(t+h) = r_i(t) + \frac{1}{\xi}hf_i(t) + \Delta r_i^B \tag{5.59}$$

In which the components $(\Delta x_i^B, \Delta y_i^B, \Delta z_i^B)$ of the random displacement Δr_i^B have to satisfy the following stochastic properties:

$$\Delta x_i^B = \Delta y_i^B = \Delta z_i^B = 0 \tag{5.60}$$

$$(\Delta x_i^B)^2 = (\Delta y_i^B)^2 = (\Delta z_i^B)^2 = \frac{2kT}{\xi}h \tag{5.61}$$

Equations similar to Eq. (5.59) hold for every particle in the system. Interactions among particles arise through the force $f_i(i=1,2,...,N)$ acting on them.

If a Brownian particle exhibits magnetic properties and has, for example, a magnetic dipole moment at the particle center, it will have a tendency to incline in the direction of an applied magnetic field. Hence, even in the case of spherical particles, the rotational motion is influenced by an external field, so that both the translational and the rotational motion of a particle are treated simultaneously in simulations.

If the unit vector of the particle direction is denoted by n_i, the equation of the change in n_i can be derived under the same conditions assumed in deriving Eq. (5.59) as:

$$n_i(t+h) = n_i(t) + \frac{1}{\xi}hT_i(t) \times n_i(t) + \Delta n_i^B \tag{5.62}$$

The friction coefficient of the rotational motion, expressed as $\xi_R = \pi \eta d^3$. The rotational displacement due to random forces, expressed as:

$$\Delta n_i^B = \Delta \phi_{\perp 1}^B n_{\perp 1} + \Delta \phi_{\perp 2}^B n_{\perp 2} \tag{5.63}$$

In which $n_{\perp 1}$ and $n_{\perp 2}$ are a set of unit vectors normal to the direction of particle i, $\Delta \phi_{\perp 1}^B$ and $\Delta \phi_{\perp 2}^B$ have the following stochastic properties:

$$\Delta \phi_{\perp 1}^B = \Delta \phi_{\perp 2}^B = 0 \tag{5.64}$$

Now consider the correspondence of quantities in the translational and rotational motion. The velocity in the translational motion corresponds to the angular velocity in the rotational motion, and the position vector corresponds to the quantity ϕ_i defined as $d\phi_i / dt = w_i$. Obviously, due to the similarity of Eqs. (5.64) and (5.65) to Eqs. (5.60) and (5.61), the components $\Delta\phi_{\perp 1}{}^B$ and $\Delta\phi_{\perp 1}{}^B$ of the vector $\Delta\phi^B$ have to satisfy Eqs. (5.64) and (5.65).

The basic Eqs. (5.59) and (5.62) for governing the translational and rotational motion of particles have been derived under the assumptions that the momentum of particles is sufficiently relaxed during the time interval h and that the force acting on the particles is substantially constant during this infinitesimally short time. This is the essence of the Ermak-McCammon method for BD simulations.

Next, the method of generating random displacements will be shown according to Eqs. (5.60) and (5.61), but, before that, the normal probability distribution needs to be briefly described. If the behavior of a stochastic variable is described by the normal distribution $\rho_{normal}(x)$ with variance σ^2, $\rho_{normal}(x)$ is written as:

$$\rho_{normal}(x) = \frac{1}{(2\pi\sigma^2)^{1/2}} \exp(-x^2 / 2\sigma^2) \tag{5.66}$$

The variance is a measure of how wide the stochastic variable x is distributed around the mean value $\langle x \rangle$, which is taken as zero for this discussion. The variance is mathematically defined as:

$$\sigma^2 = (x - x)^2 = x^2 - (x)^2 \tag{5.67}$$

If Eq. (5.66) is applied to Eqs. (5.60) and (5.61), the random displacement $\Delta x_i{}^B$ in the x-direction can be written in normal distribution form as:

$$\rho_{normal}\left(\Delta x_i{}^B\right) = \left(\frac{\xi}{4\pi kTh}\right)^{1/2} \exp\left\{\left(-\frac{\xi}{4\pi Th}\left(\Delta x_i{}^B\right)^2\right)\right\} \tag{5.68}$$

The other components also condescend a normal distribution. As seen in Eq. (5.68), larger random displacements tend to arise at a higher system temperature, due to the solvent molecules move more actively in the higher temperature case. The random displacements can therefore be generated by sampling according to the normal distributions shown in Eq. (5.68).

The main procedure for conducting the BD simulation based on Eqs. (5.59), (5.60), and (5.61) is [32, 33]:
  1. Specify the initial position of all particles.
  2. Calculate the forces acting on each particle.

3. Generate the random displacements $\Delta r_i^B = \left(\Delta x_i^B, \Delta y_i^B, \Delta z_i^B \right)(i = 1, 2, \ldots, N)$ using uniform random numbers: for example, Δx_i^B is sampled according to Eq. (5.68).
4. Calculate all the particle positions at the next time step from Eq. (5.59).
5. Return to Step 2 and repeat.

The physical quantities of interest are evaluated by the time average, similar to the molecular dynamics method.

5.5 DISSIPATIVE PARTICLE DYNAMICS METHOD

As already pointed out, it is not realist to use the MD method to simulate the motion of solvent molecules and dispersed particles simultaneously, since the characteristic time of solvent molecules is much shorter than that of dispersed particles.

Hence, in the BD method, the motion of solvent molecules is not treated, but a fluid is regarded as a continuum medium. The influence of the molecular motion is combined into the equations of motion of dispersed particles as stochastic random forces. Are there any simulation methods to simulate the motion of both the solvent molecules and the dispersed particles? As far as the motion of real solvent molecules is treated, the development of such simulation methods may be infeasible. However, if groups or clusters of solvent molecules are regarded as virtual fluid particles, such that the characteristic time of the motion of such fluid particles is not so different from that of dispersed particles, then it is possible to simulate the motion of the dispersed and the fluid particles simultaneously. These virtual fluid particles are expected to exchange their momentum, exhibit a random motion similar to Brownian particles, and interact with each other by particle-particle potentials. We call these virtual fluid particles "dissipative particles," and the simulation technique of treating the motion of dissipative particles instead of the solvent molecules is called the "dissipative particle dynamics (DPD) method" [34, 35].

The DPD method is principally applicable to simulations of colloidal dispersions that take into account the multibody hydrodynamic interactions among particles. For colloidal dispersions, the combination of the flow field solutions for a three- or four-particle system into a simulation technique enables us to address the physical situation of multibody hydrodynamic interactions as accurately as possible. However, it is really hard to solve analytically the flow field even for a three-particle system, so a solution for a nonspherical particle system is ineffectual to assay. In contrast, the DPD method does not require this type of solution of the flow field in conducting simulations of colloidal dispersions that take into account multibody hydrodynamic effects. This is because they are automatically reproduced from consideration of the interactions between the dissipative and the colloidal particles [36–38].

This approach to the hydrodynamic interactions is a great advantage of the DPD method. In addition, this method is applicable to nonspherical particle dispersions, and a good simulation technique for colloidal dispersions [39].

It will be shown the general classes of models, before proceeding to the explanation of the DPD method, applied in the modeling of a fluid for numerical simulations. Figure 5.3 schematically shows the assortment of the modeling of a fluid. Figure 5.3A shows a continuum medium model for a fluid. In this situation, a solution of a flow field can be obtained by solving the Navier-Stokes equations, which are the governing equations of the motion of a fluid. Figure 5.3C shows a microscopic model in which the solvent molecules are treated and a solution of the flow field can be obtained by pursuing the motion of the solvent molecules: this is the MD approach. Figure 5.3B shows a mesoscopic model in which a fluid is assumed to be composed of virtual fluid particles: the DPD method is classified within this category.

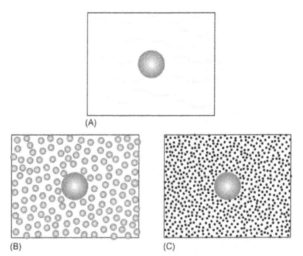

(A)

(B)   (C)

FIGURE 5.3   Modeling of a fluid: (A) the macroscopic model, (B) the mesoscopic model, and (C) the microscopic model.

In the following paragraphs, the equations of motion of the dissipative particles will be reviewed for a system composed of dissipative particles alone, without colloidal particles. For simplification's point, dissipative particles are simply called "particles" unless specifically identified.

In order that the solution of a flow field obtained from the particle motion agrees with that of the Navier-Stokes equations, the equations of motion of the particles have to be formalized in physically viable form. For example, as a physical constraint on the system behavior, the total momentum of a system should be conserved. The forces acting on particle i possibly seem to be a conservative force, exerted by other particles (particle j in this case), a dissipative force, due to the exchange of momentum and a random force, inducing the random motion of particles. The equation of motion can be written as:

$$m\frac{dv_i}{dt} = \sum_{j(\neq i)} F_{ij}^C + \sum_{j(\neq i)} F_{ij}^D + \sum_{j(\neq i)} F_{ij}^R \tag{5.69}$$

In Eq. (5.69), since F_{ij}^C is a conservative force between particles i and j, it is assumed to be dependent on the relative position $r_{ij}(=r_i - r_j)$ alone, not on velocities. This specific expression will be shown later. F_{ij}^D and F_{ij}^R have to be conserved under a Galilean transformation [40, 41] thus, they must be independent of r_i and v_i (quantities dependent on r_i and v_i are not conserved). They should be functions of the relative position vector r_{ij} and relative velocity vector $v_{ij}(=v_i - v_j)$. Furthermore, it is physically reasonable to assume that F_{ij}^R is dependent only on the relative position, and not on the relative velocity. It also has to be considered that the particle motion is isotropic and the forces between particles decrease with the particle-particle separation. The following expressions for F_{ij}^D and F_{ij}^R satisfy all the above-mentioned requirements:

$$F_{ij}^D = -\gamma w_D(r_{ij})(e_{ij}.v_{ij})e_{ij} \tag{5.70}$$

$$F_{ij}^D = \sigma w_R(r_{ij})e_{ij}\zeta_{ij} \tag{5.71}$$

In which $r_{ij} = |r_{ij}|$, and e_{ij} is the unit vector denoting the direction of a line drawn from particles j to i, expressed as $e_{ij} = r_{ij}/r_{ij}$. The stochastic variable inducing the random motion of particles and has the following characteristics:

$$\langle \zeta_{ij} \rangle = 0 \; \langle \zeta_{ij}(t)\zeta_{ij}(t') \rangle = \left(\delta_{ii}\delta_{jj} \right)\delta(t-t') \quad (5.72)$$

$\delta_{ij} = 1$ for $i = j$ and $\delta_{ij} = 0$ for the other cases. Since this variable satisfies the equation of $\zeta_{ij} = \zeta_{ji}$, the total momentum of a system is conserved. The weighting functions representing the characteristics of forces decreasing with the particle- particle separation, and γ and σ are constants specifying the strengths of the corresponding forces. As shown later, these constants are related to the system temperature and friction coefficients.

The F_{ij}^{D} acts such that the relative motion of particles i and j relaxes, and F_{ij}^{R} functions such that the thermal motion is activated. Since the action-reaction law is satisfied by F_{ij}^{R}, the conservation of the total momentum is not violated by F_{ij}^{R}

By substituting Eqs. (5.70) and (5.71) into Eq. (5.69), the equation of motion of particles can be written as:

$$m\frac{dv_i}{dt} = \sum_{j(\neq i)}F_{ij}^{C}\left(r_{ij}\right) - \sum_{j(\neq i)}\gamma w_D\left(r_{ij}\right)(e_{ij}.v_{ij})e_{ij} + \sum_{j(\neq i)}\sigma w_R\left(r_{ij}\right)e_{ij}\zeta_{ij} \quad (5.73)$$

The integral of this equation with respect to the time from t to $(t + \Delta t)$ leads to the finite difference equations specifying the motion of the simulation particles:

$$\Delta r_i = v_i \, \Delta t \quad (5.74)$$

$$\Delta v_i = \frac{1}{m}\left(\sum_{j(\neq i)}F_{ij}^{C}\left(r_{ij}\right) - \sum_{j(\neq i)}\gamma w_D\left(r_{ij}\right)(e_{ij}.v_{ij})e_{ij} \right)\Delta t \quad (5.75)$$
$$+ \frac{1}{m}\sum_{j(\neq i)}\sigma w_R\left(r_{ij}\right)e_{ij}\Delta W_{ij}$$

The ΔW_{ij} should be satisfied the following stochastic properties, which can be obtained from Eq. (5.72):

$$\Delta W_{ij} = 0 \quad (5.76)$$

$$\langle \Delta W_{ij}\Delta W_{ij} \rangle = \left(\delta_{ii}\delta_{jj} + \delta_{ij}\delta_{ij} \right) \Delta t$$

If a new stochastic variable θ_{ij} is introduced from $\Delta W_{ij} = \theta_{ij}(\Delta t)^{1/2}$, the third term in Eq. (5.75) can be written as:

$$\frac{1}{m} \sum_{j(\neq i)} \sigma w_R\left(r_{ij}\right) e_{ij} \theta_{ij} \sqrt{\Delta t} \tag{5.77}$$

In which θ_{ij} has to satisfy the following stochastic characteristics:

$$\theta_{ij} = 0$$

$$\langle \theta_{ij}\theta_{ij} \rangle = \left(\delta_{ii}\delta_{jj} + \delta_{ij}\delta_{ij}\right) \tag{5.78}$$

In simulations, values of the stochastic variable are sampled from a normal distribution with zero-mean value and unit variance or from a uniform distribution.

The constants γ and σ and the weighting functions $w_D(r_{ij})$ and $w_R(r_{ij})$, which appeared in Eq. (5.75) must be satisfied with the following relationships:

$$w_D\left(r_{ij}\right) = w_R^2\left(r_{ij}\right), \sigma^2 = 2\gamma kT \tag{5.79}$$

The second equation is called the "fluctuation-dissipation theorem." These relationships ensure a valid equilibrium distribution of particle velocities for thermodynamic equilibrium [42].

Next, the expressions for the conservative force F_{ij}^C and the weighting function $w_R(r_{ij})$ are shown. The F_{ij}^C functions as a tool for preventing particles from significantly overlapping, so that the value of $w_R(r_{ij})$ has to increase with particles i and j approaching each other. Given this consideration, these expressions may be written as:

$$F_{ij}^C = \alpha w_R\left(r_{ij}\right) e_{ij} \tag{5.80}$$

$$w_R\left(r_{ij}\right) = \begin{cases} 1 - \dfrac{r_{ij}}{r_c} & \text{for } r_{ij} \leq r_c \\ 0 & \text{for } r_{ij} > r_c \end{cases} \tag{5.81}$$

In which α is a constant representing the strength of a repulsive force. By substituting the above-mentioned expressions into Eq. (5.75) and taking into account Eq. (5.77), the final expressions for the equations of motion of particles can be obtained as:

$$\Delta r_i = v_i \, \Delta t \tag{5.82}$$

$$\Delta v_i = \frac{\alpha}{m} \left(\sum_{j(\neq i)} w_R \left(r_{ij} \right) e_{ij} \, \Delta t - \frac{\gamma}{m} \sum_{j(\neq i)} w_R^{\ 2} \left(r_{ij} \right) (e_{ij} \cdot v_{ij}) e_{ij} \, \Delta t \right) \tag{5.83}$$

$$+ \frac{(2\gamma kT)^{1/2}}{m} \sum_{j(\neq i)} w_R \left(r_{ij} \right) e_{ij} \theta_{ij} \sqrt{\Delta t}$$

As previously indicated, θ_{ij} satisfies the stochastic characteristics in Eq. (5.78) and is sampled from a normal distribution or from a uniform distribution. The DPD dynamics method simulates the motion of the dissipative particles according to Eqs. (5.82) and (5.83).

For actual simulations, the method of nondimensionalizing quantities is presented. The following delegate values are used for nondimensionalization: $(kT/m)^{1/2}$ for velocities, r_c for distances, $r_c(kT/m)^{1/2}$ for time, $(1/r_c^{\ 3})$ for number densities. Using these representative values, Eqs. (5.82) and (5.83) are nondimensionalized as

$$\Delta r_i^* = v_i^* \, \Delta t^* \tag{5.84}$$

$$\Delta v_i^* = \frac{\alpha^*}{m} \left(\sum_{j(\neq i)} w_R \left(r_{ij}^* \right) e_{ij} \, \Delta t^* - \frac{\gamma}{m} \sum_{j(\neq i)} w_R^{\ 2} \left(r_{ij}^* \right) (e_{ij} \cdot v_{ij}^*) e_{ij} \, \Delta t^* \right)$$

$$+ \frac{(2\gamma^* kT)^{1/2}}{m} \sum_{j(\neq i)} w_R \left(r_{ij}^* \right) e_{ij} \theta_{ij} \sqrt{\Delta t^*}$$

In which:

$$w_R \left(r_{ij} \right) = \begin{cases} 1 - r_{ij}^* \ \text{for} \ r_{ij}^* \leq 1 \\ 0 \ \text{for} \ r_{ij}^* > 1 \end{cases} \tag{5.86}$$

$$\alpha^* = \alpha \frac{r_c}{kT}, \gamma^* = \gamma \frac{r_c}{(mkT)^{1/2}} \tag{5.87}$$

Nondimensionalized quantities are distinguished by the superscript *. As seen in Eq. (5.85), the specification of the number density $n^*(= nr^3)$ and the number of particles with appropriate values of α^*, γ^*, and $\Delta t^* \Delta t^*$ enables us to conduct DPD simulations. If we take into account that the time is nondimensionalized by the representative time based on the average velocity $\bar{v}(\approx (kT/m)^{1/2})$ and distance r_c, the nondimensionalized time interval Δt^* has to be taken as $\Delta t^* \ll 1$.

The above-mentioned equations of motion retain flexibility and are determined by approach rather than the mathematical manipulation of certain basic key equations. These equations of motion are the revised version of the original equations, which were derived in order that the velocity distribution function of the particles converges to an equilibrium distribution for thermodynamic equilibrium. Hence, they are not the only valid equations of motion for the DPD method, and a new equation of motion may be proposed in order to enable us to conduct more accurate simulations.

The main procedure for conducting the DPD simulation is quite similar to the one we employed for BD simulations, so it is unnecessary to repeat the details here.

5.6   LATTICE BOLTZMANN METHOD

Whether or not the Lattice Boltzmann method is classified into the class of molecular simulation methods may depend on the researcher, but this method is expected to have a sufficient feasibility as a simulation technique for polymeric liquids and particle dispersions. In the Lattice Boltzmann method, a fluid is assumed to be composed of virtual fluid particles, and such fluid particles move and collide with other fluid particles in a simulation region. A simulation area is regarded as a lattice system, and fluid particles move from site to site; that is, they do not move freely in a region. The most significant difference of this method in relation to the MD method is that the Lattice Boltzmann method treats the particle distribution function of velocities rather than the positions and the velocities of the fluid particles [43–45].

Figure 5.4 illustrates the Lattice Boltzmann method for a two-dimensional system. Figure 5.4A shows that a simulation region is divided into a lattice system. Figure 5.4B is a magnification of a unit square lattice cell. Virtual fluid particles, which are regarded as groups or clusters of solvent molecules, are permitted to move only to their neighboring sites, not to other, more distant sites. That is, the fluid particles at site 0 are permitted to stay there or to move to sites 1,2,...,8 at the next time step.

This implies that fluid particles for moving to sites 1, 2, 3, and 4 have the velocity $c = (\Delta x/\Delta t)$, and those for moving to sites 5, 6, 7 and 8 have the velocity $\sqrt{2}c$, in which Δx is the lattice separation of the nearest two sites and Δt is the time interval for simulations. Since the movement speeds of fluid particles are known as c or $\sqrt{2}c$, macroscopic velocities of a fluid can be calculated by evaluating the number of particles moving to each neighboring lattice site. In the usual Lattice Boltzmann method, we treat the particle distribution function, which is defined as a quantity such that the above-mentioned number is divided by the volume and multiplied by the mass occupied by each lattice site [46–47]. This is the concept of the Lattice Boltzmann method. The two-dimensional lattice model shown in Fig. 5.4 is called the "D2Q9" model because fluid particles have nine possibilities of velocities, including the quiescent state (staying at the original site).

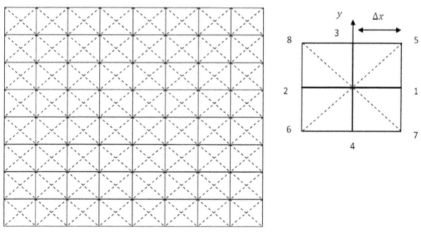

FIGURE 5.4   The Lattice Boltzmann method for a two-dimensional system.

Next, the basic equations of the particle distribution function and the method of solving these equations will be explained. The velocity vector for fluid particles moving to their neighboring site is usually denoted by c_α and, for the case of the D2Q9 model, there are nine possibilities, such as $c_0, c_1, c_2, \ldots, c_8$. For example, the velocity of the movement in the left direction in Fig. 5.4B is denoted by c_2, and c_0 is zero vector for the quiescent state ($c_0 = 0$). We consider the particle distribution function $f_\alpha(r, t)$ at the position r (at point 0 in Fig. 5.4B) at time t in the α-direction. Since $f_\alpha(r, t)$ is equal to the number density of fluid particles moving in the α-direction, multiplied by the mass of a fluid particle, the summation of the particle distribution function concerning all the directions $\alpha = 0,1,\ldots,8$ leads to the macroscopic density $\rho(r,t)$:

$$\rho(r,t) = \sum_{\alpha=0}^{8} f_\alpha(r,t) \qquad (5.88)$$

Similarly, the macroscopic velocity can be evaluated from the following relationship of the momentum per unit volume at the position r:

$$\rho(r,t)u(r,t) = \sum_{\alpha=0}^{8} f_\alpha(r,t)c_\alpha \qquad (5.89)$$

In Eqs. (5.88) and (5.89), the macroscopic density $\rho(r,t)$ and velocity $u(r, t)$ can be evaluated if the particle distribution function is known. Since fluid particles collide with the other fluid particles at each site, the rate of the number of particles moving to their neighboring sites changes. In the rarefied gas dynamics, the well-known Boltzmann equation is the basic equation specifying the velocity distribution function while taking into account the collision term due to the interactions of gaseous molecules; this collision term is a complicated integral expression. The Boltzmann equation is quite difficult to solve analytically, so an attempt has been made to simplify the collision term. One such simplified model is the Bhatnagar-Gross-Krook (BGK) collision model [48, 49]. It is well known that the BGK Boltzmann method gives rise to reasonably accurate solutions, although this collision model is expressed in quite simple form. We here show the Lattice Boltzmann equation based on the BGK model. According to this model, the particle distribution function $f_\alpha(r+c_\alpha\Delta t, t+\Delta t)$ in the α-direction at the position $(r+c_\alpha\Delta t)$ at time $(t + \Delta t)$ can be evaluated by the following equation:

$$f_\alpha(r+c_\alpha\Delta t, t+\Delta t) = f_\alpha(r,t) + \frac{1}{\tau}\left\{f_\alpha^{(0)}(r,t) - f_\alpha(r,t)\right\} \quad (5.90)$$

This equation is sometimes expressed in separate expressions indicating explicitly the two different processes of collision and transformation:

$$f_\alpha(r+c_\alpha\Delta t, t+\Delta t) = \tilde{f}_\alpha(r,t) \qquad (5.91)$$

$$\tilde{f}_\alpha(r+c_\alpha\Delta t, t+\Delta t) = \tilde{f}_\alpha(r,t)$$

$$+\frac{1}{\tau}\left\{f_\alpha^{(0)}(r,t) - f_\alpha(r,t)\right\}$$

The equilibrium distribution, expressed for the D2Q9 model as:

$$f_\alpha^{(0)} = \rho w_\alpha \left\{ 1 + 3\frac{c_\alpha.u}{c^2} - \frac{3u^2}{2c^2} + \frac{9}{2}\cdot\frac{(c_\alpha.u)^2}{c^4} \right\} \tag{5.92}$$

$$w_\alpha = \begin{array}{ll} 4/9 & \text{for } \alpha = 0 \\ 1/9 & \text{for } \alpha = 1,2,3,4 \\ 1/36 & \text{for } \alpha = 5,6,7,8 \end{array} \tag{5.93}$$

$$|c_\alpha| = \begin{array}{ll} 0 & \text{for } \alpha = 0 \\ c & \text{for } \alpha = 1,2,3,4 \\ \sqrt{2}c & \text{for } \alpha = 5,6,7,8 \end{array} \tag{5.94}$$

In these equations ρ is the local density at the position of interest, u is the fluid velocity $(u = |u|)$, $c = \Delta x / \Delta t$, and w_α is the weighting constant.

The important feature of the BGK model shown in Eq. (5.91) is that the particle distribution function in the α-direction is independent of the other directions. The particle distributions in the other directions indirectly influence $f_\alpha(r + c_\alpha \Delta t, t + \Delta t)$ through the fluid velocity u and the density ρ. The second expression in Eq. (5.91) implies that the particle distribution $f_\alpha(r,t)$ at the position r changes into $\tilde{f}_\alpha(r,t)$; Δt after the collision at the site at time t, and the first expression implies that $\tilde{f}_\alpha(r,t)$ becomes the distribution $f_\alpha(r + c_\alpha \Delta t, t + \Delta t)$ at $(r + c_\alpha \Delta t)$ after the time interval Δt.

The main procedure of the simulation is as follows [50, 51]:

1. Set appropriate fluid velocities and densities at each lattice site.
2. Calculate equilibrium particle densities $f_\alpha^{(0)}(\alpha = 0,1,\ldots,8)$ at each lattice site from Eq. (5.92) and regard these distributions as the initial distributions, $f_\alpha = f_\alpha^{(0)}(\alpha = 0,1,\ldots,8)$
3. Calculate the collision terms $f_\alpha^{(0)}(r,t)(\alpha = 0,1,\ldots,8)$ at all sites from the second expression of Eq. (5.91).
4. Evaluate the distribution at the neighboring site in the α-direction $f_\alpha(r + c_\alpha \Delta t, t + \Delta t)$ from the first expression in Eq. (5.91).
5. Calculate the macroscopic velocities and densities from Eqs. (5.88) and (5.89), and repeat the procedures from Step 3.

In addition to the above-mentioned procedures, it must be handled the treatment at the boundaries of the simulation region. For example, the periodic boundary condition, which is usually used in MD simulations, may be applicable.

For the D3Q19 model, which is applicable for three-dimensional simulations, the equilibrium distribution function is written in the same expression of Eq. (5.92), but the weighting constants are different from Eq. (5.93) and are expressed in Eq. (5.69). The basic equations for $f_\alpha \left(r + c_\alpha \Delta t, t + \Delta t \right)$ are the same as Eq. (5.90) or (5.91) and the above-mentioned simulation procedure is also directly applicable to the D3Q19 model.

5.7 CONCLUSION

Conventional molecular simulation methods perform importance sampling in configuration space (or phase space), in which a point represents a particular state of the system. All-atom molecular simulation methods treat each atom as an individual particle, which generally limits the spatial size of the simulated system to a length scale of less than tens of nanometers. MD simulations become much more complicated, if the translational and rotational motion has to be simulated for axisymmetric particle dispersion in comparison with the spherical particle system. Simulation techniques for a dispersion composed of nonspherical particles with a general shape may be obtained by generalizing the methods employed to axisymmetric particle dispersion. When treating a colloidal dispersion or a polymeric solution, the Stokesian dynamics and BD methods are usually employed as a microscopic or mesoscopic analysis tool. In these methods, dispersed particles or polymers are modeled as idealized spherical or dumbbell particles, but the base liquid is usually assumed to be a continuum medium and its effect is included in the equations of motion of the particles or the polymers only as friction terms. In the MD method, the motion of molecules (particles) is simulated according to the equations of motion and therefore it is applicable to both thermodynamic equilibrium and nonequilibrium phenomena. In contrast, the MC method generates a series of microscopic states under a certain stochastic law, irrespective of the equations of motion of particles. The BD method simulates the random motion of dispersed particles that is induced by the solvent molecules; thus, such particles are called "Brownian particles." Since the BD method with multibody hydrodynamic interactions among the particles is very complicated, an alternative method is attended that treats the friction forces between the particles and a base liquid, and the nonhydrodynamic interactions between the particles. The DPD method is principally applicable to simulations of colloidal dispersions that take into account the multibody hydrodynamic interactions among particles. But in the Lattice Boltzmann method, a fluid is assumed to be composed of virtual fluid particles, and such fluid particles move and collide with other fluid particles in a simulation region. A simulation area is regarded as a lattice system, and fluid particles move from site to site; that is, they do not move freely in a region.

KEYWORDS

- **Brownian dynamics method**
- **dissipative particle dynamics method**
- **Lattice Boltzmann methods**
- **molecular dynamics methods**
- **molecular simulation**
- **Monte Carlo methods**

REFERENCES

1. Mansoori, G. A. (2005). *Principles of Nanotechnology: Molecular-based Study of Condensed Matter in Small Systems*: World Scientific. 341.
2. Wang, Z. L. (2003). *New Developments in Transmission Electron Microscopy for Nanotechnology.* Advanced Materials. 15(18), 1497–1514.
3. Schlick, T. (2010). *Molecular Modeling and Simulation: An Interdisciplinary Guide.* 2th ed. Vol. 21, Springer. 768.
4. Turner, C. H., et al. (2008). *Simulation of Chemical Reaction Equilibria by the Reaction Ensemble Monte Carlo Method: A Review.* Molecular Simulation. 34(2), 119–146.
5. Starr, F. W., Schröder, T. B., Glotzer, S. C. (2002). *Molecular Dynamics Simulation of a Polymer Melt with a Nanoscopic Particle.* Macromolecules. 35(11), 4481–4492.
6. Allen, M. P. (2004). *Introduction to Molecular Dynamics Simulation.* Computational Soft Matter: From Synthetic Polymers to Proteins. 23, 1–28.
7. Smith, J. S., Bedrov, D., Smith, G. D. (2003). *A Molecular Dynamics Simulation Study of Nanoparticle Interactions in a Model Polymer-nanoparticle Composite.* Composites Science and Technology. 63(11), 1599–1605.
8. Phillips, J. C. et al., (2005). *Scalable Molecular Dynamics with NAMD.* Journal of Computational Chemistry. 26(16), 1781–1802.
9. Bishop, T. C., Skeel, R. D., Schulten, K. (1997). *Difficulties with Multiple Time Stepping and Fast Multipole Algorithm in Molecular Dynamics.* Journal of computational chemistry. 18(14), 1785–1791.
10. Tuckerman, M., Berne B. J., Martyna, G. J. (1992). *Reversible Multiple Time Scale Molecular Dynamics.* The Journal of Chemical Physics. 97(3), 1990–2001.

11. Humphreys, D. D., Friesner, R. A., Berne, B. J. *A Multiple-time-step Molecular Dynamics Algorithm for Macromolecules.* The Journal of Physical Chemistry, (1994). 98(27), 6885–6892.

12. Van Gunsteren, W. F., Berendsen, H. J. C. (1988). *A Leap-frog Algorithm for Stochastic Dynamics.* Molecular Simulation. 1(3), 173–185.

13. Izaguirre, J. A. et al., (2001). *Langevin Stabilization of Molecular Dynamics.* The Journal of Chemical Physics. 114, 2090–2106.

14. Wheeler, D. R., Newman, J. (2004). *Molecular Dynamics Simulations of Multicomponent Diffusion. 1. Equilibrium Method.* The Journal of Physical Chemistry B. 108(47), 18353–18361.

15. Mareschal, M., Kestemont, E. (1987). *Order and Fluctuations in Nonequilibrium Molecular Dynamics Simulations of Two-dimensional Fluids.* Journal of Statistical Physics. 48(5–6), 1187–1201.

16. Rapaport, D. C. (2004). *The Art of Molecular Dynamics Simulation.* 2 ed: Cambridge university press. 564.

17. O'Sullivan, C. (2011). *Particle-based Discrete Element Modeling: Geomechanics Perspective.* International Journal of Geomechanics. 11(6), 449–464.

18. Donev, A., Torquato, S., Stillinger, F. H. (2005). *Neighbor List Collision-driven Molecular Dynamics Simulation for Nonspherical Hard Particles. I. Algorithmic Details.* Journal of Computational Physics. 202(2), 737–764.

19. Moeendarbary, E., Ng, T. Y., Zangeneh, M. (2009). *Dissipative Particle Dynamics: Introduction, Methodology and Complex Fluid Applications-A Review.* International Journal of Applied Mechanics. 1(04), 737–763.

20. Chow, T. S. (2000). *Mesoscopic Physics of Complex Materials.* 1 ed: Springer. 196.

21. Kalos, M. H., Whitlock, P. A. (2008). *Monte Carlo Methods.* Second ed: John Wiley & Sons. 203.

22. Katsoulakis, M. A., Majda, A. J., Vlachos, D. G. (2003). *Coarse-grained Stochastic Processes for Microscopic Lattice Systems.* Proceedings of the National Academy of Sciences. 100(3), 782–787.

23. Biskos, G., Mastorakos, E., Collings, N. (2004). *Monte-Carlo Simulation of Unipolar Diffusion Charging for Spherical and Non-spherical Particles.* Journal of Aerosol Science. 35(6), 707–730.

24. Guarnieri, F., Still, W. C. (1994). *A Rapidly Convergent Simulation Method: Mixed Monte Carlo/stochastic Dynamics.* Journal of Computational Chemistry. 15(11), 1302–1310.

25. Hagerman, P. J., Zimm, B. H. (1981). *Monte Carlo Approach to the Analysis of the Rotational Diffusion of Wormlike Chains.* Biopolymers. 20(7), 1481–1502.

26. Chen, J. C., Kim, A. S. (2004). *Brownian Dynamics, Molecular Dynamics, and Monte Carlo Modeling of Colloidal Systems.* Advances in Colloid and Interface Science. 112(1), 159–173.

27. Branka, D. (1998). *Monte Carlo as Brownian Dynamics.* Molecular Physics. 94(3), 447–454.

28. Kolinski, A., Skolnick, J. (1994). *Monte Carlo Simulations of Protein Folding. I. Lattice Model and Interaction Scheme.* Proteins: Structure, Function, and Bioinformatics. 18(4), 338–352.

29. Malevanets, A., Kapral, R. (2000). *Solute Molecular Dynamics in a Mesoscale Solvent.* The Journal of Chemical Physics. 112, 7260–7269.

30. Madura, J. D., et al. (1995). *Electrostatics and Diffusion of Molecules in Solution: Simulations with the University of Houston Brownian Dynamics Program.* Computer Physics Communications. 91(1), 57–95.

31. Grassia, P. S., Hinch, E. J., Nitsche, L. C. (1995). *Computer Simulations of Brownian Motion of Complex Systems.* Journal of Fluid Mechanics. 282, 373–403.

32. Ilin, A., et al. (1995). *Parallelization of Poisson-Boltzmann and Brownian Dynamics Calculations.* in *ACS Symposium Series*: ACS Publications.

33. Elvingson, C. (1991). *A general Brownian Dynamics Simulation Program for Biopolymer Dynamics and its Implementation on a Vector Computer.* Journal of Computational Chemistry. 12(1), 71–77.

34. Kong, Y., et al. (1994). *Simulation of a Confined Polymer in Solution Using the Dissipative Particle Dynamics Method.* International Journal of Thermophysics. 15(6), 1093–1101.

35. Espafiol, P. (1995). *Hydrodynamics from Dissipative Particle Dynamics.* Physical Review E. 52, 1734–1742.

36. Satoh, A., Chantrell, R. W. (2006). *Application of the Dissipative Particle Dynamics Method to Magnetic Colloidal Dispersions.* Molecular Physics. 104(20–21), 3287–3302.

37. Jamali, S., Yamanoi M., Maia, J. (2013). *Bridging the Gap between Microstructure and Macroscopic Behavior of Monodisperse and Bimodal Colloidal Suspensions.* Soft Matter. 9(5), 1506–1515.

38. Succi, S. (2014). *Mesoscopic Particle Models of Fluid Flows*, in *Stochastic Methods in Fluid Mechanics*, Springer. 137–165.

39. T. Serrano, P. et al., (2005). *Modeling Aggregation of Colloidal Particles.* Current Opinion in Colloid & Interface Science. 10(3), 123–132.

40. Ihle, T., Kroll, D. M. (2001). *Stochastic Rotation Dynamics: A Galilean-invariant Mesoscopic Model for Fluid Flow.* Physical Review E. 63(2), 8321–8324.

41. Ihle, T., Kroll, D. M. (2003). *Stochastic Rotation Dynamics. I. Formalism, Galilean invariance, and Green-Kubo Relations.* Physical Review E. 67(6), 066705–066715.

42. Marconi, U. M. B., et al. (2008). *Fluctuation–dissipation: Response Theory in Statistical Physics.* Physics Reports. 461(4), 111–195.

43. Stickel, J. J., Powell, R. L. (2005). *Fluid Mechanics and Rheology of Dense Suspensions.* Annual Review of Fluid Mechanics. 37, 129–149.

44. Ladd, A. J. C., Verberg, R. (2001). *Lattice-Boltzmann Simulations of Particle-fluid Suspensions.* Journal of Statistical Physics. 104(5–6), 1191–1251.

45. Ahlrichs, P., Dünweg, B. (1998). *Lattice-Boltzmann Simulation of Polymer-solvent Systems.* International Journal of Modern Physics C. 9(08), 1429–1438.

46. He, X., Luo, L. S. (1997). *Lattice Boltzmann Model for the Incompressible Navier–Stokes Equation.* Journal of Statistical Physics. 88(3–4), 927–944.

47. D'Humières, D., Lallemand, P., Luo, L. S. (2001). *Lattice Boltzmann Equation on a Two-dimensional Rectangular Grid.* Journal of Computational Physics. 172(2), 704–717.

48. Skinner, J. L., Wolynes, P. G. (1979). *Derivation of Smoluchowski Equations with Corrections for Fokker-Planck and BGK Collision Models.* Physica A: Statistical Mechanics and its Applications. 96(3), 561–572.

49. d'Humières, D. (2002). *Multiple relaxation time Lattice Boltzmann Models in Three Dimensions.* Philosophical Transactions of the Royal Society of London. Series A: Mathematical, Physical and Engineering Sciences. 360(1792), 437–451.

50. Yu, D., et al. (2003). *Viscous Flow Computations with the Method of Lattice Boltzmann Equation.* Progress in Aerospace Sciences. 39(5), 329–367.

51. Nourgaliev, R. R., et al. (2003). *The Lattice Boltzmann Equation Method: Theoretical Interpretation, Numerics and Implications.* International Journal of Multiphase Flow. 29(1), 117–169.

CHAPTER 6

NUMERICAL SIMULATION OF NANOELEMENTS

CONTENTS

Abstract 168
6.1   Simulation of Nanoelements Formation and Interaction 168
6.2   Problem Statement and Modeling Technique 171
6.3   Problem Formulation for Interaction of Several Nanoelements 175
6.4   Structure and Forms of Nanoparticles 179
6.5   Nanoparticles Interaction 182
6.6   Conclusion 194
Keywords 195
References 195

ABSTRACT

The methods of numeric modeling within the framework of molecular simulation are used for calculating the interactions of nanostructural elements. A method offered is based on the pairwise static interaction potential of nanoelements that is built up with the help of the approximation of the numerical calculation results using this method. Based on the potential of the interaction of the nanostructure elements, which takes into account forces and moments of forces, the method for calculating the nanoelement interactions is applied.

6.1 SIMULATION OF NANOELEMENTS FORMATION AND INTERACTION

6.1.1 INTRODUCTION

The nanoelement structure and properties determine the properties of a nanomaterial, which form of it. One of the main factors in making nanomaterials is the basis of the material sizes, the structure and the shape of the nanoelements. Due to an increase or a decrease in the specific size of nanoelements (nanofibers, nanotubes, nanoparticles, etc.) causes the varying their physical and mechanical properties such as coefficient of elasticity, strength, deformation parameter, etc. over one order. For instance a fine-scaled material with a large ratio of the surface/interface region to the bulk, the surface/interface effect can be substantial which can be studied using a continuum model and atomistic simulation [1–3].

The calculations and experiments of the investigation into these properties show that this is primarily due to a significant rearrangement of the atomic structure and the shape of the nanoelement. The experimental survey of the parameters of the nanoelements is technically complicated and formidable because of their small sizes. In addition, the experimental results are often inconsistent. In particular, some authors have pointed to an increase in the distance between the atoms adjacent to the surface in contrast to the atoms inside the nanoelement, while others observe a decrease in the aforementioned distance [4].

Thus, further detailed systematic investigations of the problem with the use of theoretical methods and mathematical modeling, are required. The atomic structure and the shape of nanoelements depend both on their sizes and on the methods of obtaining, which can be divided into two main groups [5, 6]:

1. Obtaining nanoelements in the atomic coagulation process by "assembling" the atoms and by stopping the process when the nanoparticles grow to a desired size (the so-called "bottom-up" processes). The process of the particle growth is stopped by the change of physical or chemical conditions of the particle formation, by intercepting supplies of the substances that are necessary to form particles, or because of the limitations of the space where nanoelements form.
2. Obtaining nanoelements by breaking or demolition of more massive (coarse) formations to the segments of the desired size (the so-called "up-down" processes).

In fact, there are many publications describing the modeling of the "bottom-up" processes since past [7–8], while the "up-down" processes have been studied very little. Therefore, the objective of this work is the investigation of the regularities of the changes in the structure and shape of nanoparticles formed in the destruction ("up-down") processes depending on the nanoparticle sizes, and building up theoretical dependences describing the above parameters of nanoparticles. For calculating the characteristics of nanomaterials, the interaction of the nanoelements must be accounted since its properties significantly change because of the changes in their original shapes and sizes in the interaction process, during the formation (or usage) of the nanomaterial, in a cardinal structural rearrangement. The investigations show that using self-assembly method leads to a more organized form of a nanosystem [9–11].

There are three main processes:

1. The first process is due to the regular structure formation at the interaction of the nanostructural elements with the surface where they are situated.
2. The second one arises from the interaction of the nanostructural elements with one another.
3. The third process takes place because of the influence of the ambient medium surrounding the nanostructural elements.

The ambient medium influence can have "isotropic distribution" in the space or it can be presented by the action of separate active molecules connecting nanoelements to one another in a certain order. The external action can greatly change the original shape of the structures, which formed by the nanoelements. For example, the application of the external tensile stress leads to the "stretch" of the nanoelement system in the direction of the maximal tensile stress action and the rise in temperature, vice versa, promotes a decrease in the spatial anisotropy of the nanostructures [12].

In the self-organizing process, parallel with the linear moving, the nano-elements are in rotary movement. So the action of moment of forces caused by the asymmetry of the interaction force fields of the nanoelements, by the presence of the "attraction" and "repulsion" local regions on the nanoele-ment surface, and by the "nonisotropic" action of the ambient as well. The above phenomena are a critical key in nanotechnological processes. They al-low developing nanotechnologies for the formation of processes) and build-ing up complex spatial nanostructures consisting of different nanoelements (nanoparticles, nanotubes, fullerenes, super-molecules, etc.) [13]. However, in a number of cases, the tendency towards self-organization interferes with the formation of a desired nanostructure. Thus, the nanostructure arising from the self-organizing process is rigid and stable against external actions. For example, the "adhesion" of nanoparticles interferes with the use of separate nanoparticles in various nanotechnological processes, the uniform mixing of the nanoparticles from different materials and the formation of nanocompos-ite with desired properties. In connection with this, it is important to model the processes of static and dynamic interaction of the nanostructure elements. In this case, it is essential to take into consideration the interaction force mo-ments of the nanostructure elements, which causes the mutual rotation of the nanoelements [14–16].

The investigation of these relationships based on the mathematical model-ing methods requires the solution of the aforesaid problem on the atomic lev-el. It needs large computational aids and computational time, which makes the development of economical calculation methods necessary. So in this chapter the results of the studies of problems of numeric modeling are investigated within the framework of molecular mechanics and dynamics. The regularities of the amorphous phase formation and the nucleation and spread of the crys-talline or hypocrystalline phases are discussed over the entire nanoparticle volume depending on the process parameters, nanoparticles sizes and thermo-dynamic conditions of the ambient. Also the method for calculating the inter-actions of nanostructural elements, which is based on the potential built up is studied with the help of the approximation of the numerical calculation results using the method of molecular dynamics of the pairwise static interaction of nanoparticles. The method for calculating the ordering and self-organizing processes has been developed based on the potential of the pairwise interac-tion of the nanostructure elements, which takes into account forces and mo-ments of forces. The investigation results on the self-organization of the sys-tem consisting of two or more particles are presented and the analysis of the equilibrium stability of various types of nanostructures has been carried out.

6.2   PROBLEM STATEMENT AND MODELING TECHNIQUE

The calculating of the internal structure and the equilibrium configuration of the separate nanoparticles by the molecular mechanics and dynamics methods has two main stages:

1. The "initiation" of the task, that is, the determination of the conditions under which the process of the nanoparticle shape and structure formation begins.
2. The process of the nanoparticle formation.

Note that the original coordinates and initial velocities of the nanoparticle atoms should be determined from the calculation of the macroscopic parameters of the destructive processes at static and dynamic loadings, which occur both on the nanoscale and on the macroscale. Therefore, in the general case, the coordinates and velocities are the result of solving the problem of modeling physical-mechanical destruction processes at different structural levels [17–19].

The interaction of ordering and self-organization of the nanostructure elements calculation includes three main stages:

1. The first stage is building the internal structure and the equilibrium configuration of each separate noninteracting nanostructure element.
2. The second stage is calculating the pairwise interaction of two nanostructure elements.
3. The third stage is establishing the regularities of the spatial structure and evolution with time of the nanostructure as a whole.

6.2.1   THE CALCULATION OF THE INTERNAL STRUCTURE AND THE SHAPE OF THE NONINTERACTING NANOELEMENT

The initialization of the problem is done by giving the initial coordinates and velocities of the nanoparticle atoms:

$$\bar{x}_i = \bar{x}_{i0}, \bar{V}_i = \bar{V}_{i0}, t = 0, \bar{x}_i \subset \Omega_k \tag{6.1}$$

The molecular dynamics method can be used for calculating the structure and the equilibrium configuration of the nanoelement. The interaction of all the atoms forming the nanoelement must be considered. At the first stage of the solution, the nanoelement is not exposed to the action of external forces. It is further used for the next stage of calculations, which is taking in the equilibrium configuration time. At the first stage, the movement of the atoms forming the nanoparticle could be determined by writing the set of Langevin differential equations at the boundary conditions in Eq. (6.1) [20, 21].

$$m_i \times \frac{d\vec{V}_i}{dt} = \sum_{j=1}^{N_K} \vec{F}_{ij} + \vec{F}_i(t) - \alpha_i m_i \vec{V}_i, i = 1,2,\ldots,N_k \qquad (6.2)$$

$$\frac{d\vec{x}_i}{dt} = \vec{V}_i$$

A random set of forces at a given temperature is calculated by Gaussian distribution. The interatomic interaction forces usually are potential and determined by:

$$\vec{F}_{ij} = -\sum_1^n \frac{\partial \Phi(\vec{\rho}_{ij})}{\partial \vec{\rho}_{ij}}, i = 1,2,\ldots,N_k, j=1, 2,\ldots, N_k \qquad (6.3)$$

$$\Phi(\vec{\rho}_{ij}) = \Phi_{cb} + \Phi_{va} + \Phi_{ta} + \Phi_{pg} + \Phi_{es} + \Phi_{hb} \qquad (6.4)$$

In the general case, the potential $\Phi(\vec{\rho}_{ij})$ is given in the form of the sum of several components corresponding to different interaction types.

Here the following potentials are implied: Φ_{cb} – chemical bonds; Φ_{va} – valence angles; Φ_{ta}–torsion angles; Φ_{pg} – flat groups; Φ_{vv} – Van der Waals contacts; Φ_{es} – electrostatics; Φ_{hb} –hydrogen bonds.

The above addends have different functional forms. The parameter values for the interaction potentials are determined based on the experiments (crystallography, spectral, calorimetric, etc.) and quantum calculations [20].

Giving original coordinates (and forces of atomic interactions) and velocities of all the atoms of each nanoparticle in accordance with Eq. (6.2), at the start time, it was found the change of the coordinates and the velocities of each nanoparticle atoms with time from the equation of motion. Since the nanoparticles are not exposed to the action of external forces, they take some atomic equilibrium configuration with time that it will be used for the next calculation stage.

6.2.2 THE CALCULATION OF THE PAIRWISE INTERACTION OF THE TWO NANOSTRUCTURE ELEMENTS

At this stage of solving the problem, two interacting nanoelements are considered:

First, consider the problem statement for symmetric nanoelements, and then for arbitrary shaped nanoelements.

Now consider two symmetric nanoelements situated at the distance S from one another (Fig. 6.1) at the initial conditions:

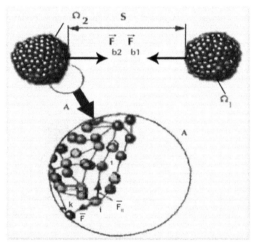

FIGURE 6.1 The scheme of the nanoparticle interaction—an enlarged view of the nanoparticle segment.

$$\bar{x}_i = \bar{x}_{i0}, \bar{V}_i = 0, t = 0, \bar{x}_i \subset \Omega_1 \cup \Omega_2 \qquad (6.5)$$

Where Ω_1, Ω_2 are the areas occupied by the first and the second nanoparticle, respectively.

The coordinates $\bar{x}_{i0}$ could be obtained from Eq. (6.2) at initial conditions given in Eq. (6.1). It allows calculating the combined interaction forces of the nanoelements:

$$\vec{F}_{b1} = -\vec{F}_{b2} = \sum_{i=1}^{N_1} \sum_{j=1}^{N_2} \vec{F}_{ij} \qquad (6.6)$$

where i, j are the atoms and N_1, N_2 are the numbers of atoms in the first and in the second nanoparticle, respectively. Forces are defined from Eq. (6.3).

In the general case, the force magnitude of the nanoparticle interaction $\left|\vec{F}_{bi}\right|$ can be written as product of functions depending on the sizes of the nanoelements and the distance between them:

$$\left|\vec{F}_{bi}\right| = \Phi_{11}(S_{\tilde{n}}) \times \Phi_{12}(D) \qquad (6.7)$$

The $\left|\vec{F}_{bi}\right|$ vector direction is determined by the direction cosines of a vector connecting the centers of the nanoelements. Now, consider two interacting asymmetric nanoelements situated at the distance S_c between their centers of mass (Fig. 6.2) and oriented at certain specified angles relative to each other.

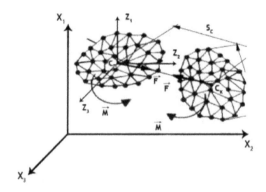

FIGURE 6.2   Two interacting nanoelements, $\vec{M}$, $\vec{F}$ are the principal moment and the principal vector of the forces, respectively.

In contrast to the previous problem, the interatomic interaction of the nanoelements leads to both the relative displacement of the nanoelements and their rotation. Consequently, in the general case, the sum of all the forces of the interatomic interactions of the nanoelements is brought to the principal vector of forces $\vec{F}_c$ and the principal moment $\overrightarrow{M}_c$.

$$\vec{F}_c = \vec{F}_{b1} = -\vec{F}_{b2} = \sum_{i=1}^{N_1}\sum_{j=1}^{N_2} \vec{F}_{ij} \qquad (6.8)$$

$$\vec{M}_c = \vec{M}_{c1} = -\vec{M}_{c2} = \sum_{i=1}^{N_1} \sum_{j=1}^{N_2} \vec{\rho}_{cj} \times \vec{F}_{ij} \tag{6.9}$$

where $\vec{\rho}_{cj}$ is a vector connecting points c and j.

The main objective of this calculation stage is making the dependences of the forces and moments of the nanostructure nanoelement interactions on the distance S_c between the centers of mass of the nanostructure nanoelements, on the angles of mutual orientation of the nanoelements (shapes of the nanoelements) and on the characteristic size of the nanoelement. In the general case, these dependences can be given in the form:

$$\vec{F}_{bi} = \vec{\Phi}_F(S_c, \Theta_1, \Theta_2, \Theta_3, D) \tag{6.10}$$

$$\vec{M}_{bi} = \vec{\Phi}_M(S_c, \Theta_1, \Theta_2, \Theta_3, D) \tag{6.11}$$

For spherical nanoelements, the angles of the mutual orientation do not influence the force of their interaction. So in Eq. (6.12), the moment is zero.

Eqs. (6.11) and (6.12) can be approximated by analogy with Eq. (6.8), as the product of functions $S_0, \Theta_1, \Theta_2, \Theta_3, D$, respectively. For the further numerical solution of the problem of the self-organization of nanoelements, it is sufficient to give the above functions in their tabular form and to use the linear (or nonlinear) interpolation of them in space.

6.3   PROBLEM FORMULATION FOR INTERACTION OF SEVERAL NANOELEMENTS

When the evolution of the nanosystem as whole (including the processes of ordering and self-organization of the nanostructure nanoelements) is investigated, the movement of each system nanoelement is considered as the movement of a single whole. In this case, the translational motion of the center of mass of each nanoelement is given in the coordinate system X_1, X_2, X_3, and the nanoelement rotation is described in the coordinate system Z_1, Z_2, Z_3, which is related to the center of mass of the nanoelement (Fig. 6.2). The system of equations describing the above processes has the form:

$$M_k \frac{d^2 X_1^K}{d\,t^2} = \sum_{J=1}^{N_e} F_{X_1}^{kj} + F_{X_1}^{ke},$$

(6.12)

$$M_k \frac{d^2 X_2^K}{d\,t^2} = \sum_{J=1}^{N_e} F_{X_2}^{kj} + F_{X_2}^{ke},$$

$$M_k \frac{d^2 X_3^K}{d\,t^2} = \sum_{J=1}^{N_e} F_{X_3}^{kj} + F_{X_3}^{ke}$$

$$J_{Z_1}^k \frac{d^2 \Theta_1^k}{d\,t^2} + \frac{d\Theta_2^k}{dt} + \frac{d\Theta_3^k}{dt}\left(J_{Z_3}^k - J_{Z_2}^k\right) = \sum_{j=1}^{N_e} M_{Z_1}^{kj} + M_{Z_1}^{ke}$$

$$J_{Z_2}^k \frac{d^2 \Theta_2^k}{d\,t^2} + \frac{d\Theta_1^k}{dt} + \frac{d\Theta_3^k}{dt}\left(J_{Z_1}^k - J_{Z_3}^k\right) = \sum_{j=1}^{N_e} M_{Z_2}^{kj} + M_{Z_2}^{ke}$$

$$J_{Z_3}^k \frac{d^2 \Theta_3^k}{d\,t^2} + \frac{d\Theta_2^k}{dt} + \frac{d\Theta_1^k}{dt}\left(J_{Z_2}^k - J_{Z_1}^k\right) = \sum_{j=1}^{N_e} M_{Z_3}^{kj} + M_{Z_3}^{ke}$$

where X_i^k, Θ_i^k are coordinates of the centers of mass and angles of the spatial orientation of the principal axis Z_1, Z_2, Z_3 of nanoelements. The initial conditions for the system of Eqs. (6.13) and (6.14) have the Form:

$$\vec{X}^k = \vec{X}_0^K; \; \Theta^k = \Theta_0^k; \; \vec{V}^k = \vec{V}_0^K; \; \frac{d\Theta^k}{dt} = \frac{d\Theta_0^k}{dt}; t = 0$$

(6.13)

6.3.1 NUMERICAL PROCEDURES AND SIMULATION TECHNIQUES

The problem, which was formulated in the previous sections generally, has no analytical solution at each stage. Therefore, the numerical methods are used. In this case, for the first stages, the numerical integration of the equation of motion of the nanoparticle atoms in the relaxation process are used in accordance with Verlet scheme [22]:

$$\vec{x}_i^{n+1} = \vec{x}_i^n + \Delta t\, \vec{V}_i^n + ((\Delta t)^2/2m_i)\left(\sum_{J=1}^{N_k} \vec{F}_{ij} + \vec{F}_i - \alpha_i m_i \vec{V}_i^n\right)^n$$

(6.14)

$$\vec{V}_i^{n+1} = (1 - \Delta t \alpha_i)\vec{V}_i^n + (\Delta t / 2m_i)\left(\sum_{J=1}^{N_k} \vec{F}_{ij} + \vec{F}_i\right)^n \quad (6.15)$$

$$+ \left(\sum_{J=1}^{N_k} \vec{F}_{ij} + \vec{F}_i\right)^{n+1})$$

Where $\vec{x}_i^n, \vec{V}_i^n$ are a coordinate and a velocity of the i-th atom at the n-th step with respect to the time and Δt is a step with respect to the time.

The solution of the Eq. (6.13) also requires the application of numerical methods of integration. Runge–Kutta method is used for solving it [23].

$$(X_i^k)_{n+1} = (X_i^k)_n + (V_i^k)_n \Delta t + \frac{1}{6}(\mu_{1i}^k + \mu_{2i}^k + \mu_{3i}^k)\Delta t \quad (6.16)$$

$$(V_i^k)_{n+1} = (V_i^k)_n + + \frac{1}{6}(\mu_{1i}^k + 2\mu_{2i}^k + 2\mu_{3i}^k + \mu_{4i}^k) \quad (6.17)$$

$$\mu_{1i}^k = \Phi_i^k\big(t_n; (X_i^k)_n, \ldots; (V_i^k)_n \ldots\big)\Delta t,$$

$$(6.17)$$

$$\mu_{2i}^k = \Phi_i^k\left(t_n + \frac{\Delta t}{2}; (X_i^k + V_i^k\frac{\Delta t}{2})_n, \ldots; (V_i^k)_n + \frac{\mu_{1i}^k}{2}, \ldots\right)\Delta t,$$

$$\mu_{3i}^k = \Phi_i^k\left(t_n + \frac{\Delta t}{2}; (X_i^k + V_i^k\frac{\Delta t}{2} + \mu_{1i}^k\frac{\Delta t}{4})_n, \ldots; (V_i^k)_n + \frac{\mu_{2i}^k}{2}, \ldots\right)\Delta t, \quad (6.18)$$

$$\mu_{4i}^k = \Phi_i^k\left(t_n + \Delta t; (X_i^k + V_i^k\Delta t + \mu_{2i}^k\frac{\Delta t}{2})_n, \ldots; (V_i^k)_n + \mu_{2i}^k, \ldots\right)\Delta t, \quad (6.19)$$

$$\Phi_i^k = \frac{1}{M_k}(\sum_{j=1}^{N_e} F_{X_3}^{kj} + F_{X_3}^{ke})$$

$$(\Phi_i^k)_{n+1} = (\Phi_i^k)_n + \left(\frac{d\Phi_i^k}{dt}\right)_n \Delta t + \frac{1}{6}(\lambda_{1i}^k + \lambda_{2i}^k + \lambda_{3i}^k)\Delta t \quad (6.20)$$

$$\left(\frac{d\Phi_i^k}{dt}\right)_{n+1} = \left(\frac{d\Phi_i^k}{dt}\right)_n \Delta t + \frac{1}{6}(\lambda_{1i}^k + 2\lambda_{2i}^k + 2\lambda_{3i}^k + \lambda_{4i}^k) \qquad (6.21)$$

$$\lambda_{1i}^k = \Psi_i^k(t_n; (\Phi_i^k)_n, \dots, \left(\frac{d\Phi_i^k}{dt}\right)_n, \dots)\Delta t$$

$$\lambda_{2i}^k = \Psi_i^k(t_n + \frac{\Delta t}{2}; \left(\Phi_i^k + \frac{d\Phi_i^k}{dt}\frac{\Delta t}{2}\right)_n, \dots, \left(\frac{d\Phi_i^k}{dt}\right)_n + \frac{\lambda_{1i}^k}{2}, \dots)\Delta t \qquad (6.22)$$

$$\lambda_{3i}^k = \Psi_i^k(t_n + \frac{\Delta t}{2}; \left(\Phi_i^k + \frac{d\Phi_i^k}{dt}\frac{\Delta t}{2} + \lambda_{1i}^k\frac{\Delta t}{4}\right)_n, \dots, \left(\frac{d\Phi_i^k}{dt}\right)_n + \frac{\lambda_{2i}^k}{2}, \dots)\Delta t$$

$$\lambda_{4i}^k = \Psi_i^k(t_n + \Delta t; \left(\Phi_i^k + \frac{d\Phi_i^k}{dt}\Delta t + \lambda_{2i}^k\frac{\Delta t}{2}\right)_n, \dots, \left(\frac{d\Phi_i^k}{dt}\right)_n$$
$$+ \lambda_{2i}^k, \dots)\Delta t$$

$$\Psi_1^k = \frac{1}{J_{Z_1}^k}\left(-\frac{d\Theta_2^k}{dt} \times \frac{d\Theta_3^k}{dt}(J_{Z_1}^k - J_{Z_2}^k) + \sum_{j=1}^{N_e} M_{Z_1}^{kj} + M_{Z_1}^{ke}\right), \qquad (6.23a)$$

$$\Psi_2^k = \frac{1}{J_{Z_2}^k}\left(-\frac{d\Theta_1^k}{dt} \times \frac{d\Theta_3^k}{dt}(J_{Z_1}^k - J_{Z_3}^k) + \sum_{j=1}^{N_e} M_{Z_2}^{kj} + M_{Z_2}^{ke}\right)$$

$$\Psi_3^k = \frac{1}{J_{Z_3}^k}\left(-\frac{d\Theta_1^k}{dt} \times \frac{d\Theta_2^k}{dt}(J_{Z_1}^k - J_{Z_2}^k) + \sum_{j=1}^{N_e} M_{Z_3}^{kj} + M_{Z_3}^{ke}\right)$$

where $i = 1, 2, 3; k = 1, 2, \dots, N_e$.

Now consider the realization of the above procedure by taking an example of the metal nanoparticle calculation.

The potentials of the atomic interaction of Morse (Eq. (6.23b)) and Lennard-Johns (Eq. (6.24)) were used in the following calculations:

$$\Phi(\vec{\rho}_{ij})_{LD} = 4\varepsilon \left((\frac{\sigma}{|\vec{\rho}_{ij}|})^{12} - (\frac{\sigma}{|\vec{\rho}_{ij}|})^{6} \right) \qquad (6.23b)$$

where $D_m, \lambda_m, \rho_0, \varepsilon, \sigma$ are the constants of the materials.

6.4   STRUCTURE AND FORMS OF NANOPARTICLES

At the first stage of the problem, the coordinates of the atoms positioned at the ordinary material lattice points (Fig. 6.3, Eq. (6.1)) were taken as the original coordinates. During the relaxation process, the initial atomic system is rearranged into a new "equilibrium" configuration in accordance with the calculations based on Eqs. (6.6)–(6.9), which satisfies the condition when the system potential energy is approaching the minimum Fig. 6.3, the plot).

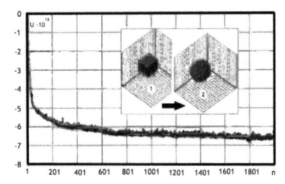

FIGURE 6.3   The initial crystalline (1) and cluster (2) structures of the nanoparticle consisting of 1331 atoms after relaxation; the plot of the potential energy U [J] variations for this atomic system in the relaxation process, where n is the number of iterations with respect to the time).

After the relaxation, the nanoparticles can have quite diverse shapes:

Globe-like, spherical centered, spherical eccentric, spherical icosahedral nanoparticles and asymmetric nanoparticles (Fig. 6.4).

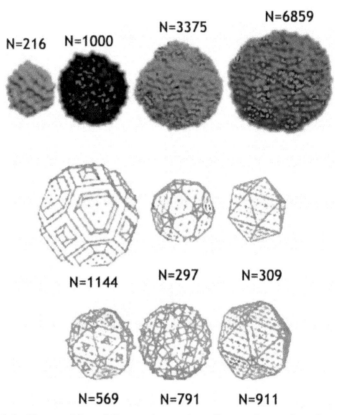

FIGURE 6.4   Nanoparticles of diverse shapes, depending on the number of atoms they contain.

In this case, the number of atoms significantly determines the shape of a nanoparticle. Note, that symmetric nanoparticles are formed only at a certain number of atoms. As a rule, in the general case, the nanoparticle deviates from the symmetric shape in the form of irregular raised portions on the surface. Besides, there are several different equilibrium shapes for the same number of atoms. The plot of the nanoparticle potential energy change in the relaxation process (Fig. 6.5) illustrates it.

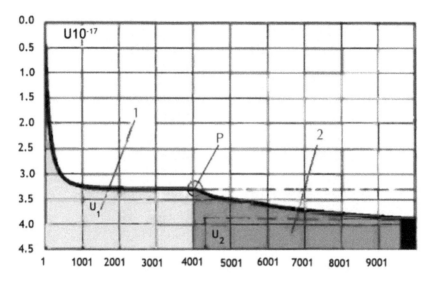

FIGURE 6.5 The plot of the potential energy change of the nanoparticle in the relaxation process. 1 – a region of the stabilization of the first nanoparticle equilibrium shape; 2 – a region of the stabilization of the second nanoparticle equilibrium shape; P – a region of the transition of the first nanoparticle equilibrium shape into the second one.

As it follows from this figure, the curve has two areas:

The area of the potential energy decrease and the area of its stabilization, which promote the formation of the first nanoparticle equilibrium shape.

Then, a repeated decrease in the nanoparticle potential energy and the stabilization area are observed corresponding to the formation of the second nanoparticle equilibrium shape. Between them, there is a region of the transition from the first shape to the second one. Because of the lesser nanoparticle potential energy, the second equilibrium shape is more stable. However, the first equilibrium shape also "exists" rather long in the calculation process. The change of the equilibrium shapes is especially characteristic of the nanoparticles with an "irregular" shape. The internal structure of the nanoparticles is of importance since their atomic structure significantly differs from the crystalline structure of the bulk materials: the distance between the atoms and the angles change, and the surface formations of different types appear. In Fig. 6.6, the change of the structure of a two dimensional nanoparticle in the relaxation process is shown.

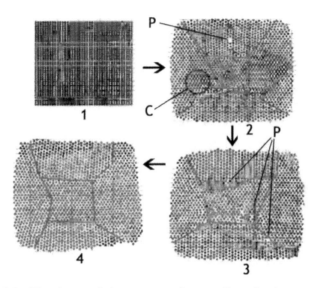

FIGURE 6.6 The change of the structure of a two-dimensional nanoparticle in the relaxation process: 1 – the initial crystalline structure; 2, 3, 4 – the nanoparticles structures which change in the relaxation process; p – pores; c – the region of compression.

The figure shows how the initial nanoparticle crystalline structure (1) is successively rearranging with time in the relaxation process (positions 2, 3, 4). Note that the resultant shape of the nanoparticle is not round. It has "remembered" the initial atomic structure. It is also of interest that in the relaxation process, in the nanoparticle, the defects in the form of pores (designation "p" in the figure) and the density fluctuation regions (designation "c" in the figure) have been formed, which are absent in the final structure.

6.5 NANOPARTICLES INTERACTION

There are some examples of nanoparticles interaction in Fig. 6.7, shows that the calculation results demonstrating the influence of the sizes of the nanoparticles on their interaction force. One can see from the plot that the larger nanoparticles are attracted stronger, the maximal interaction force increases with the size growth of the particle. The interaction force of the nanoparticles is divided by its maximal value for each nanoparticle size, respectively. The obtained plot of the "relative" (dimensionless) force (Fig. 6.8) shows that the value does not practically depend on the nanoparticle size since all the curves come close and can be approximated to one line.

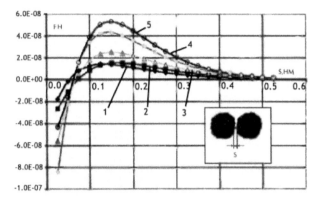

FIGURE 6.7 The dependence of the interaction force F (N) of the nanoparticles on the distance S (Nm) between them and on the nanoparticle size: 1-D = 2.04; 2-D = 2.40; 3-D = 3.05; 4-D = 3.69; 5 – d = 4.09 (Nm).

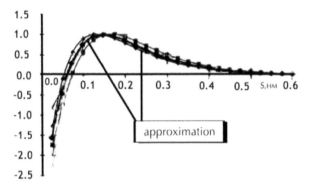

FIGURE 6.8 The dependence of the "relative" force $\bar{F}$ of the interaction of the nanoparticles on the distance S (Nm) between them.

Figure 6.9 represents the dependence of the maximal attraction force between the nanoparticles on their diameter. It is characterized by nonlinearity and a general tendency towards the growth of the maximal force with the nanoparticle size growth. The total force of the interaction between the nanoparticles is determined by multiplying of the two plots (Figs. 6.8 and 6.9).

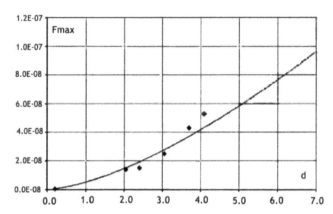

FIGURE 6.9 The dependence of the maximal attraction force F_{max}(N) the nanoparticles on the nanoparticle diameter d (Nm).

Using the polynomial approximation of the curve in Fig. 6.5 and the power mode approximation of the curve in Fig. 6.6, it could be obtained:

$$\bar{F} = (-1.13\,S^6 + 3.08S^5 - 3.41S^4 - 0.58S^3 + 0.82S - 0.00335)10^3 \quad (6.25)$$

$$F_{max} = 0.5 \times 10^{-9}.\,d^{1.499} \quad\quad (6.26)$$

$$F = F_{max}.\bar{F} \quad\quad (6.27)$$

Eqs. (6.25)–(6.27) were used for the calculation of the nanomaterial ultimate strength for different patterns of nanoparticles' "packing" in it (Fig. 6.10).

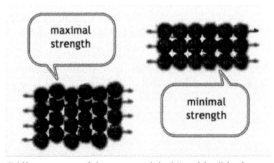

FIGURE 6.10   Different types of the nanoparticles' "packing" in the material.

Figure 6.11 shows the dependence of the ultimate strength of the nanomaterial formed by monodisperse nanoparticles on the nanoparticle sizes. One can see that with the decrease of the nanoparticle sizes, the ultimate strength of the nanomaterial increases, and vice versa. The calculations have shown that the nanomaterial strength properties are significantly influenced by the nanoparticles' "packing" type in the material.

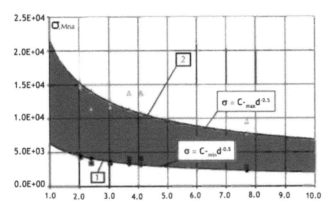

FIGURE 6.11   The dependence of the ultimate strength σ [MPa] of the nanocomposite formed by monodisperse nanoparticles on the nanoparticle sizes d (Nm).

The material strength grows when the packing density of nanoparticles increases. It should be specially noted that the material strength changes in inverse proportion to the nanoparticle diameter in the degree of 0.5, which agrees with the experimentally established law of strength change of nanomaterials (the law by Hall-petch) [24]:

$$\sigma = C.d^{-0.5} \tag{6.28}$$

where $C = C_{max} = 2.17 \times 10^4$ is for the maximal packing density; $C = C_{max} = 6.4 \times 10^3$ is for the minimal packing density.

The electrostatic forces can strongly change force of interaction of nanoparticles. For example, numerical simulation of charged sodium (NaCl) nanoparticles system (Fig. 6.12) has been carried out. Considered ensemble consists of eight separate nanoparticles. The nanoparticles interact due to Vander-Waals and electrostatic forces.

FIGURE 6.12   Nanoparticles system consists of eight nanoparticles NaCl.

Results of particles center of masses motion are introduced at Fig. 6.13 representing trajectories of all nanoparticles included into system. It shows the dependence of the modulus of displacement vector $|R|$ on time. One can see that nanoparticle moves intensively at first stage of calculation process. At the end of numerical calculation, all particles have got new stable locations, and the graphs of the radius vector $|R|$ become stationary.

However, the nanoparticles continue to "vibrate" even at the final stage of numerical calculations. Nevertheless, despite of "vibration," the system of nanoparticles occupies steady position.

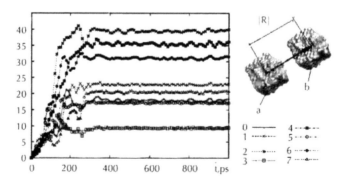

FIGURE 6.13   The dependence of nanoparticle centers of masses motion $|R|$ on the time t; a, b is the nanoparticle positions at time 0 and t, accordingly; 1–8 are the numbers of the nanoparticles.

However, one can observe a number of other situations. For example the self-organization calculation for the system consisting of 125 cubic nanoparticles is considered, the atomic interaction of which is determined by Morse potential (Fig. 6.14).

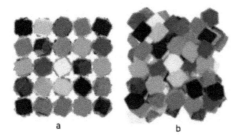

FIGURE 6.14   The positions of the 125 cubic nanoparticles: (a) initial configuration; (b) final configuration of nanoparticles.

As you see, the nanoparticles are moving and rotating in the self-organization process forming the structure with minimal potential energy.

Let us consider, for example, the calculation of the self-organization of the system consisting of two cubic nanoparticles, the atomic interaction of which is determined by Morse potential [25]. Figure 6.15 shows possible mutual positions of these nanoparticles. The positions, where the principal moment of forces is zero, corresponds to pairs of the nanoparticles 2–3; 3–4; 2–5 (Fig. 6.15) and defines the possible positions of their equilibrium.

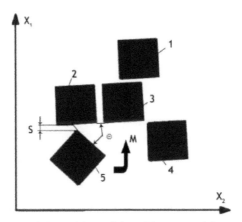

FIGURE 6.15   Characteristic positions of the cubic nanoparticles.

Figure 6.16 presents the dependence of the moment of the interaction force between the cubic nanoparticles 1–3 (Fig. 6.15) on the angle of their relative rotation. From the plot follows that when the rotation angle of particle 1 relative to particle 3 is $\pi/4$, the force moment of their interaction is zero. At an increase or a decrease in the angle the force moment appears. In the range of $\pi/8 < \theta < 3\pi/4$ the moment is small. The force moment rapidly grows outside of this range. The distance S between the nanoparticles plays a significant role in establishing their equilibrium. If $S>S_0$ (where S_0 is the distance, where the interaction forces of the nanoparticles are zero), then the particles are attracted to one another. In this case, the sign of the moment corresponds to the sign of the angle θ deviation from $\pi/4$. At $S<S_0$ (the repulsion of the nanoparticles), the sign of the moment is opposite to the sign of the angle deviation. In other words, in the first case, the increase of the angle deviation causes the increase of the moment promoting the movement of the nanoelement in the given direction, and in the second case, the angle deviation causes the increase of the moment hindering the movement of the nanoelement in the given direction. Thus, the first case corresponds to the unstable equilibrium of nanoparticles, and the second case to their stable equilibrium. The potential energy change plots for the system of the interaction of two cubic nanoparticles (Fig. 6.17) illustrate the influence of the parameter S. Here, curve 1 corresponds to the condition $S<S_0$ and it has a well-expressed minimum in the $0.3 < \theta < 1.3$ region. At $\theta < 0.3$ and $\theta < 1.3$, the interaction potential energy sharply increases, which leads to the return of the system into the initial equilibrium position. At $S > S_0$ (curves 2–5), the potential energy plot has a maximum at the $\theta = 0$ point, which corresponds to the unstable position.

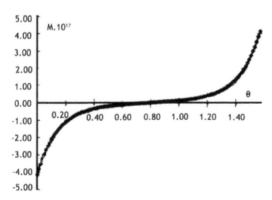

FIGURE 6.16 The dependence of the moment M (Nm) of the interaction force between cubic nanoparticles 1–3 (*see* Fig. 6.9) on the angle of their relative rotation θ [rad].

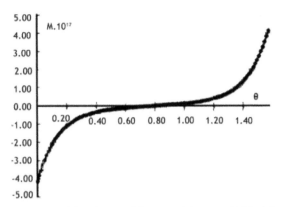

FIGURE 6.17   The plots of the change of the potential energy E (Nm) for the interaction of two cubic nanoparticles depending on the angle of their relative rotation θ [rad] and the distance between them (positions of the nanoparticles 1–3, Fig. 6.9).

The carried out theoretical analysis is confirmed by the works of the scientists from New Jersey University and California University in Berkeley who experimentally found the self-organization of the cubic microparticles of Plumbum Zirconate Titanate (PZT) [6]: the ordered groups of cubic microcrystals from PZT obtained by hydrothermal synthesis formed a flat layer of particles on the air-water interface, where the particle occupied the more stable position corresponding to position 2–3 in Fig. 6.15.

Thus, the analysis of the interaction of two cubic nanoparticles has shown that different variants of their final stationary state of equilibrium are possible, in which the principal vectors of forces and moments are zero. However, there are both stable and unstable stationary states of this system: nanoparticle positions 2–3 are stable, and positions 3–4 and 2–5 have limited stability or they are unstable depending on the distance between the nanoparticles.

Note that for the structures consisting of a large number of nanoparticles, there can be a quantity of stable stationary and unstable forms of equilibrium. Accordingly, the stable and unstable nanostructures of materials can appear. The search and analysis of the parameters determining the formation of stable nanosystems is necessary.

It is important to consider, that the method offered has limitations. This is explained by change of the nanoparticles form and accordingly variation of interaction pair potential during nanoparticles coming together at certain conditions.

The merge (accretion [4]) of two or several nanoparticles into a single whole is possible (Fig. 6.18). Change of a kind of connection cooperating nanoparticles (merging or coupling in larger particles) depending on its sizes, it is possible to explain on the basis of the analysis of the energy change graph of connection nanoparticles (Fig. 6.19). From Fig. 6.19 follows, that, though with the size increasing of a particle energy of nanoparticles connection E_{np} grows also, its size in comparison with superficial energy E_s of a particle sharply increases at reduction of the sizes nanoparticles.

Hence, for finer particles energy of connection can appear sufficient for destruction of their configuration under action of a mutual attraction and merging in larger particle.

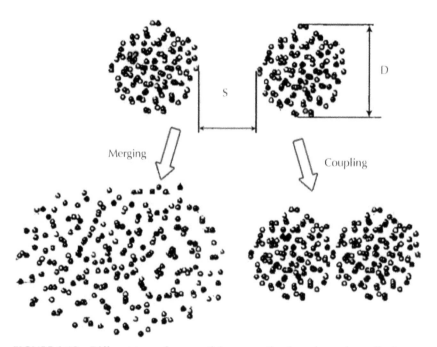

FIGURE 6.18   Different type of nanoparticles connection (merging and coupling).

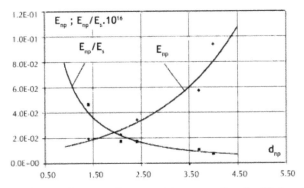

FIGURE 6.19   Change of energy of nanoparticles connection E_{np} (Nm) and E_{np} ration to superficial energy E_s depending on nanoparticles diameter d (Nm). Points designate the calculated values. Continuous lines are approximations.

Spatial distribution of particles influences on rate of the forces holding nanostructures, formed from several nanoparticles. It can see in Fig. 6.20, the chain nanoparticles formation is resulted at coupling of three nanoparticles, located in the initial moment on one line. Calculations have shown the nanoparticles form a stable chain, in this case. Thus, particles practically do not change the form and cooperate on "small platforms."

Distance between particles, which are balance, is much less than from the particles collected in group ($L_{3np}^0 < L_{2np}^0$). It indicates in the graph of forces. The maximal force of an attraction between particles, is designated by a continuous line, in some times is more than at an arrangement of particles in a chain (dashed line) $F_{3np} > F_{2np}$.

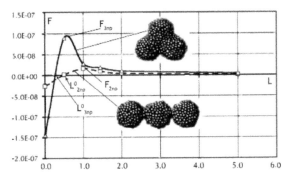

FIGURE 6.20   Change of force F (N) of three nanoparticles interaction, consisting of 512 atoms everyone, and connected among themselves on a line and on the beams missing under a corner of 120 degrees, accordingly, depending on distance between them L (Nm).

Experimental investigation of the spatial structures formed by nanoparticles [26], acknowledge that nanoparticles gather to compact objects. As the
internal nuclear structure of the connections area of nanoparticles extremely
differs from structure of a free nanoparticle.

Nanoelement interactions depend strongly on the temperature. Figure 6.21
shows the scheme of the interaction of nanoparticles at different temperatures. In Fig. 6.22, it is seen that with increasing temperature the interaction
of changes in sequence: coupling (1, 2), merging (3, 4) and with more increase
in temperature the nanoparticles dispersed.

Finally in this section the problems of nanoparticle dynamics are discussed. The analysis of interaction of nanoparticles allows drawing a conclusion on a role of initial movement energy of particles in this process. In
various processes at interaction of the nanoparticles which move with different speed, some cases can be observed such as the processes of agglomerate
formation, formation of larger particles at merge of the smaller size particles,
absorption by large particles of the smaller ones, dispersion of particles on
separate smaller ones or atoms.

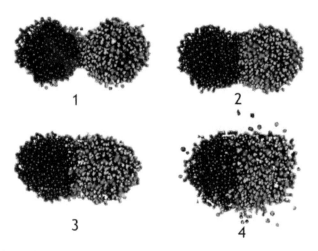

FIGURE 6.21   Change of nanoparticles connection at increase in temperature.

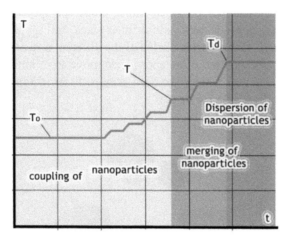

FIGURE 6.22   Curve of temperature change.

For example, in Fig. 6.23 the interactions of two particles are moving towards each other with different speed are shown. At small speed of moving, the steady agglomerate is formed (Fig. 6.23). Also Fig. 6.23 (left) is submitted interaction of two particles moving towards each other with the large speed. It is obvious that the steady formation in this case is not appearing and the particles collapse.

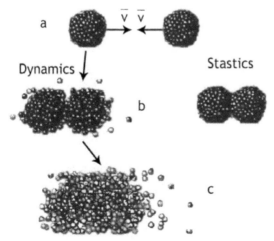

FIGURE 6.23   Pictures of dynamic interaction of two nanoparticles: a) an initial configuration nanoparticles; b) nanoparticles at dynamic interaction, c) the "cloud" of atoms formed because of dynamic destruction two nanoparticles.

The nature of nanoparticle interactions, along the speed of their movement, essentially depends on a ratio of their sizes. In Fig. 6.24, pictures of interaction of two nanoparticles of zinc of the different size are presented. As well as in the previous case of a nanoparticle move from initial situation (1) towards each other. At small initial speed, nanoparticles incorporate at contact and form a steady conglomerate (2).

FIGURE 6.24   Pictures of interaction of two nanoparticles of zinc: 1 – initial configuration of nanoparticles; 2 – connection of nanoparticles, 3, 4 – absorption by a large nanoparticle of a particle of the smaller size, 5 – destruction of nanoparticles at blow.

An increase in speed of movement makes larger nanoparticle absorbs smaller, and leads to the uniform nanoparticle formation (3, 4). At further increase in speed of nanoparticle movements, causes the smaller particle intensively takes root in big and destroys its.

The given examples show that use of dynamic processes of pressing for formation of nanomaterials demands a right choice of a mode of the loading providing integrity of nanoparticles. At big energy of dynamic loading, instead of a nanomaterial with dispersion corresponding to the initial size of nanoparticles, the nanocomposite with much larger grain will essentially change properties of a material can be obtained.

6.6   CONCLUSION

The problems of simulation of the formation and development of nanosystems with time were considered at different structural levels. Main theoretical concepts were systematized, which allowed successive modeling of a nanosystem with the use of the molecular dynamics, mesodynamics and continuum mechanics methods. Practical problems such as the calculation of dependence of the elasticity modulus of a nanoparticle on its size, the investigation of the formation and movement of nanoparticles were discussed. The calculating of the internal structure and the equilibrium configuration of the separate nanoparticles by the molecular mechanics and dynamics methods has studied. As the evolution of the nanosystem as whole (including the processes of ordering and self-organization of the nanostructure nanoelements) is investigated, the movement of each system nanoelement is considered as the movement of a single whole. The interaction force of the nanoparticles is divided by its maximal value for each nanoparticle size.

KEYWORDS

- **continuum mechanics method**
- **mesodynamics method**
- **nanoelement modeling**
- **nanoelement simulation**
- **nanoelements**

REFERENCES

1. Duan, H. L., et al. (2005). *Size-dependent Effective Elastic Constants of Solids Containing Nano-inhomogeneities with Interface Stress.* Journal of the Mechanics and Physics of Solids, 53(7), 1574–1596.
2. Diao, J., Gall, K., Dunn, M. L. (2004). *Atomistic Simulation of the Structure and Elastic Properties of Gold Nanowires.* Journal of the Mechanics and Physics of Solids, 52(9), 1935–1962.
3. Dingreville, R., Qu, J., Cherkaoui, M. (2005). *Surface Free Energy and its Effect on the Elastic Behavior of Nano-sized Particles, Wires and Films.* Journal of the Mechanics and Physics of Solids, 53(8), 1827–1854.
4. Gusev, A. I., Rempel, A. A. (2004). *Nanocrystalline Materials* Cambridge International Science Publishing. 346.
5. Shevchenko, E. V., et al. (2006). *Structural Diversity in Binary Nanoparticle Superlattices.* Nature, 439(7072), 55–59.
6. Vakhrouchev, A. V. (2006). *Simulation of Nano-elements Interactions and Self-assembling.* Modelling and Simulation in Materials Science and Engineering, 14(6), 975–991.
7. Hoare, M. R. (1979). *Structure and Dynamics of Simple Microclusters.* Advances in Chemical Physics, 40, 49–135.

8. Brooks, B. R., et al. (1983). *CHARMM: A Program for Macromolecular Energy, Minimization, and Dynamics Calculations.* Journal of computational chemistry, 4(2), 187–217.

9. Melikhov, I. V., Bozhevol'nov, V. E. (2003). *Variability and Self-organization in Nanosystems.* Journal of Nanoparticle Research, 5(5–6), 465–472.

10. Kim, D., Lu, W. (2004). *Self-organized Nanostructures in Multi-phase Epilayers.* Nanotechnology, 15(5), [667–674].

11. Friedlander, S. K. (1999). *Polymer-like Behavior of Inorganic Nanoparticle Chain Aggregates.* Journal of Nanoparticle Research, 1(1), [9–15].

12. Grzegorczyk, M., Rybaczuk, M., Maruszewski, K. (2004). *Ballistic Aggregation: An Alternative Approach to Modeling of Silica Sol–gel Structures.* Chaos, Solitons & Fractals, 19(4), 1003–1011.

13. Toma, H. E. (2000). *Supramolecular Chemistry and Technology.* Anais da Academia Brasileira de Ciências,. 72(1), 5–26.

14. Teichert, C. (2002). *Self-organization of Nanostructures in Semiconductor Heteroepitaxy.* Physics Reports. 365(5), 335–432.

15. Pohl, K., et al. (1999). *Identifying the Forces Responsible for Self-organization of Nanostructures at Crystal Surfaces.* Nature, 397(6716), 238–241.

16. Cölfen, H., Mann, S. (2003). *Higher-order Organization by Mesoscale Self-assembly and Transformation of Hybrid Nanostructures.* Angewandte Chemie International Edition, 42(21), 2350–2365.

17. Vakhrouchev, A., Lipanov, A. (1992). *A Numerical Analysis of the Rupture of Powder Materials under the Power Impact Influence.* Computers and Structures, 44(1), 481–486.

18. Schultz, A. J., Hall, C. K., Genzer, J. (2005). *Computer Simulation of Block Copolymer/ Nanoparticle Composites.* Macromolecules, 38(7), 3007–3016.

19. Brown, D., et al. (2003). *A Molecular Dynamics Study of a Model Nanoparticle Embedded in a Polymer Matrix.* Macromolecules, 36(4), 1395–1406.

20. Heermann, D. W. (1990). *Computer Simulation Methods in Theoretical Physics* Second Edition ed: Springer. 145.

21. Binder, K., Heermann, D. W. (2010). *Monte Carlo Simulation in Statistical Physics: An Introduction.* 4th ed: Springer. 216.

22. Verlet, L. (1967). *Computer "Experiments" on Classical Fluids. I. Thermodynamical Properties of Lennard-Jones Molecules.* Physical Review, 159(1), 98–103.

23. Korn, G. A., Korn, T. M. (1968). *Mathematical Handbook for Scientists and Engineers.*, New York: McGrow-Hill, 1130.

24. Glezer, A. M. (2011). *Structural Classification of Nanomaterials.* Russian Metallurgy (Metally), (4), 263–269.

25. Kang, Z. C., Wang, Z. L. (1996). *On Accretion of Nanosize Carbon Spheres.* The Journal of Physical Chemistry, 100(13), 5163–5165.

26. Shenhar, R., Norsten, T. B., Rotello, V. M. (2005). *Polymer-Mediated Nanoparticle Assembly: Structural Control and Applications.* Advanced Materials, 17(6), 657–669.

CHAPTER 7

NUMERICAL STUDY OF AXIAL AND COAXIAL ELECTROSPINNING PROCESS

CONTENTS

Abstract 198
7.1   An Introduction to Nanotechnology 199
7.2   Nanostructured Materials 201
7.3   Nanofiber Technology 209
7.4   Design Multifunctional Product by Nanostructures 218
7.5   Introduction to Theoretical Study of Electrospinning Process 243
7.6   Study of Electrospinning Jet Path 245
7.7   Electrospinning Draw Backs 248
7.8   Modelling of the Electrospinning Process 250
7.9   Electrospinning Simulation 283
7.10 Electrospinning Simulation Example 283
7.11 Applied Numerical Methods for Electrospinning 287
7.12 Concluding Remarks of Electrospinning Modeling 301
7.13 Numerical Study of Coaxial Electrospinning 302
7.14 Application of Coaxial Electrospun Nanofibers 350
7.15 Advantages and Disadvantages of Co-Electrospinning Process 360
7.16 General Assumptions in Coaxial Electrospinning Process 361
7.17 Conclusions and Future Perspectives 361
Appendixes 364

ABSTRACT

The nanostructure materials productions are most challenging and innovative processes, introducing, in the manufacturing, a new approaches such as self-assembly and self-replication. The fast growing of nanotechnology with modern computational/experimental methods gives the possibility to design multifunctional materials and products in human surroundings. Smart clothing, portable fuel cells, medical devices are some of them. Research in nanotechnology began with applications outside of everyday life and is based on discoveries in physics and chemistry. The reason for that is needed to understand the physical and chemical properties of molecules and nanostructures in order to control them.

A new approach in nanostructured materials is a computational-based material development. It is based on multiscale material and process modeling spanning, on a large spectrum of time as well as on length scales. The cost of designing and producing novel multifunctional nanomaterials can be high and the risk of investment to be significant. Computational nanomaterials research that relies on multiscale modeling has the potential to significantly reduce development costs of new nanostructured materials for demanding applications by bringing physical and microstructural information into the realm of the design engineer.

One of the most significant types of these one-dimensional nanomaterials is nanofibers which can be produced widely through electrospinning procedure. A drawback of this method however, is the unstable behavior of the liquid jet, which causes the fibers to be collected randomly. So a critical concern in this process is to achieve desirable control. Studying the dynamics of electrospinning jet would be easier and faster if it can be modeled and simulated, rather than doing experiments. By replacing the single capillary with a coaxial spinneret in electrosinning set up, it is possible to generate core-sheath and hollow nanofibers made of polymers, ceramics or composites. These nanofibers, as a kind of one dimensional nanostructure, have attracted special attention in several research groups, because these structures could further enhance material property profiles and these unique core-sheath structures offer potential in a number of applications including nanoelectronics, microfluidics, photonics, and energy storage. This study focuses on modeling and then simulating of electrospinning and coaxial electerospinning process in various views. In order to study the applicability of the electrospinning modeling equations, which discussed in detail in earlier parts of this approach, an existing mathematical model in which the jet was considered as a mechanical system, was interconnected with viscoelastic elements and used to build a numeric method. The simulation features the possibility of predicting essential parameters of electrospinning and coaxial electrospinning process and the results have good agreement with other numeric studies of electrospinning, which modeled this process based on axial direction.

7.1   AN INTRODUCTION TO NANOTECHNOLOGY

Understanding the nanoworld makes up one of the frontiers of modern science. One reason for this is that technology based on nanostructures promises to be hugely important economically [1–3]. Nanotechnology literally means any technology on a nanoscale that has applications in the real world. It includes the production and application of physical, chemical, and biological systems at scales ranging from individual atoms or molecules to submicron dimensions, as well as the integration of the resulting nanostructures into larger systems. Nanotechnology is likely to have a profound impact on our economy and society in the early twenty-first century, comparable to that of semiconductor technology, information technology, or cellular and molecular biology. Science and technology research in nanotechnology promises breakthroughs in areas such as materials and manufacturing [4], nanoelectronics [5], medicine and healthcare [6], energy [7], biotechnology [8], information technology [9], and national security [10]. It is widely felt that nanotechnology will be the next Industrial Revolution [9].

As far as "nanostructures" are concerned, one can view this as objects or structures whereby at least one of its dimensions is within nano-scale. A "nanoparticle" can be considered as a zero dimensional nano-element, which is the simplest form of nanostructure. It follows that a "nanotube" or a "nanorod" is a one-dimensional nano-element from which slightly more complex nanostructure can be constructed [11, 12].

Following this fact, a "nanoplatelet" or a "nanodisk" is a two-dimensional element, which along with its one-dimensional counterpart, is useful in the construction of nanodevices. The difference between a nanostructure and a nanodevice can be viewed upon as the analogy between a building and a machine (whether mechanical, electrical or both) [1]. It is important to know that as far as nanoscale is concerned; these nano-elements should not consider only as an element that form a structure while they can be used as a significant part of a device. For example, the use of carbon nanotube as the tip of an Atomic Force Microscope (AFM) would have it classified as a nanostructure. The same nanotube, however, can be used as a single molecule circuit, or as part of a miniaturized electronic component, thereby appearing as a nanodevice. Hence the function, along with the structure, is essential in classifying which nanotechnology subarea it belongs to. This classification will be discussed in detail in further sections [11, 13].

As long as nanostructures clearly define the solids' overall dimensions, the same cannot be said so for nanomaterials. In some instances a nanomaterial refers to a nano-sized material while in other instances a nanomaterial is a bulk material with nano-scaled structures. Nanocrystals are other groups of nanostructured materials. It is understood that a crystal is highly structured and that the repetitive unit is indeed small enough. Hence a nanocrystal refers to the size of the entire crystal itself being nano-sized, but not of the repetitive unit [14].

Nanomagnetics are the other type of nanostructured materials, which are known as highly miniaturized magnetic data storage materials with very high memory. This can be attained by taking advantage of the electron spin for memory storage, hence the term "spin-electronics," which has since been more popularly and more conveniently known as "spintronics" [1, 9, 15]. In nanobioengineering, the novel properties of nano-scale are taken advantage of for bioengineering applications. The many naturally occurring nanofibrous and nanoporous structure in the human body further adds to the impetus for research and development in this subarea. Closely related to this is molecular functionalization whereby the surface of an object is modified by attaching certain molecules to enable desired functions to be carried out such as for sensing or filtering chemicals based on molecular affinity [16, 17].

With the rapid growth of nanotechnology, nanomechanics are no longer the narrow field, which it used to be [13]. This field can be broadly categorized into the molecular mechanics and the continuum mechanics approaches which view objects as consisting of discrete many-body system and continuous media respectively. As long as the former inherently includes the size effect, it is a requirement for the latter to factor in the influence of increasing surface-to-volume ratio, molecular reorientation and other novelties as the size shrinks. As with many other fields, nanotechnology includes nanoprocessing novel materials processing techniques by which nano-scale structures and devices are designed and constructed [18, 19].

Depending on the final size and shape, a nanostructure or nanodevice can be created from the top-down or the bottom-up approach. The former refers to the act of removal or cutting down a bulk to the desired size, while the latter takes on the philosophy of using the fundamental building blocks - such as atoms and molecules, to build up nanostructures in the same manner. It is obvious that the top-down and the bottom-up nanoprocessing methodologies are suitable for the larger and two smaller ends respectively in the spectrum of nano-scale construction. The effort of nanopatterning or patterning at the nanoscale would hence fall into nanoprocessing [1, 12, 18].

7.2 NANOSTRUCTURED MATERIALS

Strictly speaking, a nanostructure is any structure with one or more dimensions measuring in the nanometer (10^{-9} m) range. Various definitions refine this further, stating that a nanostructure should have a characteristic dimension lying between 1 nm and 100 nm, putting nanostructures as intermediate in size between a molecule and a bacterium. Nanostructures are typically probed either optically (spectroscopy, photoluminescence) or in transport experiments. This field of investigation is often given the name mesoscopic transport, and the following considerations give an idea of the significance of this term [1, 2, 12, 20, 21].

What makes nanostructured materials very interesting and award them with their unique properties is that their size is smaller than critical lengths that characterize many physical phenomena. Generally, physical properties of materials can be characterized by some critical length, a thermal diffusion length, or a scattering length, for example. The electrical conductivity of a metal is strongly determined by the distance that the electrons travel between collisions with the vibrating atoms or impurities of the solid. This distance is called the mean free path or the scattering length. If the sizes of the particles are less than these characteristic lengths, it is possible that new physics or chemistry may occur [1, 9, 17].

Several computational techniques have been employed to simulate and model nanomaterials. Since the relaxation times can vary anywhere from picoseconds to hours, it becomes necessary to employ Langevin dynamics besides molecular dynamics in the calculations. Simulation of nanodevices through the optimization of various components and functions provides challenging and useful task [20, 22]. There are many examples where simulation and modeling have yielded impressive results, such as nanoscale lubrication [23]. Simulation of the molecular dynamics of DNA has been successful to some extent [24]. Quantum dots and nanotubes have been modeled satisfactorily [25, 26]. First principles calculations of nanomaterials can be problematic if the clusters are too large to be treated by Hartree–Fock methods and too small for density functional theory [1]. In the next section various classifications of these kinds of materials are considered in detail.

7.2.1 CLASSIFICATION OF NANOSTRUCTURED MATERIALS

Nanostructure materials as a subject of nanotechnology are low dimensional materials comprising of building units of a submicron or nanoscale size at least in one direction and exhibiting size effects. The first classification idea of NSMs was given by Gleiter in 1995 [3]. A modified classification scheme for these materials, in which 0D, 1D, 2D and 3D dimensions are included suggested in later researches [21]. These classifications are as follows:

7.2.1.1 0D NANOPARTICLES

A major feature that distinguishes various types of nanostructures is their dimensionality. In the past 10 years, significant progress has been made in the field of 0 dimension nanostructure materials. A rich variety of physical and chemical methods have been developed for fabricating these materials with well-controlled dimensions [3, 18]. Recently, 0D nanostructured materials such as uniform particles arrays (quantum dots), heterogeneous particles arrays, core-shell quantum dots, onions, hollow spheres and nanolenses have been synthesized by several research groups [21]. They have been extensively studied in light emitting diodes (LEDs), solar cells, single-electron transistors, and lasers.

7.2.1.2 1D NANOPARTICLES

In the last decade, 1D nanostructured materials have focused an increasing interest due to their importance in research and developments and have a wide range of potential applications [27]. It is generally accepted that these materials are ideal systems for exploring a large number of novel phenomena at the nanoscale and investigating the size and dimensionality dependence of functional properties. They are also expected to play an important role as both interconnects and the key units in fabricating electronic, optoelectronic, and EEDs with nanoscale dimensions. The most important types of this group are nanowires, nanorods, nanotubes, nanobelts, nanoribbons, hierarchical nanostructures and nanofibers [1, 18, 28].

7.2.1.3 2D NANOPARTICLES

2D nanostructures have two dimensions outside of the nanometric size range. In recent years, synthesis of 2D nanomaterial has become a focal area in materials research, owing to their many low dimensional characteristics different from the bulk properties. Considerable research attention has been focused over the past few years on the development of them. 2D nanostructured materials with certain geometries exhibit unique shape-dependent characteristics and subsequent utilization as building blocks for the key components of nanodevices [21]. In addition, these materials are particularly interesting not only for basic understanding of the mechanism of nanostructure growth, but also for investigation and developing novel applications in sensors, photocatalysts, nanocontainers, nanoreactors, and templates for 2D structures of other materials. Some of the 3D nanoparticles are junctions (continuous islands), branched structures, nanoprisms, nanoplates, nanosheets, nanowalls, and nanodisks [1].

7.2.1.4   3D NANOPARTICLES

Owing to the large specific surface area and other superior properties over their bulk counterparts arising from quantum size effect, they have attracted considerable research interest and many of them have been synthesized in the past 10 years [1, 12]. It is well known that the behaviors of NSMs strongly depend on the sizes, shapes, dimensionality and morphologies, which are thus the key factors to their ultimate performance and applications. Therefore, it is of great interest to synthesize 3D NSMs with a controlled structure and morphology. In addition, 3D nanostructures are an important material due to its wide range of applications in the area of catalysis, magnetic material and electrode material for batteries [2]. Moreover, the 3D NSMs have recently attracted intensive research interests because the nanostructures have higher surface area and supply enough absorption sites for all involved molecules in a small space [58]. On the other hand, such materials with porosity in three dimensions could lead to a better transport of the molecules. Nanoballs (dendritic structures), nanocoils, nanocones, nanopillers and nanoflowers are in this group [1, 2, 18, 29].

7.2.2   SYNTHESIS METHODS OF NANOMATERIALS

The synthesis of nanomaterials includes control of size, shape, and structure. Assembling the nanostructures into ordered arrays often becomes necessary for rendering them functional and operational. In the last decade, nanoparticles (powders) of ceramic materials have been produced in large scales by employing both physical and chemical methods. There has been considerable progress in the preparation of nanocrystals of metals, semiconductors, and magnetic materials by employing colloid chemical methods [18, 30].

The construction of ordered arrays of nanostructures by employing techniques of organic self-assembly provides alternative strategies for nanodevices. Two- and three-dimensional arrays of nanocrystals of semiconductors, metals, and magnetic materials have been assembled by using suitable organic reagents [1, 31]. Strain directed assembly of nanoparticle arrays (e.g., of semiconductors) provides the means to introduce functionality into the substrate that is coupled to that on the surface [32].

Preparation of nanoparticles is an important branch of the materials science and engineering. The study of nanoparticles relates various scientific fields, e.g., chemistry, physics, optics, electronics, magnetism and mechanism of materials. Some nanoparticles have already reached practical stage. In order to meet the nanotechnology and nano-materials development in the next century, it is necessary to review the preparation techniques of nanoparticles.

All particle synthesis techniques fall into one of the three categories: vapor-phase, solution precipitation, and solid-state processes. Although vapor-phase processes have been common during the early days of nanoparticles development, the last of the three processes mentioned above is the most widely used in the industry for production of micron-sized particles, predominantly due to cost considerations [18, 33].

Methods for preparation of nanoparticles can be divided into physical and chemical methods based on whether there exist chemical reactions [33]. On the other hand, in general, these methods can be classified into the gas phase, liquid phase and solid phase methods based on the state of the reaction system. The gas phase method includes gas-phase evaporation method (resistance heating, high frequency induction heating, plasma heating, electron beam heating, laser heating, electric heating evaporation method, vacuum deposition on the surface of flowing oil and exploding wire method), chemical vapor reaction (heating heat pipe gas reaction, laser induced chemical vapor reaction, plasma enhanced chemical vapor reaction), chemical vapor condensation and sputtering method. Liquid phase method for synthesizing nanoparticles mainly includes precipitation, hydrolysis, spray, solvent thermal method (high temperature and high pressure), solvent evaporation pyrolysis, oxidation reduction (room pressure), emulsion, radiation chemical synthesis and sol-gel processing. The solid phase method includes thermal decomposition, solid-state reaction, spark discharge, stripping and milling method [30, 33].

In other classification, there are two general approaches to the synthesis of nanomaterials and the fabrication of nanostructures, bottom-up and Top-down approach. The first one includes the miniaturization of material components (up to atomic level) with further self-assembly process leading to the formation assembly of nanostructures. During self-assembly the physical forces operating at nanoscale are used to combine basic units into larger stable structures. Typical examples are quantum dot formation during epitaxial growth and formation of nanoparticles from colloidal dispersion. The latter uses larger (macroscopic) initial structures, which can be externally controlled in the processing of nanostructures. Typical examples are etching through the mask, ball milling, and application of severe plastic deformation [3, 13].

Some of the most common methods are described in the following sub sections.

7.2.2.1 PLASMA BASED METHODS

Metallic, semiconductive and ceramic nanomaterials are widely synthesized by hot and cold plasma methods. A plasma is sometimes referred to as being "hot" if it is nearly fully ionized, or "cold" if only a small fraction, (for instance 1%), of the gas molecules are ionized, but other definitions of the terms "hot plasma" and "cold plasma" are common. Even in cold plasma, the electron temperature is still typically several thousand degrees centigrade. Generally the related equipment consists of an arc melting chamber and a collecting system. The thin films of alloys were prepared from highly pure metals by arc melting in an inert gas atmosphere. Each arc-melted ingot was flipped over and remelted three times. Then, the thin films of alloy were produced by arc melting a piece of bulk materials in a mixing gas atmosphere at a low pressure. Before the ultrafine particles were taken out from the arc-melting chamber, they were passivated with a mixture of inert gas and air to prevent the particles from burning up [34, 35].

Cold plasma method is used for producing nanowires in large scale and bulk quantity. The general equipment of this method consists of a conventional horizontal quartz tube furnace and an inductively coupled coil driven by a 13.56 MHz radio-frequency (RF or radio-frequency) power supply. This method often is called as an RF plasma method. During RF plasma method, the starting metal is contained in a pestle in an evacuated chamber. The metal is heated above its evaporation point using high voltage RF coils wrapped around the evacuated system in the vicinity of the pestle. Helium gas is then allowed to enter the system, forming high temperature plasma in the region of the coils. The metal vapor nucleates on the He gas atoms and diffuses up to a colder collector rod where nanoparticles are formed. The particles are generally passivated by the introduction of some gas such as oxygen. In the case of aluminum nanoparticles the oxygen forms a layer of aluminum oxide about the particle [1, 36].

7.2.2.2 CHEMICAL METHODS

Chemical methods have played a major role in developing materials imparting technologically important properties through structuring the materials on the nanoscale. However, the primary advantage of chemical processing is its versatility in designing and synthesizing new materials that can be refined into the final end products. The secondary most advantage that the chemical processes offer over physical methods is a good chemical homogeneity, as a chemical method offers mixing at the molecular level. On the other hand, chemical methods frequently involve toxic reagents and solvents for the synthesis of nanostructured materials. Another disadvantage of the chemical methods is the unavoidable introduction of byproducts, which require subsequent purification steps after the synthesis in other words, this process is time consuming. In spite of these facts, probably the most useful methods of synthesis in terms of their potential to be scaled up are chemical methods [33, 37]. There are a number of different chemical methods that can be used to make nanoparticles of metals, and we will give some examples. Several types of reducing agents can be used to produce nanoparticles such as $NaBEt_3H$, Li-BEt_3H, and $NaBH_4$ where Et denotes the ethyl ($-C_2H_5$) radical. For example, nanoparticles of molybdenum (Mo) can be reduced in toluene solution with $NaBEt_3H$ at room temperature, providing a high yield of Mo nanoparticles having dimensions of 1–5 nm [30].

7.2.2.3 THERMOLYSIS AND PYROLYSIS

Nanoparticles can be made by decomposing solids at high temperature having metal cations, and molecular anions or metal organic compounds. The process is called thermolysis. For example, small lithium particles can be made by decomposing lithium oxide, LiN_3. The material is placed in an evacuated quartz tube and heated to 400°C in the apparatus. At about 370°C the LiN_3 decomposes, releasing N_2 gas, which is observed by an increase in the pressure on the vacuum gauge. In a few minutes the pressure drops back to its original low value, indicating that all the N_2 has been removed. The remaining lithium atoms coalesce to form small colloidal metal particles. Particles less than 5 nm can be made by this method. Passivation can be achieved by introducing an appropriate gas [1].

Pyrolysis is commonly a solution process in which nanoparticles are directly deposited by spraying a solution on a heated substrate surface, where the constituent reacts to form a chemical compound. The chemical reactants are selected such that the products other than the desired compound are volatile at the temperature of deposition. This method represents a very simple and relatively cost-effective processing method (particularly in regard to equipment costs) as compared to many other film deposition techniques [30].

The other pyrolysis-based method that can be applied in nanostructures production is a laser pyrolysis technique, which requires the presence in the reaction medium of a molecule absorbing the CO_2 laser radiation [38, 39]. In most cases, the atoms of a molecule are rapidly heated via vibrational excitation and are dissociated. But in some cases, a sensitizer gas such as SF_6 can be directly used. The heated gas molecules transfer their energy to the reaction medium by collisions leading to dissociation of the reactive medium without, in the ideal case, dissociation of this molecule. Rapid thermalization occurs after dissociation of the reactants due to transfer collision. Nucleation and growth of NSMs can take place in the as-formed supersaturated vapor. The nucleation and growth period is very short time (0.1–10 ms). Therefore, the growth is rapidly stopped as soon as the particles leave the reaction zone. The flame-excited luminescence is observed in the reaction region where the laser beam intersects the reactant gas stream. Since there is no interaction with any walls, the purity of the desired products is limited by the purity of the reactants. However, because of the very limited size of the reaction zone with a faster cooling rate, the powders obtained in this wellness reactor present a low degree of agglomeration. The particle size is small ($\sim$ 5–50 nm range) with a narrow size distribution. Moreover, the average size can be manipulated by optimizing the flow rate, and, therefore, the residence time in the reaction zone [39, 40].

7.2.2.4   LASER BASED METHODS

The most important laser based techniques in the synthesis of nanoparticles is pulsed laser ablation. As a physical gas-phase method for preparing nanosized particles, pulsed laser ablation has become a popular method to prepare high-purity and ultra-fine nanomaterials of any composition [41, 42]. In this method, the material is evaporated using pulsed laser in a chamber filled with a known amount of a reagent gas and by controlling condensation of nanoparticles onto the support. It's possible to prepare nanoparticles of mixed molecular composition such as mixed oxides/nitrides and carbides/nitrides or mixtures of oxides of various metals by this method. This method is capable of a high rate of production of 2–3 g/min [40].

Laser chemical vapor deposition method is the next laser-based technique in which photo-induced processes are used to initiate the chemical reaction. During this method, three kinds of activation should be considered. First, if the thermalization of the laser energy is faster than the chemical reaction, pyrolytic and/or photothermal activation is responsible for the activation. Secondly, if the first chemical reaction step is faster than the thermalization, photolytical (nonthermal) processes are responsible for the excitation energy. Thirdly, combinations of the different types of activation are often encountered. During this technique a high intensity laser beam is incident on a metal rod, causing evaporation of atoms from the surface of the metal. The atoms are then swept away by a burst of helium and passed through an orifice into a vacuum where the expansion of the gas causes cooling and formation of clusters of the metal atoms. These clusters are then ionized by UV radiation and passed into a mass spectrometer that measures their mass: charge ratio [1, 41–43].

Laser-produced nanoparticles have found many applications in medicine, bio-photonics, in the development of sensors, new materials and solar cells. Laser interactions provide a possibility of chemical clean synthesis, which is difficult to achieve under more conventional NP production conditions [42]. Moreover, a careful optimization of the experimental conditions can allow a control over size distributions of the produced nanoclusters. Therefore, many studies were focused on the investigation the laser nanofabrication. In particular, many experiments were performed to demonstrate nanoparticles formation in vacuum, in the presence of a gas or a liquid. Nevertheless, it is still difficult to control the properties of the produced particles. It is believed that numerical calculations can help to explain experimental results and to better understand the mechanisms involved [43].

Despite rapid development in laser physics, one of the fundamental questions still concern the definition of proper ablation mechanisms and the processes leading to the nano particles formation. Apparently, the progress in laser systems implies several important changes in these mechanisms, which depend on both laser parameters and material properties. Among the more studied ablation mechanisms there are thermal, photochemical and photomechanical ablation processes. Frequently, however, the mechanisms are mixed, so that the existing analytical equations are hardly applicable. Therefore, numerical simulation is needed to better understand and to optimize the ablation process [44].

So far, thermal models are commonly used to describe nanosecond (and longer) laser ablation. In these models, the laser-irradiated material experiences heating, melting, boiling and evaporation. In this way, three numerical approaches were used [29, 45]:

1. **Atomistic approach** based on such methods as molecular dynamics (MD) and Direct Monte Carlo Simulation (DSMC). Typical calculation results provide detailed information about atomic positions, velocities, kinetic and potential energy;

2. **Macroscopic approach** based hydrodynamic models. These models allow the investigations of the role of the laser-induced pressure gradient, which is particularly important for ultra-short laser pulses. The models are based on a one fluid two-temperature approximation and a set of additional models (equation of state) that determines thermal properties of the target;

3. **Multi-scale approach** based on the combination of two approaches cited above was developed by several groups and was shown to be particularly suitable for laser applications.

7.3   NANOFIBER TECHNOLOGY

Nano fiber consists of two terms "Nano" and "fiber," as the latter term is looking more familiar. Anatomists observed fibers as any of the filament constituting the extracellular matrix of connective tissue, or any elongated cells or thread like structures, muscle fiber or nerve fiber. According to textile industry fiber is a natural or synthetic filament, such as cotton or nylon, capable of being spun into simply as materials made of such filaments. Physiologists and biochemists use the term fiber for indigestible plant matter consisting of polysaccharides such as cellulose, that when eaten stimulates intestinal peristalsis. Historically, the term fiber or "Fiber" in British English comes from Latin "fibra." Fiber is a slender, elongated thread like structure. Nano is originated from Greek word "nanos,, or "nannos" refer to "little old man" or "dwarf." The prefixes "nannos" or "nano" as nannoplanktons or nanoplanktons used for very small planktons measuring 2 to 20 micrometers. In modern "nano" is used for describing various physical quantities within the scale of a billionth as nanometer (length), nanosecond (time), nanogram (weight) and nanofarad (charge) [1, 4, 9, 46]. As it was mentioned before, nanotechnology refers to the science and engineering concerning materials, structures and devices which has at least one dimension is 100 nm or less. This term also refers for a fabrication technology, where molecules, specification and individual atoms, which have at least one dimension in nanometers or less, is used to design or built objects. Nano fiber, as the name suggests the fiber having a diameter range in nanometer. Fibrous structure having at least one dimension in nanometer or less is defined as Nano fiber according to National Science Foundation (NSC). The term Nano describes the diameter of the fibrous shape at anything below one micron or 1000 nm [4, 18].

Nanofiber technology is a branch of nanotechnology whose primary objective is to create materials in the form of nanoscale fibers in order to achieve superior functions [1, 2, 4]. The unique combination of high specific surface area, flexibility, and superior directional strength makes such fibers a preferred material form for many applications ranging from clothing to reinforcements for aerospace structures. Indeed, while the primary classification of nanofibers is that of nanostructure or nanomaterial, other aspects of nanofibers such as its characteristics, modeling, application and processing would enable nanofibers to penetrate into many subfields of nanotechnology [4, 46, 47].

It is obvious that nanofibers would geometrically fall into the category of one dimensional nano-scale elements that includes nanotubes and nanorods. However, the flexible nature of nanofibers would align it along with other highly flexible nano-elements such as globular molecules (assumed as zero dimensional soft matter), as well as solid and liquid films of nanothickness (two dimensional). A nanofiber is a nanomaterial in view of its diameter, and can be considered a nanostructured material if filled with nanoparticles to form composite nanofibers [1, 48].

The study of the nanofiber mechanical properties as a result of manufacturing techniques, constituent materials, processing parameters and other factors would fall into the category of nanomechanics. Indeed, while the primary classification of nanofibers is that of nanostructure or nanomaterial, other aspects of nanofibers such as its characteristics, modeling, application and processing would enable nanofibers to penetrate into many subfields of nanotechnology [1, 18].

Although the effect of fiber diameter on the performance and processibility of fibrous structures has long been recognized, the practical generation of fibers at the nanometer scale was not realized until the rediscovery and popularization of the electrospinning technology by Professor Darrell Reneker almost a decade ago [49, 50]. The ability to create nanoscale fibers from a broad range of polymeric materials in a relatively simple manner using the electrospinning process, coupled with the rapid growth of nanotechnology in recent years have greatly accelerated the growth of nanofiber technology.

Although there are several alternative methods for generating fibers in a nanometer scale, none matches the popularity of the electrospinning technology due largely to the simplicity of the electrospinning process [18]. These methods will be discussed in following sections.

7.3.1 VARIOUS NANOFIBERS PRODUCTION METHODS

As it was discussed in detail, nanofiber is defined as the fiber having at least one dimension in nanometer range which can be used for a wide range of medical applications for drug delivery systems, scaffold formation, wound healing and widely used in tissue engineering, skeletal tissue, bone tissue, cartilage tissue, ligament tissue, blood vessel tissue, neural tissue etc. It is also used in dental and orthopedic implants [4, 51, 52]. Nano fiber can be formed using different techniques including: drawing, template synthesis, phases separation, Self-assembly and electrospinning.

7.3.1.1 DRAWING

In1998, nanofibers were fabricated with citrate molecules through the process of drawing for the first time [53]. During drawing process, the fibers are fabricated by contacting a previously deposited polymer solution droplet with a sharp tip and drawing it as a liquid fiber, which is then solidified by rapid evaporation of the solvent due to the high surface area. The drawn fiber can be connected to another previously deposited polymer solution droplet thus forming a suspended fiber. Here, the predeposition of droplets significantly limits the ability to extend this technique, especially in three-dimensional configurations and hard to access spatial geometries. Furthermore, there is a specific time in which the fibers can be pulled. The viscosity of the droplet continuously increases with time due to solvent evaporation from the deposited droplet. The continual shrinkage in the volume of the polymer solution droplet affects the diameter of the fiber drawn and limits the continuous drawing of fibers [54].

To overcome the above-mentioned limitation is appropriate to use hollow glass micropipettes with a continuous polymer dosage. It provides greater flexibility in drawing continuous fibers in any configuration. Moreover, this method offers increased flexibility in the control of key parameters of drawing such as waiting time before drawing (due to the required viscosity of the polymer edge drops), the drawing speed or viscosity, thus enabling repeatability and control on the dimensions of the fabricated fibers. Thus, drawing process requires a viscoelastic material that can undergo strong deformations while being cohesive enough to support the stresses developed during pulling [54, 55].

7.3.1.2 TEMPLATE SYNTHESIS

Template synthesis implies the use of a template or mold to obtain a desired material or structure. Hence the casting method and DNA replication can be considered as template-based synthesis. In the case of nanofiber creation by [56], the template refers to a metal oxide membrane with through-thickness pores of nano-scale diameter. Under the application of water pressure on one side and restrain from the porous membrane causes extrusion of the polymer which, upon coming into contact with a solidifying solution, gives rise to nanofibers whose diameters are determined by the pores [1, 57].

This method is an effective route to synthesize nanofibrils and nanotubes of various polymers. The advantage of the template synthesis method is that the length and diameter of the polymer fibers and tubes can be controlled by the selected porous membrane, which results in more regular nanostructures. General feature of the conventional template method is that the membrane should be soluble so that it can be removed after synthesis in order to obtain single fibers or tubes. This restricts practical application of this method and gives rise to a need for other techniques [1, 56, 57].

7.3.1.3 PHASE SEPARATION METHOD

This method consists of five basic steps: polymer dissolution, gelation, solvent extraction, freezing and freeze-drying. In this process, it is observed that gelatin is the most difficult step to control the porous morphology of nano fiber. Duration of gelation varied with polymer concentration and gelation temperature. At low gelation temperature, nano-scale fiber network is formed, whereas, high gelation temperature led to the formation of platelet-like structure. Uniform nanofiber can be produced as the cooling rate is increased, polymer concentration affects the properties of nano fiber, as polymer concentration is increased porosity of fiber decreased and mechanical properties of fiber are increased [1, 58].

7.3.1.4 SELF-ASSEMBLY

Self–assembly refers to the build-up of nanoscale fibers using smaller molecules. In this technique, a small molecule is arranged in a concentric manner so that they can form bonds among the concentrically arranged small molecules, which upon extension in the plane's normal gives the longitudinal axis of a nanofiber. The main mechanism for a generic self-assembly is the intra-molecular forces that bring the smaller unit together. A hydrophobic core of alkyl residues and a hydrophilic exterior lined by peptide residues was found in obtained fiber. It is observed that the nano fibers produced with this technique have diameter range 5–8 mm approximately and several microns in length [1, 59].

Although there are a number of techniques used for the synthesis of nanofiber but Electrospinning represents an attractive technique to fabricate polymeric biomaterial into nanofibers. Electrospinning is one of the most commonly used methods for the production of nanofiber. It has a wide advantage over the previously available fiber formation techniques because here electrostatic force is used instead of conventionally used mechanical force for the formation of fibers. This method will be debated comprehensively in following section.

7.3.1.5   ELECTROSPINNING OF NANOFIBERS

Electrospinning is a straightforward and cost-effective method to produce novel fibers with diameters in the range of from less than 3 nm to over 1 mm, which overlaps contemporary textile fiber technology. During this process, an electrostatic force is applied to a polymeric solution to produce nanofiber [60, 61] with diameter ranging from 50 nm to 1000 nm or greater [49, 62, 63]; Due to surface tension the solution is held at the tip of syringe. Polymer solution is charged due to applied electric force. In the polymer solution, a force is induced due to mutual charge repulsion that is directly opposite to the surface tension of the polymer solution. Further increases in the electrical potential led to the elongation of the hemispherical surface of the solution at the tip of the syringe to form a conical shape known as "Taylor cone" [50, 64]. The electric potential is increased to overcome the surface tension forces to cause the formation of a jet, ejects from the tip of the Taylor cone. Due to elongation and solvent evaporation, charged jet instable and gradually thins in air primarily [62, 65–67]. The charged jet forms randomly oriented nanofibers that can be collected on a stationary or rotating grounded metallic collector [50]. Electrospinning provides a good method and a practical way to produce polymer fibers with diameters ranging from 40–2000 nm [49, 50].

7.3.1.5.1   THE HISTORY OF ELECTROSPINNING METHODOLOGY

William Gilbert discovered the first record of the electrostatic attraction of a liquid in 1600 [68]. The first electrospinning patent was submitted by John

Francis Cooley in 1900 [69]. After that in 1914 John Zeleny studied on the behavior of fluid droplets at the end of metal capillaries which caused the beginning of the mathematical model the behavior of fluids under electrostatic forces [65]. Between 1931 and 1944 Anton Formhals took out at least 22 patents on electrospinning [69]. In 1938, N. D. Rozenblum and I. V. Petryanov-Sokolov generated electrospun fibers, which they developed into filter materials [70]. Between 1964 and 1969 Sir Geoffrey Ingram Taylor produced the beginnings of a theoretical foundation of electrospinning by mathematically modeling the shape of the (Taylor) cone formed by the fluid droplet under the effect of an electric field [71, 72]. In the early 1990s several research groups (such as Reneker) demonstrated electrospun nano-fibers. Since 1995, the number of publications about electrospinning has been increasing exponentially every year [69].

7.3.1.5.2   ELECTROSPINNING PROCESS

Electrospinning process can be explained in five significant steps including [48, 73–75]:

a) Charging of the polymer fluid

The syringe is filled with a polymer solution, the polymer solution is charged with a very high potential around 10–30 kV. The nature of the fluid and polarity of the applied potential free electrons, ions or ion-pairs are generated as the charge carriers form an electrical double layer. This charging induction is suitable for conducting fluid, but for nonconducting fluid charge directly injected into the fluid by the application of electrostatic field.

b) Formation of the cone jet (Taylor cone)

The polarity of the fluid depends upon the voltage generator. The repulsion between the similar charges at the free electrical double layer works against the surface tension and fluid elasticity in the polymer solution to deform the droplet into a conical shaped structure i.e., known as a Taylor cone. Beyond a critical charge density Taylor-cone becomes unstable and a jet of fluid is ejected from the tip of the cone.

c) Thinning of the jet in the presence of an electric field

The jet travels a path to the ground; this fluid jet forms a slender continuous liquid filament. The charged fluid is accelerated in the presence of an electrical field. This region of fluid is generally linear and thin.

d) Instability of the jet

Fluid elements accelerated under electric field and thus stretched and succumbed to one or more fluid instabilities, which distort as they grow following many spiral and distort the path before collected on the collector electrode. This region of instability is also known as whipping region.

e) Collection of the jet

Charged electro spun fibers travel downfield until its impact with a lower potential collector plate. The orientation of the collector affects the alignment of the fibers. Different type of collector also affects the morphology and the properties of producing nanofiber. Different type of collectors are used: Rotating drum collector, moving belt collector, rotating wheel with beveled edge, multifilament thread, parallel bars, simple mesh collector etc.

7.3.1.5.3   ELECTROSPINNING SET-UPS

Electrospinning is conducted at room temperature with atmospheric conditions. The typical set up of electrospinning apparatus is shown in Fig. 7.1. Basically, an electrospinning system consists of three major components: a high voltage power supply, a spinneret (such as a pipette tip) and a grounded collecting plate (usually a metal screen, plate, or rotating mandrel) and uses a high voltage source to inject charge of a certain polarity into a polymer solution or melt, which is then accelerated towards a collector of opposite polarity [73, 76, 77]. Most of the polymers are dissolved in some solvents before electrospinning, and when it completely dissolves, forms polymer solution. The polymer fluid is then introduced into the capillary tube for electrospinning. However, some polymers may emit unpleasant or even harmful smells, so the processes should be conducted within chambers having a ventilation system. In the electrospinning process, a polymer solution held by its surface tension at the end of a capillary tube is subjected to an electric field and an electric charge is induced on the liquid surface due to this electric field. When the electric field applied reaches a critical value, the repulsive electrical forces overcome the surface tension forces. Eventually, a charged jet of the solution is ejected from the tip of the Taylor cone and an unstable and a rapid whipping of the jet occurs in the space between the capillary tip and collector which leads to evaporation of the solvent, leaving a polymer behind. The jet is only stable at the tip of the spinneret and after that instability starts. Thus, the electrospinning process offers a simplified technique for fiber formation [50, 73, 78, 79].

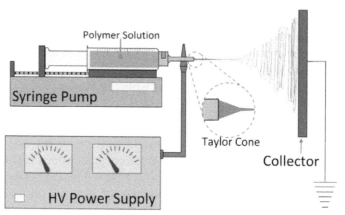

FIGURE 7.1   Scheme of a conventional electrospinning set-up.

7.3.1.5.4   *THE EFFECTIVE PARAMETERS ON ELECTROSPINNING*

The electrospinning process is generally governed by many parameters, which can be classified broadly into solution parameters, process parameters, and ambient parameters. Each of these parameters significantly affects the fiber morphology obtained as a result of electrospinning, and by proper manipulation of these parameters we can get nanofibers of desired morphology and diameters. These effective parameters are sorted as below [63, 67, 73, 76]:

   a) Polymer solution parameters which includes: molecular weight and solution viscosity, surface tension, solution conductivity and dielectric effect of solvent.

   b) Processing parameters which include voltage, feed rate, temperature, effect of collector, the diameter of the orifice of the needle.

a) Polymer solution parameters
1) Molecular weight and solution viscosity
Higher the molecular weight of the polymer, increases molecular entanglement in the solution, hence there is an increase in viscosity. The electro spun jet eject with high viscosity during it is stretched to a collector electrode leading to formation of continuous fiber with higher diameter, but very high viscosity makes difficult to pump the solution and also lead to the drying of the solution at the needle tip. As a very low viscosity lead in bead formation in the resultant electro spun fiber, so the molecular weight and viscosity should be acceptable to form nanofiber [48, 80].

2) Surface tension

Lower viscosity leads to decrease in surface tension resulting bead formation along the fiber length because the surface area is decreased, but at the higher viscosity effect of surface tension is nullified because of the uniform distribution of the polymer solution over the entangled polymer molecules. So, lower surface tension is required to obtain smooth fiber and lower surface tension can be achieved by adding of surfactants in polymer solution [80, 81].

3) Solution conductivity

Higher conductivity of the solution followed a higher charge distribution on the electrospinning jet, which leads to increase in stretching of the solution during fiber formation. Increased conductivity of the polymer solution lowers the critical voltage for the electro spinning. Increased charge leads to the higher bending instability leading to the higher deposition area of the fiber being formed, as a result jet path is increased and finer fiber is formed. Solution conductivity can be increased by the addition of salt or polyelectrolyte or increased by the addition of drugs and proteins, which dissociate into ions when dissolved in the solvent formation of smaller diameter fiber [67, 80].

4) Dielectric effect of solvent

Higher the dielectric property of the solution lesser is the chance of bead formation and smaller is the diameter of electro spun fiber. As the dielectric property is increased, there is increase in the bending instability of the jet and the deposition area of the fiber is increased. As jet path length is increased fine fiber deposit on the collector [67, 80].

b) Processing condition parameters

1) Voltage

Taylor cone stability depends on the applied voltage; at the higher voltage greater amount of charge causes the jet to accelerate faster leading to smaller and unstable Taylor cone. Higher voltage leads to greater stretching of the solution due to fiber with small diameter formed. At lower voltage the flight time of the fiber to a collector plate increases that led to the formation of fine fibers. There is greater tendency to bead formation at high voltage because of increased instability of the Taylor cone, and theses beads join to form thick diameter fibers. It is observed that the better crystallinity in the fiber obtained at higher voltage, because with very high voltage acceleration of fiber increased that reduced flight time and polymer molecules do not have much time to align them and fiber with less crystallinity formed. Instead of DC if AC voltage is provided for electro spinning it forms thicker fibers [48, 80].

2) Feed rate

As the feed rate is increased, there is an increase in the fiber diameter because greater volume of solution is drawn from the needle tip [80].

3) Temperature

At high temperature, the viscosity of the solution is decreased and there is increase in higher evaporation rate, which allows greater stretching of the solution, and a uniform fiber is formed [82].

4) Effect of collector

In electro spinning, collector material should be conductive. The collector is grounded to create stable potential difference between needle and collector. A nonconducting material collector reduces the amount of fiber being deposited with lower packing density. But in case of conducting collector there is accumulation of closely packed fibers with higher packing density. Porous collector yields fibers with lower packing density as compared to nonporous collector plate. In porous collector plate the surface area is increased so residual solvent molecules gets evaporated fast as compared to nonporous. Rotating collector is useful in getting dry fibers as it provides more time to the solvents to evaporate. It also increases fiber morphology [83]. The specific hat target with proper parameters has a uniform surface electric field distribution, the target can collect the fiber mats of uniform thickness and thinner diameters with even size distribution [80].

5) Diameter of pipette orifice

Orifice with small diameter reduces the clogging effect due to less exposure of the solution to the atmosphere and leads to the formation of fibers with smaller diameter. However, very small orifice has the disadvantage that it creates problems in extruding droplets of solution from the tip of the orifice [80].

7.4   DESIGN MULTIFUNCTIONAL PRODUCT BY NANOSTRUCTURES

The largest variety of efficient and elegant multifunctional materials is seen in natural biological systems, which occur sometimes in the simple geometrical forms in man-made materials. The multifunctionality of a material could be achieved by designing the material from the micro to macroscales (bottom up design approach), mimicking the structural formations created by nature [84]. Biological materials present around us have a large number of ingenious solutions and serve as a source of inspiration. There are different ways of producing multifunctional materials that depend largely on whether these materials are structural composites, smart materials, or nanostructured materials. The nanostructure materials are most challenging and innovative processes, introducing, in the manufacturing, a new approaches such as self-assembly and self-replication. For bio-materials involved in surface-interface related processes, common geometries involve capillaries, dendrites, hair, or fin-like attachments supported on larger substrates. It may be useful to incorporate similar hierarchical structures in the design and fabrication of multifunctional synthetic products that include surface sensitive functions such as sensing, reactivity, charge storage, transport property or stress transfer. Significant effort is being directed in order to fabricate and understand materials involving multiple length scales and functionalities. Porous fibrous structures can behave like lightweight solids providing significantly higher surface area compared to compact ones. Depending on what is attached on their surfaces, or what matrix is infiltrated in them, these core structures can be envisioned in a wide variety of surface active components or net-shape composites. If nanoelements can be attached in the pores, the surface area within the given space can be increased by several orders of magnitude, thereby increasing the potency of any desired surface functionality. Recent developments in electrospinning have made these possible, thanks to a coelectrospinning polymer suspension [85]. This opens up the possibility of taking a functional material of any shape and size, and attaching nanoelements on them for added surface functionality. The fast growing nanotechnology with modern computational/experimental methods gives the possibility to design multifunctional materials and products in human surroundings. Smart clothing, portable fuel cells, medical devices are some of them. Research in nanotechnology began with applications outside of everyday life and is based on discoveries in physics and chemistry. The reason for that is need to understand the physical and chemical properties of molecules and nanostructures in order to control them. For example, nanoscale manipulation results in new functionalities for textile structures, including self-cleaning, sensing, actuating, and communicating. Development of precisely controlled or programmable medical nanomachines and nanorobots is great promise for nanomedicine. Once nanomachines are available, the ultimate dream of every medical man becomes reality. The miniaturization of instruments on micro and nano-dimensions promises to make our future lives safer with more humanity. A new approach in material synthesis is a computational-based material development. It is based on multiscale material and process modeling spanning, on a large spectrum of time as well as on length scales. Multi-scale materials design means to design materials from a molecular scale up to a macro scale. The ability to manipulate at atomic and molecular level is also creating materials and structures that have unique functionalities and characteristics. Therefore, it will be and revolutionizing next-generation technology ranging from structural materials to nano-electro-mechanical systems (NEMs), for medicine and bioengineering applications. Recent research development in nanomaterials has been progressing at a tremendous speed for it can totally change the ways in which materials can be made with unusual properties. Such research includes the synthetic of nanomaterials, manufacturing processes, in terms of the controls of their nano-structural and geometrical properties, moldability and mixability with other matrix for nanocomposites. The cost of designing and producing a novel multifunctional material can be high and the risk of investment to be significant [12, 22].

Computational materials research that relies on multiscale modeling has the potential to significantly reduce development costs of new nanostructured materials for demanding applications by bringing physical and microstructural information into the realm of the design engineer. As there are various potential applications of nanotechnology in design multifunctional product, only some of the well-known properties come from by nano-treatment are critically highlighted [12, 22, 30]. This section review current research in nanotechnology application of the electrospinning nanofiber, from fibber production and development to end uses as multifunctional nanostructure device and product. The electrospinning phenomena are described from experimental point of view to it simulation as multiscale problem.

7.4.1 THE MULTIFUNCTIONAL MATERIALS AND PRODUCTS

7.4.1.1 RESPONSIVE NANOPARTICLES

There are several directions in the research and development of the responsive nanoparticle (RNP) applications. Development of particles that respond by changing stability of colloidal dispersions is the first directions. Stimuli-responsive emulsions and foams could be very attractive for various technologies in coating industries, cosmetic, and personal care (Fig. 7.2) The RNPs compete with surfactants and, hence, the costs for the particle production will play a key role. The main challenge is the development of robust and simple methods for the synthesis of RNPs from inexpensive colloidal particles and suspensions. That is indeed not a simple job since most of commercially available NPs are more expensive than surfactants. Another important application of RNPs for tunable colloidal stability of the particle suspensions is a very broad area of biosensors [86, 87].

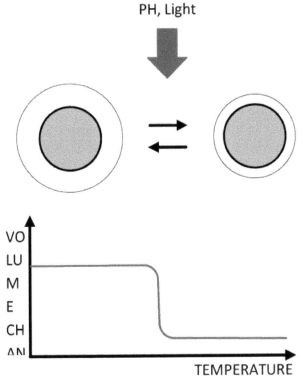

FIGURE 7.2   Stimuli-responsive nanoparticles.

The second direction is stimuli-responsive capsules that can release the cargo upon external stimuli (See Fig. 7.1). The capsules are interesting for biomedical applications (drugs delivery agents) and for composite materials (release of chemicals for self-healing). The most challenging task in many cases is to engineering systems capable to work with demanded stimuli. It is not a simple job for many biomedical applications where signaling biomolecules are present in very small concentrations and a range of changes of many properties is limited by physiological conditions. A well-known challenge is related to the acceptable size production of capsules. Many medical applications need capsules less than 50 nm in diameter. Fabrication of capsules with a narrow pore size distribution and tunable sizes could dramatically improve the mass transport control [86, 88].

A hierarchically organized multi-compartment RNPs are in the focus. These particles could respond to weak signals, to multiple signals, and could demonstrate a multiple response. They can perform logical operations with multiple signals, store energy, absorb and consume chemicals, and synthesize and release chemicals. In other words, they could operate as an autonomous intelligent minidevice. The development of such RNPs can be considered as a part of biomimetics inspired by living cells or logic extension of the bottom up approach in nanotechnology. The development of the intelligent RNPs faces numerous challenges related to the coupling of many functional building blocks in a single hierarchically structured RNP. These particles could find applications for intelligent drug delivery, removal of toxic substances, diagnostics in medicine, intelligent catalysis, microreactors for chemical synthesis and biotechnology, new generation of smart products for personal use, and others [88, 89].

7.4.1.2　NANOCOATINGS

In general, the coating's thickness is at least an order of magnitude lower than the size of the geometry to be coated. The coating's thickness less than 10 nm is called nanocoating. Nanocoatings are materials that are produced by shrinking the material at the molecular level to form a denser product. Nanostructure coatings have an excellent toughness, good corrosion resistance, wear and adhesion properties. These coatings can be used to repair component parts instead of replacing them, resulting in significant reductions in maintenance costs. Additionally, the nanostructure coatings will extend the service life of the component due to the improved properties over conventional coatings [90, 91].

7.4.1.3　FIBROUS NANOSTRUCTURE

The nanofibers are basic building block for plants and animals. From the structural point of view, a uniaxial structure is able to transmit forces along its length and reducing required mass of materials. Nanofibers serves as the another platform for multifunctional hierarchical example. The successful design concepts of nature, the nanofiber become an attractive basic building component in the construction of hierarchically organized nanostructures. To follow nature's design, a process that is able to fabricate nanofiber from a variety materials and mixtures is a prerequisite [92, 93].

Control of the nanofibers arrangement is also necessary to optimize structural requirements. Finally, incorporation of other components into the nanofibers is required to form a complex, hierarchically organized composite. A nanofiber fabrication technique known as electrospinning process has the potential to play a vital role in the construction of a multilevels nanostructure [94]. In this paper, we will introduce electrospinning as a potential technology for use as a platform for multifunctional, hierarchically organized nanostructures. Electrospinning is a method of producing superfine fibers with diameters ranging from 10 nm to 100 nm. Electrospinning occurs when the electrical forces at the surface of a polymer solution overcome the surface tension and cause an electrically charged jet of polymer solution to be ejected. A schematic drawing of the electrospinning process is shown in Fig. 7.3. The electrically charged jet undergoes a series of electrically induced instabilities during its passage to the collection surface, which results in complicated stretching and looping of the jet [50, 60]. This stretching process is accompanied by the rapid evaporation of the solvent molecules, further reducing the jet diameter. Dry fibers are accumulated on the surface of the collector, resulting in a nonwoven mesh of nanofibers.

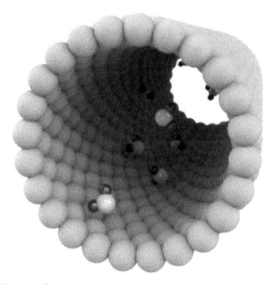

FIGURE 7.3   Nanocoatings.

Basically, an electrospinning system consists of three major components: a high voltage power supply, an emitter (e.g., a syringe) and a grounded collecting plate (usually a metal screen, plate, or rotating mandrel). There are wide ranges of polymers that used in electrospinning and are able to form fine nanofibers within the submicron range and used for varied applications. Electrospun nanofibers have been reported as being from various synthetic polymers, natural polymers or a blend of both including proteins, nucleic acids [74]. The electrospinning process is solely governed by many parameters, classified broadly into rheological, processing, and ambient parameters. Rheological parameters include viscosity, conductivity, molecular weight, and surface tension and process parameters include applied electric field, tip to collector distance and flow rate. Each of these parameters significantly affect the fibers morphology obtained as a result of electrospinning, and by proper manipulation of these parameters we can get nanofibers fabrics of desired structure and properties on multiple material scale.

Among these variables, ambient parameters encompass the humidity and temperature of the surroundings which play a significant role in determining the morphology and topology of electrospun fabrics. Nanofibrous assemblies such as nonwoven fibrous sheet, aligned fibrous fabric, continuous yarn and 3D structure have been fabricated using electrospinning [51]. Physical characteristics of the electrospun nanofibers can also be manipulated by selecting the electrospinning conditions and solution. Structure organization on a few hierarchical levels (See Fig. 7.4) has been developed using electrospinning (Fig. 7.5). Such hierarchy and multifunctionality potential will be described in the following sections. Finally, we will describe how electrospun multifunctional, hierarchically organized nanostructure can be used in applications such as healthcare, defense and security, and environmental.

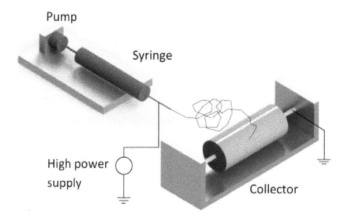

FIGURE 7.4   The electrospinning process.

FIGURE 7.5   Multiscale electrospun fabric.

The slender-body approximation is widely used in electrospinning analysis of common fluids [51]. The presence of nanoelements (nanoparticles, carbon nanotube, clay) in suspension jet complicate replacement 3D axisymmetric with 1D equivalent jet problem under solid-fluid interaction force on nanolevel domain. The applied electric field induced dipole moment, while torque on the dipole rotate and align the nanoelement with electric field. The theories developed to describe the behavior of the suspension jet fall into two levels macroscopic and microscopic. The macroscopic governing equations of the electrospinning are equation of continuity, conservation of the charge, balance of momentum and electric field equation. Conservation of mass for the jet requires that [61, 95] (Fig. 7.6).

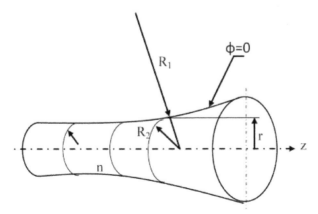

FIGURE 7.6   Geometry of the Jet flow.

For polymer suspension stress tensor τ_{ij} come from polymeric τ_{ij} and solvent contribution tensor via constitutive equation:

$$\tau_v = \hat{\tau}_v + n_s.\dot{y}_v \qquad (7.4.1)$$

where n_s is solvent viscosity, and γ_{ij} strain rate tensor. The polymer contribution tensor $\tau^\wedge_{ij}$ depend on microscopic models of the suspension. Microscopic approach represents the microstructural features of material by means of a large number of micromechanical elements (beads, platelet, rods) obeying stochastic differential equations. The evolution equations of the microelements arise from a balance of momentum on the elementary level. For example, rheological behavior of the dilute suspension of the carbon nanotube (CNTs) in polymer matrix can be described as FENE dumbbell model [96].

$$\langle Q.Q \rangle^{\nabla} = \delta_v - \frac{c\langle Q.Q \rangle}{1 - tr\langle Q.Q \rangle / b_{mAx}} \qquad (7.4.2)$$

where $<Q.Q>$ is the suspension configuration tensor (See Fig. 7.7), c is a spring constant, and max b is maximum CNT extensibility. Subscript ∇ represent the upper convected derivative,

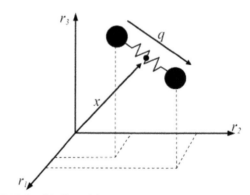

FIGURE 7.7   FENE dumbbell model.

and λ denote a relaxation time. The polymeric stress can be obtained from the following relation:

$$\frac{\hat{\tau}_v}{nkT} = \delta_v - \frac{c\langle Q.Q \rangle}{1 - tr\langle Q.Q \rangle / b_{mAx}} \qquad (7.4.3)$$

where k is Boltzmann's constant, T is temperature, and n is dumbbells density. Orientation probability distribution function ψ of the dumbbell vector Q can be described by the Fokker-Planck equation, neglecting rotary diffusivity.

$$\frac{\partial \psi}{\partial t} + \frac{\partial}{\partial Q}(\psi . Q) = 0 \qquad (7.4.4)$$

Solution equations (the Eqs. (4.3) and (4.4)) with supposition that flow in orifice is Hamel flow [97], give value orientation probability distribution function ψ along streamline of the jet. Rotation motion of a nanoelement (CNTs for example) in a Newtonian flow can be described as short fiber suspension model as another rheological model [8].

$$\frac{dp}{dt} = \frac{1}{2}\omega_v P_i + \frac{1}{2}\Theta\left[\frac{d\gamma_v}{dt}P_j - \frac{d\gamma_{kl}}{dt}P_k P_t P_l\right] - D_r \frac{1}{\psi}\frac{\partial \psi}{\partial t} \qquad (7.4.5)$$

Where p is a unit vector in nanoelement axis direction, ω_{ij} is the rotation rate tensor, γ_{ij} is the deformation tensor, D_r is the rotary diffusivity and θ is shape factor. Microscopic models for evolution of suspension microstructure can be coupled to macroscopic transport equations of mass and momentum to yield micromacro multiscale flow models. The presence of the CNTs in the solution contributes to new form of instability with influences on the formation of the electrospun mat. The high strain rate on the nanoscale with complicated microstructure requires innovative research approach from the computational modeling point of view [98].

Figure 7.8 illustrate multiscale treatment the CNTs suspension in the jet, one time as short flexible cylinder in solution (microscale), and second time as coarse grain system with polymer chain particles and CNT(nanoscale level).

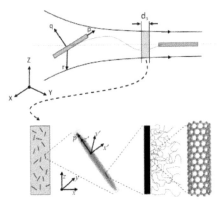

FIGURE 7.8   The CNTs alignment in jet flow.

7.4.2 MULTIFUNCTIONAL NANOFIBER-BASED STRUCTURE

The variety of materials and fibrous structures that can be electrospun allow for the incorporation and optimization of various functions to the nanofiber, either during spinning or through postspinning modifications. A schematic of the multilevel organization of an electrospun fiber based composite is shown in Fig. 7.9.

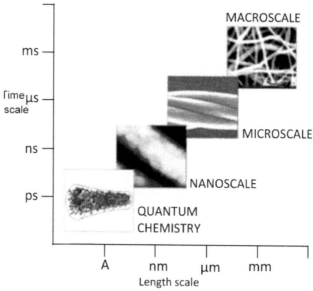

FIGURE 7.9 Multiscale electrospun fabrics.

Based on current technology, at least four different levels of organization can be put together to form a nanofiber based hierarchically organized structure. At the first level, nanoparticles or a second polymer can be mixed into the primary polymer solution and electrospun to form composite nanofiber. Using a dual-orifice spinneret design, a second layer of material can be coated over an inner core material during electrospinning to give rise to the second level organization. Two solution reservoirs, one leading to the inner orifice and the other to the outer orifice will extrude the solutions simultaneously. Otherwise, other conditions for electrospinning remain the same. Rapid evaporation of the solvents during the spinning process reduces mixing of the two solutions therefore forming core-shell nanofiber. At the same level, various surface coating or functionalization techniques may be used to introduce additional property to the fabricated nanofiber surface. Chemical functionality is a vital component in advance multifunctional composite material to detect and respond to changes in its environment. Thus various surface modifications techniques have been used to construct the preferred arrangement of chemically active molecules on the surface with the nanofiber as a supporting base. The third level organization will see the fibers oriented and organized to optimize its performance. A multilayered nanofiber membrane or mixed materials nanofibers membrane can be fabricated *in situ* through selective spinning or using a multiple orifice spinneret design, respectively. Finally, the nanofibrous assembly may be embedded within a matrix to give the fourth-level organization. The resultant structure will have various properties and functionality due its hierarchical organization. Nanofiber structure at various levels have been constructed and tested for various applications and will be covered in the following sections. To follow surface functionality and modification, jet flow must be solved on multiple scale level. All above scale (nanoscale) can be solved by use particle method together with coarse grain method on supramolecular level [50, 51].

7.4.3 NANOFIBER EFFECTIVE PROPERTIES

The effective properties of the nanofiber can be determined by homogenization procedure using representative volume element (RVE). There is need for incorporating more physical information on microscale in order to precise determine material behavior model. For electrospun suspension with nano-elements (CNTs,), a concentric composite cylinder embedded with a caped carbon nanotube represents RVE as shown by Fig. 7.10. A carbon nanotube with a length 2l, radii 2a is embedded at the center of matrix materials with a radii R and length 2L.

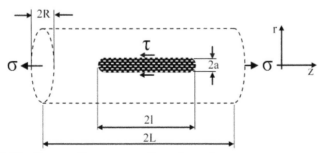

FIGURE 7.10   The nanofiber representative volume element.

The discrete atomic nanotube structure replaced the effective (solid) fiber having the same length and outer diameter as a discrete nanotube with effective Young's nanotube modulus determined from atomic structure. The stress and strain distribution in RVE was determined using modified shear-lag model [99]. For the known stress and strain distribution under RVE we can calculate elastic effective properties quantificators. The effective axial module E_{33}, and the transverse module $E_{11} = E_{22}$, can be calculated as follow

$$E_{33} = \frac{\langle \sigma_{zz} \rangle}{\langle \varepsilon_{zz} \rangle}$$

$$E_{11} = \frac{\langle \sigma_{xx} \rangle}{\langle \varepsilon_{xx} \rangle} \qquad (7.4.6)$$

where denotes a volume average under volume V as defined by:

$$\langle \Xi \rangle = \frac{1}{V} \int_V \Xi\,(x, y, z).\,dV. \qquad (7.4.7)$$

The three-phase concentric cylindrical shell model has been proposed to predict effective modulus of nanotube reinforced nanofibers. The modulus of nanofiber depends strongly upon the thickness of the interphase and CNTs diameter [12].

7.4.4   NETWORK MACROSCOPIC PROPERTIES

Macroscopic properties of the multifunctional structure determine final value of the any engineering product. The major objective in the determination of macroscopic properties is the link between atomic and continuum types of modeling and simulation approaches. The multiscale method such as quasi-continuum, bridge method, coarse-grain method, and dissipative particle dynamics are some popular methods of solution [98, 100]. The main advantage of the mesoscopic model is its higher computational efficiency than the molecular modeling without a loss of detailed properties at molecular level. Peri-dynamic modeling of fibrous network is another promising method, which allows damage, fracture and long-range forces to be treated as natural components of the deformation of materials [101]. In the first stage, effective fiber properties are determined by homogenization procedure, while in the second stage the point-bonded stochastic fibrous network at mesoscale is replaced by continuum plane stress model. Effective mechanical properties of nanofiber sheets at the macro scale level can be determined using the 2D Timoshenko beam-network. The critical parameters are the mean number of crossings per nanofiber, total nanofiber crossing in sheet and mean segment length [102]. Let as first consider a general planar fiber network characterized by fiber concentration n and fiber angular and length distribution ψ (φ, 1), where φ and l are fiber orientation angle and fiber length, respectively. The fiber radius r is considered uniform and the fiber concentration n is defined as the number of fiber per unite area (Fig. 7.11).

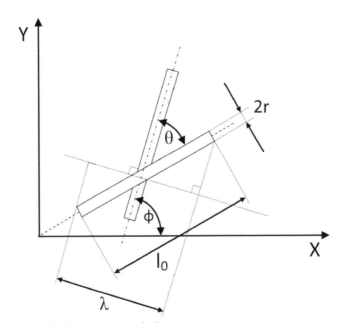

FIGURE 7.11   The fiber contact analysis.

The Poisson probability distribution can be used to describe the fiber segment length distribution for electrospun fabrics, a portion of the fiber between two neighboring contacts:

$$f(\ell) = \frac{1}{\ell}\exp(-\ell/\overline{\ell})$$ (7.4.8)

where l is the mean segment length. The total number fiber segments $N^{\wedge}$ in the rectangular region $b^* h$:

$$\overline{N} = \{n.\ell_0(\langle\lambda\rangle + 2r) - 1\}.n.b.h$$ (7.4.9)

with

$$\langle \lambda \rangle = \int_0^\phi \int_0^\infty \psi(\,9.\ell).\lambda(9).d\ell.d9 \qquad (7.4.10)$$

where the dangled segments at fiber ends have been excluded. The fiber network will be deformed in several ways. The strain energy in fiber segments come from bending, stretching and shearing modes of deformation (see Fig. 7.12).

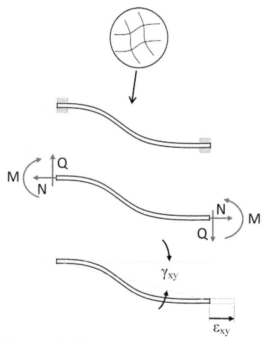

FIGURE 7.12   Fiber network 2D model.

$$U = N.\ell_0.b.h\frac{1}{2}\iint \frac{E.A}{\ell}\varepsilon_{XX}^2.\psi(\varphi,\ell).\ell.d\ell.d\varphi$$
$$+n.\ell_0.\{\,(\langle\lambda\rangle+2r)-1\,\}.b.h.\frac{1}{2}\Big\{\iint \frac{G.A}{\ell}..\gamma_{xy}^2.\psi(\varphi,\ell).\ell.d\ell.d\varphi$$
$$+\iint \frac{3.E.I}{\ell^3}\gamma_{xy}^2.\psi(\varphi,\ell).\ell.d\ell.d\varphi\,\} \qquad (7.4.11)$$

where A and I are beam cross-section area and moment of inertia, respectively. The first term on right side is stretching mode, while second and last term are shear bending modes respectively.

The effective material constants for fiber network can be determined using homogenization procedure concept for fiber network. The strain energy fiber network for representative volume element is equal to strain energy continuum element with effective material constant. The strain energy of the representative volume element under plane stress conditions are:

$$U = \frac{1}{2} \cdot \langle \varepsilon_v \rangle \cdot C_{vkl} \langle \varepsilon_{kl} \rangle \cdot V \qquad (7.4.12)$$

where is $V.b.h.2.r$ representative volume element, C_{ijkl} are effective elasticity tensor. The square bracket $<\,>$ means macroscopic strain value. Microscopic deformation tensor was assume of a fiber segments ε_{ij} is compatible with effective macroscopic strain $<\varepsilon_{ij}>$ of effective continuum (affine transformation). This is bridge relations between fiber segment microstrain ε_{ij} and macroscopic strain $<\varepsilon_{ij}>$ in the effective medium. Properties of this nanofibrous structure on the macro scale depend on the 3D joint morphology. The joints can be modeled as contact torsional elements with spring and dashpot [102]. The elastic energy of the whole random fiber network can be calculated numerically, from the local deformation state of the each segment by finite element method [103]. The elastic energy of the network is then the sum of the elastic energies of all segments. We consider here tensile stress, and the fibers are rigidly bonded to each other at every fiber-fiber crossing points. To mimic the microstructure of electrospun mats, we generated fibrous structures with fibers positioned in horizontal planes, and stacked the planes on top of one another to form a 2D or 3D structure. The representative volume element dimensions are considered to be an input parameter that can be used among other parameters to control the solid volume fraction of the structure, density number of fiber in the simulations. The number of intersections/unit area and mean lengths are obtained from image analysis of electrospun sheets. For the random point field the stochastic fiber network was generated. Using polar coordinates and having the centerline equation of each fiber, the relevant parameters confined in the simulation box is obtained. The procedure is repeated until reaching the desired parameters is achieved [22, 95, 104]. The nonload bearing fiber segments were removed and trimmed to keep dimensions $b*h$ of the representative window (see Fig. 7.13).

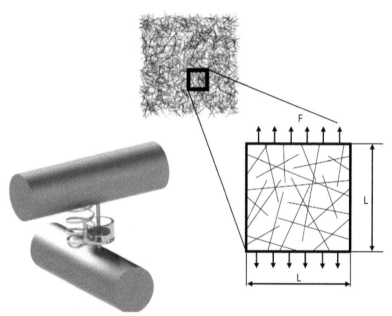

FIGURE 7.13   Representative volume element of the network.

A line representative network model is replaced by finite element beam mesh. The finite element analyzes were performed in a network of 100 fibers, for some CNTs volume fractions values. Nanofibers were modeled as equivalent cylindrical beam as mentioned above. Effective mechanical properties of nanofiber sheets at the macro scale level can be determined using the 2D Timoshenko beam-network.

For a displacement-based form of beam element, the principle of virtual work is assumed valid. For a beam system, a necessary and sufficient condition for equilibrium is that the virtual work done by sum of the external forces and internal forces vanish for any virtual displacement $\delta W = 0$. *The W is the virtual work, which the work is done by imaginary or virtual displacements.*

$$W = \oiiint_v v\delta\varepsilon_v + \oiiint_v F_j\delta u_s dV + \oiint_A T\delta u_l dA \qquad (7.4.13)$$

where, ε is the strain, σ is the stress, F is the body force, δu is the virtual displacement, and T is the traction on surface A. The symbol δ is the variational operator designating the virtual quantity. Finite element interpolation for displacement field [15]:

$$\{u\} = [N]\{\hat{u}\} \tag{7.4.14}$$

where is $\{u\}$ u displacement vector of arbitrary point and $\{u\hat{\ }\}$ is nodal displacement point's vector. (N) is shape function matrix. After FEM procedure the problem is reduced to the solution of the linear system of equations.

$$[K_e].\{u\} = \{f\} \tag{7.4.15}$$

where are $\{u\}$ is global displacement vector, $\{f\}$ global nodal force vector, and $[K_e]$ global stiffness matrix. Finite element analyzes were performed for computer generated network of 100 fibers. The comparison of calculated data with experimental data [99] for nanotube sheet shows some discrepancies (Fig. 7.14). A rough morphological network model for the sheets can explain this on the one hand and simple joint morphology on the other hand [103].

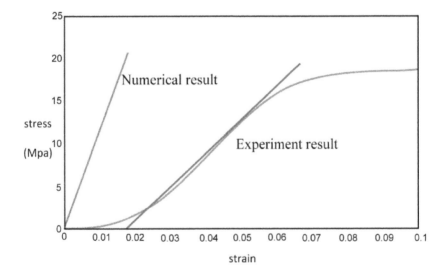

Figure 14. The stress-strain curve

FIGURE 7.14   The stress-strain curve.

7.4.5   FLOW IN FIBER NETWORK

Electrospun nanofiber materials are becoming an integral part of many recent applications and products. Such materials are currently being used in tissue engineering, air and liquid filtration, protective clothing's, drug delivery, and many others. Permeability of fibrous media is important in many applications, therefore during the past few decades, there have been many original studies dedicated to this subject. Depending on the fiber diameter and the air thermal conditions, there are four different regimes of flow around a fiber:

a) Continuum regime ($K_N \leq 10^{-3}$),
b) Slip-flow regime ($10^{-3} \leq K_N \leq 0.25$),
c) Transient regime ($0.25 \leq K_N \leq 10$).
d) Free molecule regime ($N_K \geq 10$),

Here, $K_N = 2\lambda/d$ is the fiber Knudson number, where $\lambda = RT//\sqrt{2}N^{\hat{}}\pi. d^2 p$ is the mean free path of gas molecules, d is fiber diameter, $N^{\hat{}}$ is Avogadro number. Airflow around most electrospun nanofibers is typically in the slip or transition flow regimes. In the context of air filtration, the 2D analytical work of Kuwabara [105] has long been used for predicting the permeability across fibrous filters. The analytical expression has been modified by Brown [106] to develop an expression for predicting the permeability across filter media operating in the slip flow regime. The ratio of the slip to no-slip pressure drops obtained from the simplified 2D models may be used to modify the more realistic, and so more accurate, existing 3D permeability models in such a way that they could be used to predict the permeability of nanofiber structure. To test this supposition, for above developed 3D virtual nanofibrous structure, the Stokes flow equations solved numerically inside these virtual structures with an appropriate slip boundary condition that is developed for accounting the gas slip at fiber surface.

7.4.5.1   FLOW FIELD CALCULATION

A steady state, laminar, incompressible model has been adopted for the flow regime inside our virtual media. Implemented in the Fluent code is used to solve continuity and conservation of linear momentum in the absence of inertial effects [107]:

$$\nabla . v = 0 \qquad\qquad (7.4.16)$$

$$\nabla P = \mu . \Delta^2 v \qquad\qquad (7.4.17)$$

The grid size required to mesh the gap between two fibers around their crossover point is often too small. The computational grid used for computational fluid dynamics (CFD) simulations needs to be fine enough to resolve the flow field in the narrow gaps, and at the same time coarse enough to cover the whole domain without requiring infinite computational power. Permeability of a fibrous material is often presented as a function of fiber radius, r, and solid volume fraction α, of the medium. Here, we use the continuum regime analytical expressions of Jackson and James [108], developed for 3D isotropic fibrous structures given as

$$\frac{k}{r^2} = \frac{3r^2}{20a}[-\ell n(a) - 0.931] \qquad (7.4.18)$$

Brown [106] has proposed an expression for the pressure drop across a fibrous medium based on the 2D cell model of Kuwabara [105] with the slip boundary condition:

$$\Delta P_{SLIP} = \frac{4\mu a. hV. (1 + 1.996K_N}{r^2[\hat{K} + 1.996K_N(-0.5. \ell na - 0.25 + 0.25a^2]}$$

$$\hat{K} = -0.5. \ell na - 0.25 + 0.25a^2 \qquad (7.4.19)$$

where $K^\wedge = -0.5. \, ln\alpha - 0.75 + \alpha - 0.25\alpha^2$, Kuwabara hydrodynamic factor, h is fabric thickness, and V is velocity. As discussed in the some reference [*], permeability (or pressure drop) models obtained using ordered 2D fiber arrangements are known for under predicting the permeability of a fibrous medium. In order to overcome this problem, if a correction factor can be derived based on the above 2D expression, and used with the realistic expressions developed for realistic 3D fibrous structures. From Eq. (19) we have for the case of no-slip boundary condition ($K_N = 0$):

$$\Delta P_{NONSLIP} = \frac{4\mu a. hV}{r^2\hat{K}} \qquad (7.4.20)$$

The correction factor is defined as:

$$\Xi = \frac{\Delta P_{NONSLIP}}{\Delta P_{SLIP}} \qquad (7.4.21)$$

to be used in modifying the original permeability expressions of Jackson and James [109], and/or any other expression based on the no-slip boundary condition, in order to incorporate the slip effect. For instance, the modified expression of Jackson and James can be presented as:

$$k_z = \frac{3r^2}{20\alpha}[-\ln(\alpha) - 0.931].\,\Xi \qquad (7.4.22)$$

Operating pressure has no influence on the pressure drop in the continuum region, while pressure drop in the free molecular region is linearly proportional to the operating pressure. While there are many equations available for predicting the permeability of fibrous materials made up of coarse fibers, there are no accurate "easy-to-use" permeability expressions that can be used for nanofiber media. Figure 7.15 is drown corrected Jackson and James data (blue line). Points on figure are CFD numerical data.

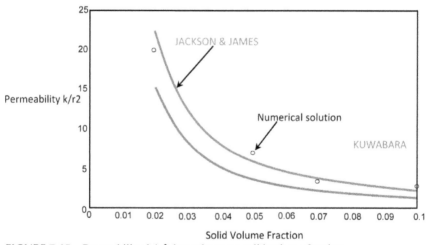

FIGURE 7.15   Permeability $k\,/\,r^2$ dependence on solid volume fraction.

7.4.6   SOME ILLUSTRATIVE EXAMPLES

7.4.6.1   FUEL CELL EXAMPLE

Fuel cells are electrochemical devices capable of converting hydrogen or hydrogen-rich fuels into electrical current by a metal catalyst. There are many kinds of fuel cells, such as proton exchange mat (PEM) fuel cells, direct methanol fuel cells, alkaline fuel cells and solid oxide fuel cells [110]. PEM fuel cells are the most important one among them because of high power density and low operating temperature. Pt nanoparticle catalyst is a main component in fuel cells. The price of Pt has driven up the cell cost and limited the commercialization. Electrospun materials have been prepared as alternative catalyst with high catalytic efficiency, good durability and affordable cost. Binary PtRh and PtRu nanowires were synthesized by electrospinning, and they had better catalytic performance than commercial nanoparticle catalyst because of the one-dimensional features [111]. Pt nanowires also showed higher catalytic activities in a polymer electrolyte membrane fuel cell [112]. Instead of direct use as catalyst, catalyst supporting material is another important application area for electrospun nanofibers. Pt clusters were electrodeposited on a carbon nanofiber mat for methanol oxidation, and the catalytic peak current of the composite catalyst reached 420 mA/mg compared with 185 mA/mg of a commercial Pt catalyst [113]. Pt nanoparticles were immobilized on polyimide-based nanofibers using a hydrolysis process and Pt nanoparticles were also loaded on the carbon nanotube containing polyamic acid nanofibers to achieve high catalytic current with long-term stability [114]. Proton exchange mat is the essential element of PEM fuel cells and normally made of a Nafion film for proton conduction. Because pure Nafion is not suitable for electrospinning due to its low viscosity in solution, it is normally mixed with other polymers to make blend nanofibers. Blend Nafion/PEO nanofibers were embedded in an inert polymer matrix to make a proton conducting mat [115], and a high proton conductivity of 0.06–0.08 S/cm at 15 °C in water and low water swelling of 12–23 wt% at 25 °C were achieved [116].

7.4.6.2 PROTECTIVE CLOTHING EXAMPLE

The development of smart nanotextiles has the potential to revolutionize the functionality of our clothing and the fabrics in our surroundings. This is made possible by such developments as new materials, fibbers, and finishing; inherently conducting polymers; carbon nanotubes; an antimicrobial nanocoatings. These additional functionalities have numerous applications in healthcare, sports, military applications, fashion, and etc. Smart textiles become a critical part of the emerging area of body sensor networks incorporating sensing, actuation, control and wireless data transmission [51, 52, 117] (Fig. 7.16).

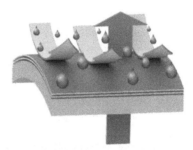

FIGURE 7.16   Ultrathin layer for selective transport.

7.4.6.3 *MEDICAL DEVICE*

Basic engineered nanomaterial and biotechnology products will be enormously useful in future medical applications. We know nanomedicine as the monitoring, repair, construction and control of biological systems at the nanoscale level, using engineered nanodevices and nanostructures. The upper portion of the dress contains cotton coated with silver nanoparticles. Silver possesses natural antibacterial qualities that are strengthened at the nanoscale, thus giving the ability to deactivate many harmful bacteria and viruses. The silver infusion also reduces the need to wash the garment, since it destroys bacteria, and the small size of the particles prevents soiling and stains [16, 118] (Fig. 7.17).

FIGURE 7.17   Cotton coated with silver.

7.4.6.3.1   DRUG DELIVERY AND RELEASE CONTROL

Controlled release is an efficient process of delivering drugs in medical therapy. It can balance the delivery kinetics, immunize the toxicity and side effects, and improve patient convenience [119]. In a controlled release system, the active substance is loaded into a carrier or device first, and then released at a predictable rate in vivo when administered by an injected or noninjected route. As a potential drug delivery carrier, electrospun nanofibers have exhibited many advantages. The drug loading is very easy to implement via electrospinning process, and the high-applied voltage used in electrospinning process had little influence on the drug activity. The high specific surface area and short diffusion passage length give the nanofiber drug system higher overall release rate than the bulk material (e.g., film). The release profile can be finely controlled by modulation of nanofiber morphology, porosity and composition.

Nanofibers for drug release systems mainly come from biodegradable polymers and hydrophilic polymers. Model drugs that have been studied include water soluble [120], poor-water soluble [121] and water insoluble drugs [122]. The release of macromolecules, such as DNA [123] and bioactive proteins from nanofibers was also investigated. In most cases, water-soluble drugs, including DNA and proteins, exhibited an early stage burst [124]. For some applications, preventing postsurgery induced adhesion for instance, and such an early burst release will be an ideal profile because most infections occur within the first few hours after surgery. A recent study also found that when a poorly water-soluble drug was loaded into PVP nanofibers [125], 84.9% of the drug can be released in the first 20 seconds when the drug-to-PVP ratio was kept as 1:4, which can be used for fast drug delivery systems. However, for a long-lasting release process, it would be essential to maintain the release at an even and stable pace, and any early burst release should be avoided. For a water insoluble drug, the drug release from hydrophobic nanofibers into buffer solution is difficult. However, when an enzyme capable of degrading nanofibers exists in the buffer solution, the drug can be released at a constant rate because of the degradation of nanofibers [122]. For example, when rifampin was encapsulated in PLA nanofibers, no drug release was detected from the nanofibers. However, when the buffer solution contained proteinase K, the drug release took place nearly in zero-order kinetics, and no early burst release happened. Similarly, initial burst release did not occur for poor-water soluble drugs, but the release from a nonbiodegradable nanofiber could follow different kinetics [126]. In another example, blending a hydrophilic but water-insoluble polymer (PEG-g-CHN) with PLGA could assist in the release of a poor-water soluble drug Iburprofen [127]. However, when a water-soluble polymer was used, the poorly soluble drug was released accompanied with dissolving of the nanofibers, leading to a low burst release [128]. The early burst release can be reduced when the drug is encapsulated within the nanofiber matrix. When an amphiphilic block copolymer, PEG-b-PLA was added into Mefoxin/PLGA nanofibers, the cumulative amount of the released drug at earlier time points was reduced and the drug release was prolonged [129]. The reason for the reduced burst release was attributed to the encapsulation of some drug molecules within the hydrophilic block of the PEG-b-PLA. Amphiphilic block copolymer also assisted the dispersion and encapsulation of water-soluble drug into nanofibers when the polymer solution used an oleophilic solvent, such as chloroform, during electrospinning [130]. In this case, a water-in-oil emulsion can be electrospun into uniform nanofibers, and drug molecules are trapped by hydrophilic chains. The swelling of the hydrophilic chains during releasing assists the diffusion of drug from nanofibers to the buffer. Coating nanofibers with a shell could be an effective way to control the release profile.

When a thin layer of hydrophobic polymer, such as poly (p-xylylene) (PPX), was coated on PVA nanofibers loaded with bovine serum albumin (BSA)/luciferase, the early burst release of the enzyme was prevented [131]. Fluorination treatment [132] on PVA nanofibers introduced functional C-F groups and made the fiber surface hydrophobic, which dramatically decreased the initial drug burst and prolonged the total release time. The polymer shell can also be directly applied via a coaxial co-electrospinning process, and the nanofibers produced are normally named "core-sheath" bicomponent nanofibers. In this case, even a pure drug can be entrapped into nanofiber as the core, and the release profile was less dependent on the solubility of drug released [133]. A research has compared the release behavior of two drug-loaded PLLA nanofibers prepared using blend and coaxial electrospinning [134]. It was found that the blend fibers still showed an early burst release, while the threads made of core-sheath fibers provided a stable release of growth factor and other therapeutic drugs. In addition, the early burst release can also be lowered via encapsulating drugs into nanomaterial, followed by incorporating the drug-loaded nanomaterials into nanofibers. For example, halloysite nanotubes loaded with tetracycline hydrochloride were incorporated into PLGA nanofibers and showed greatly reduced initial burst release [135].

7.4.7   CONCLUDING REMARKS OF MULTIFUNCTIONAL NANOSTRUCTURES DESIGN

Electrospinning is a simple, versatile, and cost-effective technology, which generates nonwoven fibers with high surface area to volume ratio, porosity and tunable porosity. Because of these properties this process seems to be a promising candidate for various applications especially nanostructure applications. Electrospun fibers are increasingly being used in a variety of applications such as, tissue engineering scaffolds, wound healing, drug delivery, immobilization of enzymes, as membrane in biosensors, protective clothing, cosmetics, affinity membranes, filtration applications etc. In summary, mother Nature has always used hierarchical structures such as capillaries and dendrites to increase multifunctional of living organs. Material scientists are at beginning to use this concept and create multiscale structures where nanotubes, nanofillers can be attached to larger surfaces and subsequently functionalized. In principle, many more applications can be envisioned and created. Despite of several advantages and success of electrospinning there are some critical limitations in this process such as small pore size inside the fibers. Several attempts in these directions are being made to improve the design through multi-layering, inclusion of nanoelements and blending with polymers with different degradation behavior. As new architectures develop, a new wave of surface-sensitive devices related to sensing, catalysis, photovoltaic, cell scaffolding, and gas storage applications is bound to follow.

7.5   INTRODUCTION TO THEORETICAL STUDY OF ELECTROSPINNING PROCESS

Electrospinning is a procedure in which an electrical charge to draw very fine (typically on the micro or nano scale) fibers from polymer solution or molten. Electrospinning shares characteristics of both electrospraying and conventional solution dry spinning of fibers. The process does not require the use of coagulation chemistry or high temperatures to produce solid threads from solution. This makes the process more efficient to produce the fibers using large and complex molecules. Recently, various polymers have been successfully electrospun into ultrafine fibers mostly in solvent solution and some in melt form [79, 136]. Optimization of the alignment and morphology of the fibers is produced by fitting the composition of the solution and the configuration of the electrospinning apparatus such as voltage, flow rate, and etc. As a result, the efficiency of this method can be improved [137]. Mathematical and theoretical modeling and simulating procedure will assist to offer an in-depth insight into the physical understanding of complex phenomena during electrospinningand might be very useful to manage contributing factors toward increasing production rate [75, 138].

Despite the simplicity of the electrospinning technology, industrial applications of it are still relatively rare, mainly due to the notable problems with very low fiber production rate and difficulties in controlling the process [67].

Modeling and simulation (M&S) give information about how something will act without actually testing it in real. The model is a representation of a real object or system of objects for purposes of visualizing its appearance or analyzing its behavior. Simulation is a transition from a mathematical or computational model for description of the system behavior based on sets of input parameters [104, 139]. Simulation is often the only means for accurately predicting the performance of the modeled system [140]. Using simulation is generally cheaper and safer than conducting experiments with a prototype of the final product. Also simulation can often be even more realistic than traditional experiments, as they allow the free configuration of environmental and operational parameters and can often be run faster than in real time. In a situation with different alternatives analysis, simulation can improve the efficiency, in particular when the necessary data to initialize can easily be obtained from operational data. Applying simulation adds decision support systems to the tool box of traditional decision support systems [141].

Simulation permits set up a coherent synthetic environment that allows for integration of systems in the early analysis phase for a virtual test environment in the final system. If managed correctly, the environment can be migrated from the development and test domain to the training and education domain in real systems under realistic constraints [142].

A collection of experimental data and their confrontation with simple physical models appears as an effective approach towards the development of practical tools for controlling and optimizing the electrospinning process. On the other hand, it is necessary to develop theoretical and numerical models of electrospinning because of demanding a different optimization procedure for each material [143]. Utilizing a model to express the effect of electrospinning parameters will assist researchers to make an easy and systematic way of presenting the influence of variables and by means of that, the process can be controlled. Additionally, it causes to predict the results under a new combination of parameters. Therefore, without conducting any experiments, one can easily estimate features of the product under unknown conditions [95].

7.6   STUDY OF ELECTROSPINNING JET PATH

To yield individual fibers, most, if not all of the solvents must be evaporated by the time the electrospinning jet reaches the collection plate. As a result, volatile solvents are often used to dissolve the polymer. However, clogging of the polymer may occur when the solvent evaporates before the formation of the Taylor cone during the extrusion of the solution from several needles. In order to maintain a stable jet while still using a volatile solvent, an effective method is to use a gas jacket around the Taylor cone through two coaxial capillary tubes. The outer tube which surrounds the inner tube will provide a controlled flow of inert gas which is saturated with the solvent used to dissolve the polymer. The inner tube is then used to deliver the polymer solution. For 10 wt% poly (L-lactic acid) (PLLA) solution in dichloromethane, electrospinning was not possible due to clogging of the needle. However, when N2 gas was used to create a flowing gas jacket, a stable Taylor cone was formed and electrospinning was carried out smoothly.

7.6.1   THE THINNING JET (JET STABILITY)

The conical meniscus eventually gives rise to a slender jet that emerges from the apex of the meniscus and propagates downstream. Hohman et al. [60] first reported this approach for the relatively simple case of Newtonian fluids. This suggests that the shape of the thinning jet depends significantly on the evolution of the surface charge density and the local electric field. As the jet thins down and the charges relax to the surface of the jet, the charge density and local field quickly pass through a maximum, and the current due to advection of surface charge begins to dominate over that due to bulk conduction.

The crossover occurs on the length scale given by [6]:

$$L_N = \left(K^4 Q^7 \rho^3 (\ln X)^2 / 8\pi^2 E_\infty I^5 \varepsilon^{-2} \right)^{1/5} \qquad (7.6.1)$$

This length scale defines the 'nozzle regime' over which the transition from the meniscus to the steady jet occurs. Sufficiently far from the nozzle regime, the jet thins less rapidly and finally enters the asymptotic regime, where all forces except inertial and electrostatic forces ceases to influence the jet. In this regime, the radius of the jet decreases as follows:

$$h = \left(\frac{Q^3 \rho}{2\pi^2 E_\infty I} \right)^{1/4} z^{-1/4} \qquad (7.6.2)$$

Here z is the distance along the centerline of the jet. Between the 'nozzle regime' and the 'asymptotic regime,' the evolution of the diameter of the thinning jet can be affected by the viscous response of the fluid. Indeed by balancing the viscous and the electrostatic terms in the force balance equation it can be shown that the diameter of the jet decreases as:

$$h = \left(\frac{6\mu Q^2}{\pi E_\infty I} \right)^{1/2} z^{-1} \tag{7.6.3}$$

In fact, the straight jet section has been studied extensively to understand the influence of viscoelastic behavior on the axisymmetric instabilities [93] and crystallization [60] and has even been used to extract extensional viscosity of polymeric fluids at very high strain rates.

For highly strain-hardening fluids, Yu et al. [144] demonstrated that the diameter of the jet decreased with a power-law exponent of −1/2, rather than −1/4 or −1, as discussed earlier for Newtonian fluids. This −1/2 power-law scaling for jet thinning in viscoelastic fluids has been explained in terms of a balance between electromechanical stresses acting on the surface of the jet and the viscoelastic stress associated with extensional strain hardening of the fluid. Additionally, theoretical studies of viscoelastic fluids predict a change in the shape of the jet due to non-Newtonian fluid behavior. Both Yu et al. [144] and Han et al. [145] have demonstrated that substantial elastic stresses can be accumulated in the fluid as a result of the highstrain rate in the transition from the meniscus into the jetting region. This elastic stress stabilizes the jet against external perturbations. Further downstream the rate of stretching slows down, and the longitudinal stresses relax through viscoelastic processes. The relaxation of stresses following an extensional deformation, such as those encountered in electrospinning, has been studied in isolation for viscoelastic fluids [146]. Interestingly, Yu et al. [144] also observed that, elastic behavior notwithstanding, the straight jet transitions into the whipping region when the jet diameter becomes of the order of 10 mm.

7.6.2   THE WHIPPING JET (JET INSTABILITY)

While it is in principle possible to draw out the fibers of small diameter by electrospinning in the cone-jet mode alone, the jet does not typically solidify enough en route to the collector and succumbs to the effect of force imbalances that lead to one or more types of instability. These instabilities distort the jet as they grow. A family of these instabilities exists, and can be analyzed in the context of various symmetries (axisymmetric or nonaxisymmetric) of the growing perturbation to the jet.

Some of the lower modes of this instability observed in electrospinning have been discussed in a separate review [81]. The 'whipping instability' occurs when the jet becomes convectively unstable and its centerline bends. In this region, small perturbations grow exponentially, and the jet is stretched out laterally. Shin et al. [62] and Fridrikh et al. [63] have demonstrated how the whipping instability can be largely responsible for the formation of solid fiber in electrospinning. This is significant, since as recently as the late 1990s the bifurcation of the jet into two more or less equal parts (also called 'splitting' or 'splaying') were thought to be the mechanism through which the diameter of the jet is reduced, leading to the fine fibers formed in electrospinning. In contrast to 'splitting' or 'splaying,' the appearance of secondary, smaller jets from the primary jet have been observed more frequently and in situ [64, 147]. These secondary jets occur when the conditions on the surface of the jet are such that perturbations in the local field, for example, due to the onset of the slight bending of the jet, is enough to overcome the surface tension forces and invoke a local jetting phenomenon.

The conditions necessary for the transition of the straight jet to the whipping jet has been discussed in the works of Ganan-Calvo [148], Yarin et al. [64], Reneker et al. [66] and Hohman et al. [60].

During this whipping instability, the surface charge repulsion, surface tension, and inertia were considered to have more influence on the jet path than Maxwell's stress, which arises due to the electric field and finite conductivity of the fluid. Using the equations reported by Hohman et al. [60] and Fridrikh et al. [63] obtained an equation for the lateral growth of the jet excursions arising from the whipping instability far from the onset and deep into the nonlinear regime. These developments have been summarized in the review article of Rutledge and Fridrikh.

The whipping instability is postulated to impose the stretch necessary to draw out the jet into fine fibers. As discussed previously, the stretch imposed can make an elastic response in the fluid, especially if the fluid is polymeric in nature. An empirical rheological model was used to explore the consequences of nonlinear behavior of the fluid on the growth of the amplitude of the whipping instability in numerical calculations [63, 79]. There it was observed that the elasticity of the fluid significantly reduces the amplitude of oscillation of the whipping jet. The elastic response also stabilizes the jet against the effect of surface tension. In the absence of any elasticity, the jet eventually breaks up and forms an aerosol. However, the presence of a polymer in the fluid can stop this breakup if:

$$\tau / \left(\frac{\rho h^3}{\gamma} \right)^{1/2} \geq 1 \qquad (7.6.4)$$

Where τ is the relaxation time of the polymer, ρ is the density of the fluid, h is a characteristic radius, and γ is the surface tension of the fluid.

7.7   ELECTROSPINNING DRAW BACKS

Electrospinning has attracted much attention both to academic research and industry application because electrospinning (1) can fabricate continuous fibers with diameters down to a few nanometers, (2) is applicable to a wide range of materials such as synthetic and natural polymers, metals as well as ceramics and composite systems, (3) can prepare nanofibers with low cost and high yielding [47].

Despite the simplicity of the electrospinning technology, industrial applications of it are still relatively rare, mainly due to the notable problems of very low fiber production rate and difficulties in controlling the process [50, 67]. The usual federate for electrospinning is about 1.5 ml/hr. Given a solution concentration of 0.2 g/ml, the mass of nanofiber collected from a single needle after an hour is only 0.3 g. In order for electrospinning to be commercially viable, it is necessary to increase the production rate of the nanoflbers. To do so, multiple-spinning setup is necessary to increase the yield while at the same time maintaining the uniformity of the nanofiber mesh [48].

Optimization of the alignment and morphology of the fibers which is produced by fitting the composition of the solution and the configuration of the electrospinning apparatus such as voltage, flow rate, and etc., can be useful to improve the efficiency of this method [137]. Mathematical and theoretical modeling and simulating procedure will assist to offer an in-depth insight into the physical understanding of complex phenomena during electrospinningand might be very useful to manage contributing factors toward increasing production rate [75, 138].

Presently, nanofibers have attracted the attention of researchers due to their remarkable micro and nano structural characteristics, high surface area, small pore size, and the possibility of their producing three dimensional structure that enable the development of advanced materials with sophisticated applications [73].

Controlling the property, geometry, and mass production of the nanofibers, is essential to comprehend quantitatively how the electrospinning process transforms the fluid solution through a millimeter diameter capillary tube into solid fibers which are four to five orders smaller in diameter [74].

As mentioned above, the electrospinning gives us the impression of being a very simple and easily controlled technique for the production of nanofibers. But, actually the process is very intricate. Thus, electrospinning is usually described as the interaction of several physical instability processes. The bending and stretching of the jet are mainly caused by the rapidly whipping which is an essential element of the process induced by these instabilities. Until now, little is known about the detailed mechanisms of the instabilities and the splaying process of the primary jet into multiple filaments. It is thought to be responsible that the electrostatic forces overcome surface tensions of the droplet during undergoing the electrostatic field and the deposition of jets formed nanofibers [47].

Though electrospinning has been become an indispensable technique for generating functional nanostructures, many technical issues still need to be resolved. For example, it is not easy to prepare nanofibers with a same scale in diameters by electrospinning; it is still necessary to investigate systematically the correlation between the secondary structure of nanofiber and the processing parameters; the mechanical properties, photoelectric properties and other property of single fiber should be systematically studied and optimized; the production of nanofiber is still in laboratory level, and it is extremely important to make efforts for scaled-up commercialization; nanofiber from electrospinning has a the low production rate and low mechanical strength which hindered it's commercialization; in addition, another more important issue should be resolved is how to treat the solvents volatilized in the process.

Until now, lots of efforts are putted on the improvement of electrospinning installation, such as the shape of collectors, modified spinnerets and so on. The application of multijets electrospinning provides a possibility to produce nanofibers in industrial scale. The development of equipments, which can collect the poisonous solvents and the application of melt electrospinning, which would largely reduce the environment problem, create a possibility of the industrialization of electrospinning. The application of water as the solvent for elelctrospinning provide another approach to reduce environmental pollution, which is the main fact hindered the commercialization of electrospinning. In summary, electrospinning is an attractive and promising approach for the preparation of functional nanofibers due to its wide applicability to materials, low cost and high production rate [47].

7.8   MODELLING OF THE ELECTROSPINNING PROCESS

The electrospinning process is a fluid dynamics related problem. Controlling the property, geometry, and mass production of the nanofibers, is essential to comprehend quantitatively how the electrospinning process transforms the fluid solution through a millimeter diameter capillary tube into solid fibers, which are four to five orders smaller in diameter [74]. Although information on the effect of various processing parameters and constituent material properties can be obtained experimentally, theoretical models offer in-depth scientific understanding which can be useful to clarify the affecting factors that cannot be exactly measured experimentally. Results from modeling also explained how processing parameters and fluid behavior lead to the nanofiber of appropriate properties. The term "properties" refers to basic properties (such as fiber diameter, surface roughness, fiber connectivity, etc.), physical properties (such as stiffness, toughness, thermal conductivity, electrical resistivity, thermal expansion coefficient, density, etc.) and specialized properties (such as biocompatibility, degradation curve, etc. for biomedical applications) [48, 73].

For example, the developed models can be used for the analysis of mechanisms of jet deposition and alignment on various collecting devices in arbitrary electric fields [149].

The various method formulated by researchers are prompted by several applications of nanofibers. It would be sufficient to briefly describe some of these methods to observed similarities and disadvantages of these approaches. An abbreviated literature review of these models will be discussed in the following sections.

7.8.1   MODELING ASSUMPTIONS

Just as in any other process modeling, a set of assumptions are required for the following reasons:

  a. To furnish industry – based applications whereby speed of calculation, but not accuracy, is critical,

  b. To simplify – hence enabling checkpoints to be made before more detailed models can proceed.

  c. For enabling the formulations to be practically traceable.

The first assumption to be considered as far as electrospinning is concerned is conceptualizing the jet itself. Even though the most appropriate view of a jet flow is that of a liquid continuum, the use of nodes connected in series

by certain elements that constitute rheological properties has proven successful [64, 66]. The second assumption is the fluid constitutive properties. In the
discrete node model [66], the nodes are connected in series by a Maxwell unit,
i.e., a spring and dashpot in series, for quantifying the viscoelastic properties.

In analyzing viscoelastic models, we apply two types of elements: the
dashpot element which describes the force as being in proportion to the velocity (recall friction), and the spring element which describes the force as being
in proportion to elongation. One can then develop viscoelastic models using
combinations of these elements. Among all possible viscoelastic models, the
Maxwell model was selected by [66] due to its suitability for liquid jet as
well as its simplicity. Other models are either unsuitable for liquid jet or too
detailed.

In the continuum models a power law can be used for describing the liquid
behavior under shear flow for describing the jet flow [150]. At this juncture,
we note that the power law is characterized from a shear flow, while the jet
flow in electrospinning undergoes elongational flow. This assumption will be
discussed in detail in following sections.

The other assumption which should be applied in electrospinning modeling is about the coordinate system. The method for coordinate system selection in electrospinning process is similar to other process modeling, the
system that best bring out the results by (i) allowing the computation to be
performed in the most convenient manner and, more importantly, (ii) enabling
the computation results to be accurate. In view of the linear jet portion during
the initial first stage of the jet, the spherical coordinate system is eliminated.
Assuming the second stage of the jet to be perfectly spiraling, due to bending
of the jet, the cylindrical coordinate system would be ideal. However, experimental results have shown that the bending instability portion of the jet is not
perfectly expanding spiral. Hence the Cartesian coordinate system, which is
the most common of all coordinate system, is adopted.

Depending on the processing parameters (such as applied voltage, volume
flow rate, etc.) and the fluid properties (such as surface tension, viscosity, etc.)
as many as 10 modes of electrohydrodynamically driven liquid jet have been
identified [151]. The scope of jet modes is highly abbreviated in this chapter
because most electrospinning processes that lead to nanofibers consist of only
two modes, the straight jet portion and the spiraling (or whipping) jet portion.
Insofar as electrospinning process modeling is involved, the following classification indicates the considered modes or portion of the electrospinning jet.

1. Modeling the criteria for jet initiation from the droplet [64, 152];
2. Modeling the straight jet portion [150, 153–155];

3.   Modeling the entire jet [60, 61, 66, 156].

7.8.2   CONSERVATION RELATIONS

Balance of the producing accumulation is, particularly, a basic source of quantitative models of phenomena or processes. Differential balance equations are formulated for momentum, mass and energy through the contribution of local rates of transport expressed by the principle of Newton's, Fick's and Fourier laws. For a description of more complex systems like electrospinning that involved strong turbulence of the fluid flow, characterization of the product property is necessary and various balances are required [157].

The basic principle used in modeling of chemical engineering process is a concept of balance of momentum, mass and energy, which can be expressed in a general form as:

$$A = I + G - O - C \qquad (7.8.1)$$

where

A: Accumulation built up within the system
I: Input entering through the system surface
G: Generation produced in system volume
O: Output leaving through system boundary
C: Consumption used in system volume

The form of expression depends on the level of the process phenomenon description [157, 158]

According to the electrospinning models, the jet dynamics are governed by a set of three equations representing mass, energy and momentum conservation for the electrically charge jet [159].

In electrospinning modeling for simplification of describing the process, researchers consider an element of the jet and the jet variation versus time is neglected.

7.8.2.1   MASS CONSERVATION

The concept of mass conservation is widely used in many fields such as chemistry, mechanics, and fluid dynamics. Historically, mass conservation was discovered in chemical reactions by Antoine Lavoisier in the late eighteenth century, and was of decisive importance in the progress from alchemy to the modern natural science of chemistry. The concept of matter conservation is useful and sufficiently accurate for most chemical calculations, even in modern practice [160].

The equations for the jet follow from Newton's Law and the conservation laws obey, namely, conservation of mass and conservation of charge [60].

According to the conservation of mass equation

$$\pi R^2 v = Q \qquad (7.8.2)$$

$$I = I_{conduction} + I_{convection} \qquad (7.8.3)$$

For incompressible jets, by increasing the velocity the radius of the jet decreases. At the maximum level of the velocity, the radius of the jet reduces. The macromolecules of the polymers are compacted together closer while the jet becomes thinner as it shown in Fig. 7.3. When the radius of the jet reaches the minimum value and its speed becomes maximum to keep the conservation of mass equation, the jet dilates by decreasing its density, which called electrospinning dilation (Fig. 7.18) [161, 162].

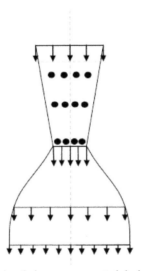

FIGURE 7.18 Macromolecular chains are compacted during the electrospinning.

7.8.2.2 ELECTRIC CHARGE CONSERVATION

An electric current is a flow of electric charge. Electric charge flows when there is voltage present across a conductor. In physics, charge conservation is the principle that electric charge can neither be created nor destroyed. The net quantity of electric charge, the amount of positive charge minus the amount of negative charge in the universe, is always conserved. The first written statement of the principle was by American scientist and statesman Benjamin Franklin in 1747 [163]. Charge conservation is a physical law, which states that the change in the amount of electric charge in any volume of space is exactly equal to the amount of charge in a region and the flow of charge into and out of that region [164].

During the electrospinning process, the electrostatic repulsion between excess charges in the solution stretches the jet. This stretching also decreases the jet diameter that this leads to the law of charge conservation as the second governing equation [165].

In electrospinning process, the electric current, which induced by electric field included two parts, conduction and convection.

The conventional symbol for current is I:

$$I = I_{conduction} + I_{convection} \qquad (7.8.4)$$

Electrical conduction is the movement of electrically charged particles through a transmission medium. The movement can form an electric current in response to an electric field. The underlying mechanism for this movement depends on the material.

$$I_{conduction} = J_{cond} \times S = KE \times \pi R^2 \qquad (7.8.5)$$

$$J = \frac{I}{A(s)} \qquad (7.8.6)$$

$$I = J \times S \qquad (7.8.7)$$

Convection current is the flow of current with the absence of an electric field.

$$I_{convection} = J_{conv} \times S = 2\pi R(L) \times \sigma v \qquad (7.8.8)$$

$$J_{conv} = \sigma v \qquad (7.8.9)$$

So, the total current can be calculated as:

$$\pi R^2 KE + 2\pi Rv\sigma = I \qquad (7.8.10)$$

$$\frac{\partial}{\partial t}(2\pi R\sigma) + \frac{\partial}{\partial z}\left(\pi R^2 KE + 2\pi Rv\sigma\right) = 0 \qquad (7.8.11)$$

7.8.3 MOMENTUM BALANCE

In classical mechanics, linear momentum or translational momentum is the product of the mass and velocity of an object. Like velocity, linear momentum is a vector quantity, possessing a direction as well as a magnitude:

$$P = mv \tag{7.8.12}$$

Linear momentum is also a conserved quantity, meaning that if a closed system (one that does not exchange any matter with the outside and is not acted on by outside forces) is not affected by external forces, its total linear momentum cannot change. In classical mechanics, conservation of linear momentum is implied by Newton's laws of motion; but it also holds in special relativity (with a modified formula) and, with appropriate definitions, a (generalized) linear momentum conservation law holds in electrodynamics, quantum mechanics, quantum field theory, and general relativity [163]. For example, according to the third law, the forces between two particles are equal and opposite. If the particles are numbered 1 and 2, the second law states:

$$F_1 = \frac{dP_1}{dt} \tag{7.8.13}$$

$$F_2 = \frac{dP_2}{dt} \tag{7.8.14}$$

Therefore:

$$\frac{dP_1}{dt} = -\frac{dP_2}{dt} \tag{7.8.15}$$

$$\frac{d}{dt}(P_1 + P_2) = 0 \tag{7.8.16}$$

If the velocities of the particles are v_{11} and $_{12}$ before the interaction, and afterwards they are v_{21} and v_{22}, then

$$m_1 v_{11} + m_2 v_{12} = m_1 v_{21} + m_2 v_{22} \tag{7.8.17}$$

This law holds no matter how complicated the force is between the particles. Similarly, if there are several particles, the momentum exchanged between each pair of particles adds up to zero, so the total change in momentum is zero. This conservation law applies to all interactions, including collisions and separations caused by explosive forces. It can also be generalized to situations where Newton's laws do not hold, for example in the theory of relativity and in electrodynamics [153, 166]. The momentum equation for the fluid can be derived as follow:

$$\rho(\frac{dv}{dt} + v\frac{dv}{dz}) = \rho g + \frac{d}{dz}[\tau_{zz} - \tau_{rr}] + \frac{\gamma}{R^2} \cdot \frac{dr}{dz} + \frac{\sigma}{\varepsilon_0} \frac{d\sigma}{dz} + (\varepsilon - \varepsilon_0)(E\frac{dE}{dz}) + \frac{2\sigma E}{r} \quad (7.8.18)$$

But commonly the momentum equation for electrospinning modeling is formulated by considering the forces on a short segment of the jet [153, 166].

$$\frac{d}{dz}(\pi R^2 \rho v^2) = \pi R^2 \rho g + \frac{d}{dz}[\pi R^2(-p + \tau_{zz})] + \frac{\gamma}{R}.2\pi RR' + 2\pi R(t_t^e - t_n^e R') \quad (7.8.19)$$

As it is shown in the Fig. 7.19, the element's angels could be defined as α and β. According to the mathematical relationships, it is obvious that:

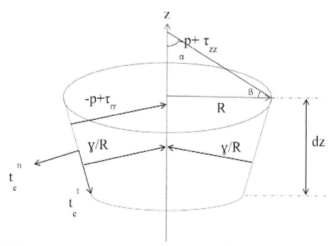

FIGURE 7.19   Momentum balance on a short section of the jet.

$$\alpha + \beta = \frac{\pi}{2} \qquad (7.8.20)$$

$$\sin \alpha = \tan \alpha$$
$$\qquad (7.8.21)$$
$$\cos \alpha = 1$$

Due to the Figure, relationships between these electrical forces are as below:

$$t_n^e \sin \alpha \cong t_n^e \tan \alpha \cong -t_n^e \tan \beta \cong -\frac{dR}{dz}t_n^e = -R't_n^e \qquad (7.8.22)$$

$$t_t^e \cos \alpha \cong t_t^e \qquad (7.8.23)$$

So the effect of the electric forces in the momentum balance equation can be presented as:

$$2\pi RL(t_t^e - R't_n^e)dz \tag{7.8.24}$$

(Notation: In the main momentum equation, final formula is obtained by dividing into dz)

Generally, the normal electric force is defined as:

$$t_n^e \cong \frac{1}{2}\bar{\varepsilon}E_n^2 = \frac{1}{2}\bar{\varepsilon}(\frac{\sigma}{\bar{\varepsilon}})^2 = \frac{\sigma^2}{2\bar{\varepsilon}} \tag{7.8.25}$$

A little amount of electric forces is perished in the vicinity of the air.

$$E_n = \frac{\sigma}{\bar{\varepsilon}} \tag{7.8.26}$$

The electric force can be presented by:

$$F = \frac{\Delta We}{\Delta l} = \frac{1}{2}(\varepsilon - \bar{\varepsilon})E^2 \times \Delta S \tag{7.8.27}$$

The force per surface unit is:

$$\frac{F}{\Delta S} = \frac{1}{2}(\varepsilon - \bar{\varepsilon})E^2 \tag{7.8.28}$$

Generally the electric potential energy is obtained by:

$$Ue = -We = -\int F.ds \tag{7.8.29}$$

$$\Delta We = \frac{1}{2}(\varepsilon - \bar{\varepsilon})E^2 \times \Delta V = \frac{1}{2}(\varepsilon - \bar{\varepsilon})E^2 \times \Delta S.\Delta l \tag{7.8.30}$$

So, finally it could be resulted:

$$t_n^e = \frac{\sigma^2}{2\bar{\varepsilon}} - \frac{1}{2}(\varepsilon - \bar{\varepsilon})E^2 \tag{7.8.31}$$

$$t_t^e = \sigma E \tag{7.8.32}$$

7.8.4 COULOMB'S LAW

Coulomb's law is a mathematical description of the electric force between charged objects which is formulated by the 18th-century French physicist

Charles-Augustin de Coulomb. It is analogous to Isaac Newton's law of gravity. Both gravitational and electric forces decrease with the square of the distance between the objects, and both forces act along a line between them [167]. In Coulomb's law, the magnitude and sign of the electric force are determined by the electric charge, more than the mass of an object. Thus, a charge which is a basic property matter determines how electromagnetism affects the motion of charged targets [163].

Coulomb force is thought to be the main cause for the instability of the jet in the electrospinning process [168]. This statement is based on the Earnshaw's theorem, named after Samuel Earnshaw [169], which claims that "A charged body placed in an electric field of force cannot rest in stable equilibrium under the influence of the electric forces alone." This theorem can be notably adapted to the electrospinning process [168]. The instability of charged jet influences on jet deposition and as a consequence on nanofiber formation. Therefore, some researchers applied developed models to the analysis of mechanisms of jet deposition and alignment on various collecting devices in arbitrary electric fields [66].

The equation for the potential along the centerline of the jet can be derived from Coulomb's law. Polarized charge density is obtained:

$$\rho_{p'} = -\vec{\nabla}.\vec{P}'$$

(7.8.33)

where P' is polarization:

$$\vec{P}' = (\varepsilon - \overline{\varepsilon})\vec{E}$$

(7.8.34)

By substituting P' in the last equation:

$$\rho_{p'} = -(\overline{\varepsilon} - \varepsilon)\frac{dE}{dz'}$$

(7.8.35)

Beneficial charge per surface unit can be calculated as below:

$$\rho_{p'} = \frac{Q_b}{\pi R^2}$$

(7.8.36)

$$Q_b = \rho_b.\pi R^2 = -(\overline{\varepsilon} - \varepsilon)\pi R^2 \frac{dE}{dz'}$$

(7.8.37)

$$Q_b = -(\overline{\varepsilon} - \varepsilon)\pi \frac{d(ER^2)}{dz'}$$

(7.8.38)

$$\rho_{sb} = Q_b.dz' = -(\overline{\varepsilon} - \varepsilon)\pi\frac{d}{dz'}(ER^2)dz' \tag{7.8.39}$$

The main equation of Coulomb's law:

$$F = \frac{1}{4\pi\varepsilon_0}\frac{qq_0}{r^2} \tag{7.8.40}$$

The electric field is:

$$E = \frac{1}{4\pi\varepsilon_0}\frac{q}{r^2} \tag{7.8.41}$$

The electric potential can be measured:

$$\Delta V = -\int E.dL \tag{7.8.42}$$

$$V = \frac{1}{4\pi\varepsilon_0}\frac{Q_b}{r} \tag{7.8.43}$$

According to the beneficial charge equation, the electric potential could be rewritten as:

$$\Delta V = Q(z) - Q_\infty(z) = \frac{1}{4\pi\overline{\varepsilon}}\int\frac{(q - Q_b)}{r}dz' \tag{7.8.44}$$

$$Q(z) = Q_\infty(z) + \frac{1}{4\pi\overline{\varepsilon}}\int\frac{q}{r}dz' - \frac{1}{4\pi\overline{\varepsilon}}\int\frac{Q_b}{r}dz' \tag{7.8.45}$$

$$Q_b = -(\overline{\varepsilon} - \varepsilon)\pi\frac{d(ER^2)}{dz'} \tag{7.8.46}$$

The surface charge density's equation is:

$$q = \sigma.2\pi RL \tag{7.8.47}$$

$$r^2 = R^2 + (z - z')^2 \tag{7.8.48}$$

$$r = \sqrt{R^2 + (z - z')^2} \tag{7.8.49}$$

The final equation, which obtained by substituting the mentioned equations is:

$$Q(z) = Q_\infty(z) + \frac{1}{4\pi\bar{\varepsilon}} \int \frac{\sigma.2\pi R}{\sqrt{(z-z')^2 + R^2}} dz' - \frac{1}{4\pi\bar{\varepsilon}} \int \frac{(\bar{\varepsilon}-\varepsilon)\pi}{\sqrt{(z-z')^2 + R^2}} \frac{d(ER^2)}{dz'} \qquad (7.8.50)$$

It is assumed that β is defined:

$$\beta = \frac{\varepsilon}{\bar{\varepsilon}} - 1 = -\frac{(\bar{\varepsilon}-\varepsilon)}{\bar{\varepsilon}} \qquad (7.8.51)$$

So, the potential equation becomes:

$$Q(z) = Q_\infty(z) + \frac{1}{2\bar{\varepsilon}} \int \frac{\sigma.R}{\sqrt{(z-z')^2 + R^2}} dz' - \frac{\beta}{4} \int \frac{1}{\sqrt{(z-z')^2 + R^2}} \frac{d(ER^2)}{dz'} \qquad (7.8.52)$$

The asymptotic approximation of χ is used to evaluate the integrals mentioned above:

$$\chi = \left(-z + \xi + \sqrt{z^2 - 2z\xi + \xi^2 + R^2}\right) \qquad (7.8.53)$$

Where χ is "aspect ratio" of the jet (L = length, R_0 = Initial radius)
This leads to the final relation to the axial electric field:

$$E(z) = E_\infty(z) - \ln \chi \left(\frac{1}{\bar{\varepsilon}} \frac{d(\sigma R)}{dz} - \frac{\beta}{2} \frac{d^2 (ER^2)}{dz^2} \right) \qquad (7.8.54)$$

7.8.5 FORCES CONSERVATION

There exists a force, as a result of charge build-up, acting upon the droplet coming out of the syringe needle pointing toward the collecting plate, which can be either grounded or oppositely charged. Furthermore, similar charges within the droplet promote jet initiation due to their repulsive forces. Nevertheless, surface tension and other hydrostatic forces inhibit the jet initiation because the total energy of a droplet is lower than that of a thin jet of equal volume upon consideration of surface energy. When the forces that aid jet initiation (such as electric field and Coulombic) overcome the opposing forces (such as surface tension and gravitational), the droplet accelerates toward the collecting plate. This forms a jet of very small diameter. Other than initiating jet flow, the electric field and Coulombic forces tend to stretch the jet, thereby contributing towards the thinning effect of the resulting nanofibers.

   In the flow path modeling, we recall the Newton's Second Law of motion

$$m\frac{d^2P}{dt^2} = \Sigma f \qquad (7.8.55)$$

Where, m (equivalent mass) and the various forces are summed as

$$\Sigma f = f_C + f_E + f_V + f_S + f_A + f_G + \ldots \qquad (7.8.56)$$

In which subscripts C, E, V, S, A and G correspond to the Coulombic, electric field, viscoelastic, surface tension, air drag and gravitational forces respectively. A description of each of these forces based on the literature [66] is summarized in Table 7.1, Here, V_0 = applied voltage; h = distance from pendent drop to ground collector; σ_V= viscoelastic stress; v = kinematic viscosity.

TABLE 7.1   Description of Itemized Forces or Terms Related to Them

Forces	Equations
Coulombic	$f_C = \dfrac{q^2}{l^2}$
Electric field	$f_E = -\dfrac{qV_0}{h}$
Viscoelastic	$f_V = \dfrac{d\sigma_V}{dt} = \dfrac{G}{l}\dfrac{dl}{dt} - \dfrac{G}{\eta}\sigma_V$
Surface Tension	$f_S = \dfrac{\alpha\pi R^2 k}{\sqrt{x_i^2 + y_i^2}}\left[i\lvert x\rvert Sin(x) + i\lvert y\rvert Sin(y)\right]$
Air drag	$f_A = 0.65\pi R\rho_{air} v^2\left(\dfrac{2vR}{V_{air}}\right)^{-0.81}$
Gravitational	$f_G = \rho g\pi R^2$

7.8.6   CONSTITUTIVE EQUATIONS

In modern condensed matter physics, the constitutive equation plays a major role. In physics and engineering, a constitutive equation or relation is a relation between two physical quantities that is specific to a material or substance, and approximates the response of that material to external stimulus, usually as applied fields or forces [170]. There are a sort of mechanical equation of state, and describe how the material is constituted mechanically. With these constitutive relations, the vital role of the material is reasserted [171]. There are two groups of constitutive equations: Linear and nonlinear constitutive equations [172]. These equations are combined with other governing physical laws to solve problems; for example in fluid mechanics the flow of a fluid in a pipe, in solid state physics the response of a crystal to an electric field, or in structural analysis, the connection between applied stresses or forces to strains or deformations [170].

The first constitutive equation (constitutive law) was developed by Robert Hooke and is known as Hooke's law. It deals with the case of linear elastic materials. Following this discovery, this type of equation, often called a "stress-strain relation" in this example, but also called a "constitutive assumption" or an "equation of state" was commonly used [173]. Walter Noll advanced the use of constitutive equations, clarifying their classification and the role of invariance requirements, constraints, and definitions of terms like "material," "isotropic," "aeolotropic," etc. The class of "constitutive relations" of the form stress rate = f (velocity gradient, stress, density) was the subject of Walter Noll's dissertation in 1954 under Clifford Truesdell [170]. There are several kinds of constitutive equations, which are applied commonly in electrospinning. Some of these applicable equations are discussed as following:

7.8.6.1   OSTWALD-DE WAELE POWER LAW

Rheological behavior of many polymer fluids can be described by power law constitutive equations [172]. The equations that describe the dynamics in electrospinning constitute, at a minimum, those describing the conservation of mass, momentum and charge, and the electric field equation. Additionally a constitutive equation for the fluid behavior is also required [76]. A Power-law fluid, or the Ostwald-De Waele relationship, is a type of generalized Newtonian fluid for which the shear stress, τ, is given by:

$$\tau = K'\left(\frac{\partial v}{\partial y}\right)^{m} \tag{7.8.57}$$

Which $\partial v/\partial y$ is the shear rate or the velocity gradient perpendicular to the plane of shear. The power law is only a good description of fluid behavior across the range of shear rates to which the coefficients are fitted. There are a number of other models that better describe the entire flow behavior of shear-dependent fluids, but they do so at the expense of simplicity, so the power law is still used to describe fluid behavior, permit mathematical predictions, and correlate experimental data [166, 174].

Nonlinear rheological constitutive equations applicable for polymer fluids (Ostwald-De Waele power law) were applied to the electrospinning process by Spivak and Dzenis [77, 150, 175].

$$\hat{\tau}^c = \mu \left[tr \left(\dot{\gamma}^2 \right) \right]^{(m-1)/2} \dot{\gamma} \qquad (7.8.58)$$

$$\mu = K \left(\frac{\partial v}{\partial y} \right)^{m-1} \qquad (7.8.59)$$

Viscous Newtonian fluids are described by a special case of equation above with the flow index $m = 1$. Pseudoplastic (shear thinning) fluids are described by flow indices $0 \le m \le 1$. Dilatant (shear thickening) fluids are described by the flow indices $m > 1$ [150].

7.8.6.2 GIESEKUS EQUATION

In 1966, Giesekus established the concept of anisotropic forces and motions into polymer kinetic theory. With particular choices for the tensors describing the anisotropy, one can obtained Giesekus constitutive equation from elastic dumbbell kinetic theory [176, 177]. The Giesekus equation is known to predict, both qualitatively and quantitavely, material functions for steady and nonsteady shear and elongational flows. However, the equation sustains two drawbacks: it predicts that the viscosity is inversely proportional to the shear rate in the limit of infinite shear rate and it is unable to predict any decrease in the elongational viscosity with increasing elongation rates in uniaxial elongational flow. The first one is not serious because of the retardation time, which is included in the constitutive equation, but the second one is more critical because the elongational viscosity of some polymers decreases with increasing of elongation rate [178, 179].

In the main Giesekus equation, the tensor of excess stresses depending on the motion of polymer units relative to their surroundings was connected to a sequence of tensors characterizing the configurational state of the different kinds of network structures present in the concentrated solution or melt. The respective set of constitutive equations indicates [180–181]:

$$S_k + \eta \frac{\partial C_k}{\partial t} = 0 \qquad (7.8.60)$$

The equation below indicates the upper convected time derivative (Oldroyd derivative):

$$\frac{\partial C_k}{\partial t} = \frac{DC_k}{Dt} - \left[C_k \nabla v + (\nabla v)^T C_k \right] \qquad (7.8.61)$$

(Note: The upper convective derivative is the rate of change of any tensor property of a small parcel of fluid that is written in the coordinate system rotating and stretching with the fluid.)

C_k also can be measured as follows:

$$C_k = 1 + 2E_k \qquad (7.8.62)$$

According to the concept of "recoverable strain" S_k may be understood as a function of E_k and vice versa. If linear relations corresponding to Hooke's law are adopted.

$$S_k = 2\mu_k E_k \qquad (7.8.63)$$

So:

$$S_k = \mu_k (C_k - 1) \qquad (7.8.64)$$

The Eq. (7.8.60) becomes:

$$S_k + \lambda_k \frac{\partial S_k}{\partial t} = 2\eta D \qquad (7.8.65)$$

$$\lambda_k = \frac{\eta}{\mu_k} \qquad (7.8.66)$$

As a second step in order to rid the model of the shortcomings is the scalar mobility constants B_k, which are contained in the constants η. This mobility constant can be represented as:

$$\tfrac{1}{2}(\beta_k S_k + S_k \beta_k) + \tilde{\eta}\frac{\partial C_k}{\partial t} = 0 \qquad (7.8.67)$$

The two parts of Eq. (7.8.67) reduces to the single constitutive equation:

$$\beta_k + \tilde{\eta}\frac{\partial C_k}{\partial t} = 0 \tag{7.8.68}$$

The excess tension tensor in the deformed network structure where the well-known constitutive equation of a so-called Neo-Hookean material is proposed [180, 182]:

Neo-Hookean equation: $\quad S_k = 2\mu_k E_k = \mu_k(C_k - 1) \tag{7.8.69}$

$$\mu_k = NKT$$
$$\beta_k = 1 + \alpha(C_k - 1) = (1 - \alpha) + \alpha C_k \tag{7.8.70}$$

where K is Boltzmann's constant.

By substitution Eqs. (7.8.69) and (7.8.70) in the Eq. (7.8.64), it can obtained where the condition $0 \le \alpha \le 1$ must be fulfilled, the limiting case $\alpha = 0$ corresponds to an isotropic mobility [183].

$$0 \le \alpha \le 1 \qquad [1 + \alpha(C_k - 1)](C_k - 1) + \lambda_k\frac{\partial C_k}{\partial t} = 0 \tag{7.8.71}$$

$$\alpha = 1 \qquad C_k(C_k - 1) + \lambda_k\frac{\partial C_k}{\partial t} = 0 \tag{7.8.72}$$

$$0 \le \alpha \le 1 \qquad C_k = \frac{S_k}{\mu_k} + 1 \tag{7.8.73}$$

By substituting equations above in Eq. (7.8.64), it becomes:

$$\left[1 + \frac{\alpha S_k}{\mu_k}\right]\frac{S_k}{\mu_k} + \lambda_k\frac{\partial C_k}{\partial t} = 0 \tag{7.8.74}$$

$$\frac{S_k}{\mu_k} + \frac{\alpha S_k^2}{\mu_k^2} + \lambda_k\frac{\partial(S_k/\mu_k + 1)}{\partial t} = 0 \tag{7.8.75}$$

$$\frac{S_k}{\mu_k} + \frac{\alpha S_k^2}{\mu_k^2} + \frac{\lambda_k}{\mu_k}\frac{\partial S_k}{\partial t} = 0 \tag{7.8.76}$$

$$S_k + \frac{\alpha S_k^2}{\mu_k} + \lambda_k \frac{\partial S_k}{\partial t} = 0 \qquad (7.8.77)$$

D means the rate of strain tensor of the material continuum [180].

$$D = \frac{1}{2}\left[\nabla v + (\nabla v)^T\right] \qquad (7.8.78)$$

The equation of the upper convected time derivative for all fluid properties can be calculated as:

$$\frac{\partial \otimes}{\partial t} = \frac{D \otimes}{Dt} - \left[\otimes . \nabla v + (\nabla v)^T . \otimes\right] \qquad (7.8.79)$$

$$\frac{D \otimes}{Dt} = \frac{\partial \otimes}{\partial t} + \left[(v.\nabla).\otimes\right] \qquad (7.8.80)$$

By replacing S_k instead of the symbol:

$$\lambda_k \frac{\partial S_k}{\partial t} = \lambda_k \frac{DS_k}{Dt} - \lambda_k\left[S_k \nabla v + (\nabla v)^T S_k\right] = \lambda_k \frac{DS_k}{Dt} - \lambda_k (v.\nabla)S_k \qquad (7.8.81)$$

By simplification the equation above:

$$S_k + \frac{\alpha S_k^2}{\mu_k} + \lambda_k \frac{DS_k}{Dt} = \lambda_k (v.\nabla)S_k \qquad (7.8.82)$$

$$S_k = 2\mu_k E_k \qquad (7.8.83)$$

The assumption of $E_k = 1$ would lead to the next equation:

$$S_k + \frac{\alpha \lambda_k S_k^2}{\eta} + \lambda_k \frac{DS_k}{Dt} = \frac{\eta}{\mu_k}(2\mu_k)D = 2\eta D = \eta\left[\nabla v + (\nabla v)^T\right] \qquad (7.8.84)$$

In electrospinning modeling articles τ is used commonly instead of S_k [154, 159, 161].

$$S_k \leftrightarrow \tau$$

$$\tau + \frac{\alpha \lambda_k \tau^2}{\eta} + \lambda_k \tau_{(1)} = \eta\left[\nabla v + (\nabla v)^T\right] \qquad (7.8.85)$$

7.8.6.3   MAXWELL EQUATION

Maxwell's equations are a set of partial differential equations that, together with the Lorentz force law, form the foundation of classical electrodynamics, classical optics, and electric circuits. These fields are the bases of modern electrical and communications technologies. Maxwell's equations describe how electric and magnetic fields are generated and altered by each other and by charges and currents. They are named after the Scottish physicist and mathematician James Clerk Maxwell who published an early form of those equations between 1861 and 1862 [184, 185]. It will be discussed in the next section in detail.

7.8.7 MICROSCOPIC MODELS

One of the aims of computer simulation is to reproduce experiment to elucidate the invisible microscopic details and further explain the experiments. Physical phenomena occurring in complex materials cannot be encapsulated within a single numerical paradigm. In fact, they should be described within hierarchical, multilevel numerical models in which each submodel is responsible for different spatial-temporal behavior and passes out the averaged parameters of the model, which is next in the hierarchy. The understanding of the nonequilibrium properties of complex fluids such as the viscoelastic behavior of polymeric liquids, the rheological properties of ferrofluids and liquid crystals subjected to magnetic fields, based on the architecture of their molecular constituents is useful to get a comprehensive view of the process. The analysis of simple physical particle models for complex fluids has developed from the molecular computation of basic systems (atoms, rigid molecules) to the simulation of macromolecular 'complex' system with a large number of internal degrees of freedom exposed to external forces [186, 187].

The most widely used simulation methods for molecular systems are Monte Carlo, Brownian dynamics and molecular dynamics. The microscopic approach represents the microstructural features of material by means of a large number of micromechanical elements (beads, platelet, rods) obeying stochastic differential equations. The evolution equations of the microelements arise from a balance of momentum at the elementary level. The Monte Carlo method is a stochastic strategy that relies on probabilities. The Monte Carlo sampling technique generates large numbers of configurations or microstates of equilibrated systems by stepping from one microstate to the next in a particular statistical ensemble. Random changes are made to the positions of the species present, together with their orientations and conformations where appropriate. Brownian dynamics are an efficient approach for simulations of large polymer molecules or colloidal particles in a small molecule solvent. Molecular dynamics is the most detailed molecular simulation method, which computes the motions of individual molecules. Molecular dynamics efficiently evaluates different configurational properties and dynamic quantities which cannot generally be obtained by Monte Carlo [188, 189].

The first computer simulation of liquids was carried out in 1953. The model was an idealized two-dimensional representation of molecules as rigid disks. For macromolecular systems, the coarse-grained approach is widely used as the modeling process is simplified, hence becomes more efficient, and the characteristic topological features of the molecule can still be maintained. The level of detail for a coarse-grained model varies in different cases. The whole molecule can be represented by a single particle in a simulation and interactions between particles incorporate average properties of the whole molecule. With this approach, the number of degrees of freedom is greatly reduced [190].

On the other hand, a segment of a polymer molecule can also be represented by a particle (bead). The first coarse-grained model, called the 'dumbbell' model, was introduced in the 1930s (Fig. 7.20). Molecules are treated as a pair of beads interacting via a harmonic potential. However, by using this model, it is possible to perform kinetic theory derivations and calculations for nonlinear rheological properties and solve some flow problems. The analytical results for the dumbbell models can also be used to check computer simulation procedures in molecular dynamics and Brownian dynamics [191, 192].

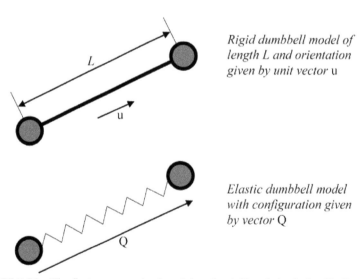

Rigid dumbbell model of length L and orientation given by unit vector u

Elastic dumbbell model with configuration given by vector Q

FIGURE 7.20   The first coarse-grained models – the rigid and elastic dumbbell models.

The bead-rod and bead-spring model (Fig. 7.21) were introduced to model chainlike macromolecules. Beads in the bead-rod model do not represent the atoms of the polymer chain backbone, but some portion of the chain, normally 10 to 20 monomer units. These beads are connected by rigid and massless rods. While in the bead-spring model, a portion of the chain containing several hundreds of backbone atoms are replaced by a "spring" and the masses of the atoms are concentrated on the mass of beads [193].

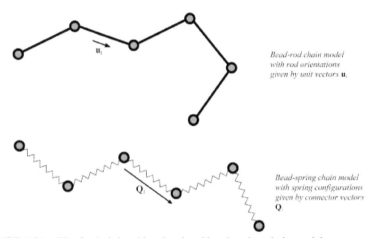

Bead-rod chain model
with rod orientations
given by unit vectors $\mathbf{u}_i$

Bead-spring chain model
with spring configurations
given by connector vectors
$\mathbf{Q}_i$

FIGURE 7.21   The freely jointed bead-rod and bead-spring chain models.

If the springs are taken to be Hookean springs, the bead-spring chain is referred to as a Rouse chain or a Rouse-Zimm chain. This approach has been applied widely as it has a large number of internal degrees of freedom and exhibits orientability and stretchability. However, the disadvantage of this model is that it does not have a constant contour length and can be stretched out to any length. Therefore, in many cases finitely extensible springs with two more parameters, the spring constant and the maximum extensibility of an individual spring, can be included so the contour length of the chain model cannot exceed a certain limit [194, 195].

The understanding of the nonequilibrium properties of complex fluids (Fig. 7.22) such as the viscoelastic behavior of polymeric liquids, the rheological properties of ferrofluids and liquid crystals subjected to magnetic fields, based on the architecture of their molecular constituents [186].

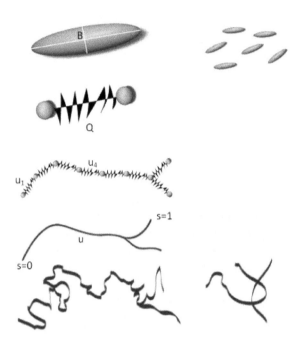

FIGURE 7.22   Simple microscopic models for complex fluids by using dumbbell model.

Dumbbell models are very crude representations of polymer molecules. Too crude to be of much interest to a polymer chemist, since it in no way accounts for the details of the molecular architecture. It certainly does not have enough internal degrees of freedom to describe the very rapid motions that contribute, for example, to the complex viscosity at high frequencies. On the other hand, the elastic dumbbell model is orientable and stretchable, and these two properties are essential for the qualitative description of steady-state rheological properties and those involving slow changes with time. For dumbbell models one can go through the entire program of endeavor—from molecular model for fluid dynamics—for illustrative purposes, in order to point the way towards the task that has ultimately to be performed for more realistic models. According to the researches, dumbbell models must, to some extend then, be regarded as mechanical playthings, somewhat disconnected from the real world of polymers. When used intelligently, however, they can be useful pedagogically and very helpful in developing a qualitative understanding of rheological phenomena [186, 196].

The simplest model of flexible macromolecules in a dilute solution is the elastic dumbbell (or bead-spring) model. This has been widely used for purely mechanical theories of the stress in electrospinning modeling [197].

A Maxwell constitutive equation was first applied by Reneker et al. in 2000. Consider an electrified liquid jet in an electric field parallel to its axis. They modeled a segment of the jet by a viscoelastic dumbbell. They used a Gaussian electrostatic system of units. According to this model (Fig. 7.23), each particle in the electric field exerts repulsive force on another particle [66].

He had three main assumptions [66, 198]:

α) The background electric field created by the generator is Considered static;

β) The fiber is a perfect insulator;

χ) The polymer solution is a viscoelastic medium with constant elastic modulus, viscosity and surface tension.

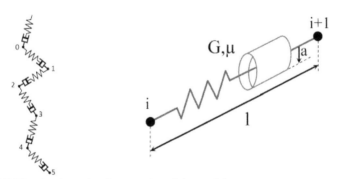

FIGURE 7.23 A schematic of one section of the model.

The researcher considered the governing equations for each bead as [198]:

$$\frac{d}{dt}\left(\pi a^2 l\right) = 0 \tag{7.8.86}$$

Therefore, the stress between these particles can be measured by [66]:

$$\frac{d\sigma}{dt} = G\frac{dl}{ldt} - \frac{G}{\eta}\sigma \tag{7.8.87}$$

The stress can be calculated by a Maxwell viscoelastic constitutive equation [199]:

$$\dot{\tau} = G\left(\varepsilon' - \frac{\tau}{\eta}\right)$$

(7.8.88)

Where ε' is the Lagrangian axial strain [199]:

$$\varepsilon' \equiv \frac{\partial \dot{x}}{\partial \xi}.\hat{i}.$$

(7.8.89)

Equation of motion for beads can be written as [200]:

mass × acceleration = viscous drag + Brownian motion force   (7.8.90)

... + force of one bead on another through the connector

The momentum balance for a bead is [198]:

$$m\frac{dv}{dt} = \underbrace{-\frac{q^2}{l^2}}_{Coulomb\ forces} - \underbrace{qE}_{Electric\ force} + \underbrace{\pi a^2 \sigma}_{Mechanical\ forces}$$

(7.8.91)

So the momentum conservation for model charges can be calculated as [201]:

$$m_i\frac{dv_i}{dt} = \underbrace{q_i\sum_{i \neq j} q_j K \frac{r_i - r_j}{|r_i - r_j|^3}}_{Coulomb\ forces} + \underbrace{q_i E}_{Electric\ force} + \underbrace{\pi a_{i,i+1}^2 \sigma_{i,i+1}\frac{r_{i+1} - r_i}{|r_{i+1} - r_i|} - \pi a_{i-1,i}^2 \sigma_{i-1,i}\frac{r_i - r_{i-1}}{|r_i - r_{i-1}|}}_{Mechanical\ forces}$$

(7.8.92)

Boundary condition assumptions: A small initial perturbation is added to the position of the first bead, the background electric field is axial and uniform and the first bead is described by a stationary equation. For solving these equations some dimensionless parameters are defined then by simplifying equations, the equations are solved by using boundary conditions [198, 201].

Now, an example for using this model for the polymer structure is mentioned. For a dumbbell consists of two (Fig. 7.24), which are connected with a nonlinear spring, the spring force law is given by [96]:

$$F = -\frac{HQ}{1 - Q^2/Q_0^2}$$

(7.8.93)

Now if we considered the model for the polymer matrix such as carbon nanotube, the rheological behavior can be obtained as [96, 202]:

$$\tau_{ij} = \tau_p + \tau_s$$

(7.8.94)

$$\tau_p = \underbrace{n_a \langle Q_a F_a \rangle}_{aggregated \ \ dumbbells} + \underbrace{n_f \langle Q_f F_f \rangle}_{free \ \ dumbbells} - nkT\delta_{ij} \tag{7.8.95}$$

$$\tau_s = \eta \dot{\gamma} \tag{7.8.96}$$

$$\lambda \langle Q.Q \rangle^{\nabla} = \delta_{ij} - \frac{c \langle Q.Q \rangle}{1 - tr \langle Q.Q \rangle / b_{max}} \tag{7.8.97}$$

The polymeric stress can be obtained from the following relation [96]:

$$\frac{\hat{\tau}_{ij}}{n_d kT} = \delta_{ij} - \frac{c \langle Q.Q \rangle}{1 - tr \langle Q.Q \rangle / b_{max}} \tag{7.8.98}$$

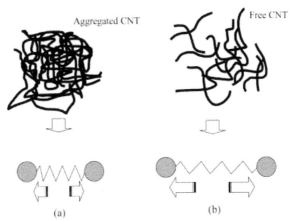

(a)   (b)

FIGURE 7.24 Modeling of two kinds of dumbbell sets, (a) aggregate FENE dumbbell, which has lower mobility and (b) free FENE dumbbell which has higher mobility.

7.8.8 SCALING

The physical aspect of a phenomenon can use the language of differential equation which represents the structure of the system by selecting the variables that characterize the state of it and certain mathematical constraint on the values of those variables can take on. These equations can predict the behavior of the system over a quantity such as time. For an instance, a set of continuous functions of time that describe the way the variables of the system developed over time starting from a given initial state [208]. In general, the renormalization group theory, scaling and fractal geometry, are applied to the understanding of the complex phenomena in physics, economics and medicine [209].

In more recent times, in statistical mechanics, the expression "scaling laws" has referred to the homogeneity of form of the thermodynamic and correlation functions near critical points, and to the resulting relations among the exponents that occur in those functions. From the viewpoint of scaling, electrospinning modeling can be studied in two ways, allometric and dimensionless analysis. Scaling and dimensional analysis actually started with Newton, and allometry exists everywhere in our daily life and scientific activity [209, 210].

7.8.8.1 ALLOMETRIC SCALING

Electrospinning applies electrically generated motion to spin fibers. So, it is difficult to predict the size of the produced fibers, which depends on the applied voltage in principal. Therefore, the relationship between the radius of the jet and the axial distance from the nozzle is always the subject of investigation [211, 212]. It can be described as an allometric equation by using the values of the scaling exponent for the initial steady, instability and terminal stages [213].

The relationship between r and z can be expressed as an allometric equation of the form:

$$r \approx z^b \qquad\qquad (7.8.99)$$

When the power exponent, b = 1 the relationship is isometric and when b ≠ 1 the relationship is allometric [211, 214]. In another view, b = −1/2 is considered for the straight jet, b = −1/4 for instability jet and b = 0 for final stage [172, 212].

Due to high electrical force acting on the jet, it can be illustrated [211]:

$$\frac{d}{dz}\left(\frac{v^2}{2}\right) = \frac{2\sigma E}{\rho r} \qquad\qquad (7.8.100)$$

Equations of mass and charge conservations applied here as mentioned before [211, 214–215]

From the above equations it can be seen that [161, 211].

$$r \approx z^b, \sigma \approx r, E \approx r^{-2}, \frac{d v^2}{dz} \approx r^{-2}$$
(7.8.101)

So it is obtained for the initial part of jet $r \approx z^{-1/2}$, $r \approx z^{-1/4}$ for the instable stage and $r \approx z^0$ for the final stage.

The charged jet can be considered as a one-dimensional flow as mentioned. If the conservation equations modified, they would change as [211]:

$$2\pi r \sigma^\alpha v + K \pi r^2 E = I$$
(7.8.102)

$$r \approx z^{-\alpha/(\alpha+1)}$$
(7.8.103)

where α is a surface charge parameter, the value of α depends on the surface charge in the jet. When $\alpha = 0$ no charge in jet surface, and in $\alpha = 1$ use for full surface charge.

Allometric scaling equations are more widely investigated by different researchers. Some of the most important allometric relationships for electrospinning are presented in Table 7.2.

TABLE 7.2   Investigated Scaling Laws Applied in Electrospinning Model

Parameters	Equation	Ref.
The conductance and polymer concentration	$g \approx c^\beta$	[161]
The fiber diameters and the solution viscosity	$d \approx \eta^\alpha$	[212]
The mechanical strength and threshold voltage	$\bar{\sigma} \approx E_{threshold}^{-\alpha}$	[216]
The threshold voltage and the solution viscosity	$E_{threshold} \approx \eta^{1/4}$	[216]
The viscosity and the oscillating frequency	$\eta \approx \omega^{-0.4}$	[216]
The volume flow rate and the current	$I \approx Q^b$	[215]
The current and the fiber radius	$I \approx r^2$	[217]
The surface charge density and the fiber radius	$\sigma \approx r^3$	[217]
The induction surface current and the fiber radius	$\varphi \approx r^2$	[217]
The fiber radius and AC frequency	$r \approx \Omega^{1/4}$	[172]

where β, α and b = scaling exponent.

7.8.8.2   *DIMENSIONLESS ANALYSIS*

One of the simplest, yet most powerful, tools in the physics is dimensional analysis in which there are two kinds of quantities: dimensionless and dimensional.

In physics and all science, dimensional analysis is the analysis of the relationships between different physical quantities by identifying their dimensions. The dimension of any physical quantity is the combination of the basic physical dimensions that compose it, although the definitions of basic physical dimensions may vary. Some fundamental physical dimensions, based on the SI system of units, are length, mass, time, and electric charge. (The SI unit of electric charge is, however, defined in terms of units of length, mass and time, and, for example, the time unit and the length unit are not independent but can be linked by the speed of light c.) Other physical quantities can be expressed in terms of these fundamental physical dimensions. Dimensional analysis is based on the fact that a physical law must be independent of the units used to measure the physical variables. A straightforward practical consequence is that any meaningful equation (and any inequality and inequation) must have the same dimensions on the left and right sides. Dimensional analysis is routinely used as a check on the plausibility of derived equations and computations. It is also used to categorize types of physical quantities and units based on their relationship to or dependence on other units.

Dimensionless quantities which are without associated physical dimensions, are widely used in mathematics, physics, engineering, economics, and in everyday life (such as in counting). Numerous well-known quantities, such as π, e, and φ, are dimensionless. They are "pure" numbers, and as such always have a dimension of 1 [218, 219].

Dimensionless quantities are often defined as products or ratios of quantities that are not dimensionless, but whose dimensions cancel out when their powers are multiplied [220].

The basic principle of dimensional analysis was known to Isaac Newton (1686) who referred to it as the "Great Principle of Similitude." James Clerk Maxwell played a major role in establishing modern use of dimensional analysis by distinguishing mass, length, and time as fundamental units, while referring to other units as derived. The 19th-century French mathematician Joseph Fourier made important contributions based on the idea that physical laws like F = ma should be independent of the units employed to measure the physical variables. This led to the conclusion that meaningful laws must be homogeneous equations in their various units of measurement, a result which was eventually formalized in the Buckingham π theorem. This theorem describes how every physically meaningful equation involving n variables can be equivalently rewritten as an equation of n − m dimensionless parameters, where m is the rank of the dimensional matrix. Furthermore, and most importantly, it provides a method for computing these dimensionless parameters from the given variables.

A dimensional equation can have the dimensions reduced or eliminated through nondimensionalization, which begins with dimensional analysis, and involves scaling quantities by characteristic units of a system or natural units of nature. This gives insight into the fundamental properties of the system, as illustrated in the examples below.

In nondimensional scaling, there are two key steps:

(a) Identify a set of physically relevant dimensionless groups, and
(b) Determine the scaling exponent for each one.

Dimensional analysis will help you with step (a), but it cannot be applicable possibly for step (b).

A good approach to systematically getting to grips with such problems is through the tools of dimensional analysis (Bridgman, 1963). The dominant balance of forces controlling the dynamics of any process depends on the relative magnitudes of each underlying physical effect entering the set of governing equations [221]. Now, the most general characteristics parameters, which used in dimensionless analysis in electrospinning are introduced in Table 7.3.

TABLE 7.3 Characteristics Parameters Employed and Their Definitions

Parameter	Definition
Length	R_0
Velocity	$v_0 = \dfrac{Q}{\pi R_0^2 K}$
Electric field	$E_0 = \dfrac{I}{\pi R_0^2 K}$
Surface charge density	$\sigma_0 = \bar{\varepsilon} E_0$
Viscose stress	$\tau_0 = \dfrac{\eta_0 v_0}{R_0}$

For achievement of a simplified form of equations and reduction a number of unknown variables, the parameters should be subdivided into characteristic scales in order to become dimensionless. Electrospinning dimensionless groups are shown in Table 7.4 [222].

TABLE 7.4   Dimensionless Groups Employed and Their Definitions

Name	Definition	Field of application
Froude number	$Fr = \dfrac{v_0^2}{gR_0}$	The ratio of inertial to gravitational forces
Reynolds number	$Re = \dfrac{\rho v_0 R_0}{\eta_0}$	The ratio of the inertia forces of the viscous forces
Weber number	$We = \dfrac{\rho v_0^2 R_0}{\gamma}$	The ratio of the surface tension forces to the inertia forces
Deborah number	$De = \dfrac{\lambda v_0}{R_0}$	The ratio of the fluid relaxation time to the instability growth time
Electric Peclet number	$Pe = \dfrac{2\bar{\varepsilon} v_0}{KR_0}$	The ratio of the characteristic time for flow to that for electrical conduction
Euler number	$Eu = \dfrac{\varepsilon_0 E^2}{\rho v_0^2}$	The ratio of electrostatic forces to inertia forces
Capillary number	$Ca = \dfrac{\eta v_0}{\gamma}$	The ratio of inertia forces of viscous forces
Ohnesorge number	$oh = \dfrac{\eta}{\left(\rho\gamma R_0\right)^{1/2}}$	The ratio of viscous force to surface force
Viscosity ratio	$r_\eta = \dfrac{\eta_p}{\eta_0}$	The ratio of the polymer viscosity to total viscosity

TABLE 7.4 *(Continued)*

Name	Definition	Field of application
Aspect ratio	$\chi = \dfrac{L}{R_0}$	The ratio of the length of the primary radius of jet
Electrostatic force parameter	$\varepsilon = \dfrac{\overline{\varepsilon} E_0^2}{\rho v_0^2}$	The relative importance of the electrostatic and hydrodynamic forces
Dielectric constant ratio	$\beta = \dfrac{\varepsilon}{\overline{\varepsilon}} - 1$	The ratio of the field without the dielectric to the net field with the dielectric

The governing and constitutive equations can be transformed into a dimensionless form using the dimensionless parameters and groups.

7.8.9 SOME OF ELECTROSPINNING MODELS

The most important mathematical models for electrospinning process are classified in the Table 7.5. According to the year, advantages and disadvantages of the models:

TABLE 7.5 The Most Important Mathematical Models for Electrospinning

Researchers	Model	Year	Ref.
Taylor, G. I. Melcher, J. R.	Leaky dielectric model Dielectric fluid Bulk charge in the fluid jet considered to be zero Only axial motion Steady state part of jet	1969	[223]
Ramos	Slender body Incompressible and axi-symmetric and viscous jet under gravity force No electrical force Jet radius decreases near zero Velocity and pressure of jet only change during axial direction Mass and volume control equations and Taylor expansion were applied to predict jet radius	1996	[224]

TABLE 7.5   *(Continued)*

Researchers	Model	Year	Ref.
Saville, D. A.	Electrohydrodynamic model The hydrodynamic equations of dielectric model was modified Using dielectric assumption This model can predict drop formation Considering jet as a cylinder (ignoring the diameter reduction) Only for steady state part of the jet	1997	[225]
Spivak, A. Dzenis, Y.	Spivak and Dzenis model The motion of a viscose fluid jet with lower conductivity were surveyed in an external electric field Single Newtonian Fluid jet The electric field assumed to be uniform and constant, unaffected by the charges carried by the jet Use asymptotic approximation was applied in a long distance from the nozzle Tangential electric force assumed to be zero Using nonlinear rheological constitutive equation (Ostwald-de-wale law), nonlinear behavior of fluid jet were investigated	1998	[150]
Jong Wook	Droplet formation model Droplet formation of charged fluid jet was studied in this model The ratio of mass, energy and electric charge transition are the most important parameters on droplet formation Deformation and break-up of droplets were investigated too Newtonian and Non-Newtonian fluids Only for high conductivety and viscouse fluids	2000	[226]

TABLE 7.5 *(Continued)*

Researchers	Model	Year	Ref.
Reneker, D. H. Yarin, A. L.	Reneker model	2000	[227]
	For description of instabilities in viscoelastic jets		
	Using molecular chain theory, behavior of polymer chain of spring-bead model in electric field was studied		
	Electric force based on electric field cause instability of fluid jet while repulsion force between surface charges make perturbation and bending instability		
	The motion paths of these two cases were studied		
	Governing equations momentum balance, motion equations for each bead, Maxwell tension and columbic equations		
Hohman, M. Shin, M.	Stability theory	2001	[60]
	This model is based on a dielectric model with some modification for Newtonian fluids.		
	This model can describe whipping, bending and Rayleigh instabilities and introduced new ballooning instability.		
	Four motion regions were introduced: dipping mode, spindle mode, oscillating mode, precession mode.		
	Surface charge density introduced as the most effective parameter on instability formation.		
	Effect of fluid conductivity and viscosity on nanofibers diameter were discussed.		
	Steady solutions may be obtained only if the surface charge density at the nozzle is set to zero or a very low value		
Feng, J. J	Modifying Hohman model	2002	[153]
	For both Newtonian and non-Newtonian fluids		
	Unlike Hohman model, the initial surface charge density was not zero, so the "ballooning instability" did not accrue.		
	Only for steady state part of the jet		
	Simplifying the electric field equation,which Hohman used in order to eliminate Ballooning instability		

TABLE 7.5   *(Continued)*

Researchers	Model	Year	Ref.
Wan-Guo-Pan	Wan-Guo-Pan model	2004	[175]
	They introduced thermo-electro-hydro dynamics model in electrospinning process		
	This model is a modification on Spivak model which mentioned before		
	The governing equations in this model: Modified Maxwell equation, Navier-Stocks eqs., and several rheological constitutive equation		
Ji- Haun	AC-electrospinning model	2005	[172]
	Whipping instability in this model was distinguished as the most effective parameter on uncontrollable deposition of nanofibers		
	Applying AC current can reduce this instability so make oriented nanofibers		
	This model found a relationship between axial distance from nozzle and jet diameter		
	This model also connected AC frequency and jet diameter		
Roozemond (Eindhoven University and Technology)	Combination of slender body and dielectric model	2007	[228]
	In this model, a new model for viscoelastic jets in electrospinning were presented by combining these two models		
	All variables were assumed uniform in cross section of the jet but they changed in during z direction		
	Nanofiber diameter can be predicted		
Wan	Electromagnetic model	2012	[229]
	Results indicated that the electromagnetic field which made because of electrical field in charged polymeric jet is the most important reason of helix motion of jet during the process		
Dasri	Dasri model	2012	[230]
	This model was presented for description of unstable behavior of fluid jet during electrospinning		
	This instability causes random deposition of nanofiber on surface of the collector		
	This model described dynamic behavior of fluid by combining assumption of Reneker and Spivak models		

The most frequent numeric mathematical methods, which were used in different models, are listed in Table 7.6.

TABLE 7.6   Applied Numerical Methods for Electrospinning

Method	Ref.
Relaxation method	[153, 159, 231]
Boundary integral method (boundary element method)	[199, 226]
Semi-inverse method	[159, 172]
(Integral) control-volume formulation	[224]
Finite element method	[223]
Kutta-Merson method	[232]
Lattice Boltzmann method with finite difference method	[233]

7.9   ELECTROSPINNING SIMULATION

Electrospun polymer nanofibers demonstrate outstanding mechanical and thermodynamic properties as compared to macroscopic-scale structures. These features are attributed to nanofiber microstructure [234, 235]. Theoretical modeling predicts the nanostructure formations during electrospinning. This prediction could be verified by various experimental condition and analysis methods, which called simulation. Numerical simulations can be compared with experimental observations as the last evidence [149, 236].

Parametric analysis and accounting complex geometries in simulation of electrospinning are extremely difficult due to the nonlinearity nature in the problem. Therefore, a lot of researches have done to develop an existing electrospinning simulation of viscoelastic liquids [231].

7.10   ELECTROSPINNING SIMULATION EXAMPLE

In order to survey of electrospinning modeling application, the main equations were applied for simulating the process according to the constants, which summarized in Table 7.7.

Mass and charge conservations allow v and σ to be expressed in terms of R and E, and the momentum and E-field equations can be recast into two second-order ordinary differential equations for R and E. The slender-body theory (the straight part of the jet) was assumed to investigate the jet behavior during the spinning distance. The slope of the jet surface (R') is maximum at the origin of the nozzle. The same assumption has been used in most previous models concerning jets or drops. The initial and boundary conditions, which govern the process, are introduced as:

Initial values ($z=0$):

$$R(0) = 1$$
$$E(0) = E_0$$
$$\tau_{prr} = 2r_\eta \frac{R_0^{'}}{R_0^3}$$
$$\tau_{pzz} = -2\tau_{prr}$$

(7.9.1)

Feng [153] indicated that E(0) effect is limited to a tiny layer below the nozzle which its thickness is a few percent of R_0. It was assumed that the shear inside the nozzle is effective in stretching of polymer molecules as compared with the following elongation.

Boundary values ($z=\chi$):

$$R(\chi) + 4\chi R'(\chi) = 0$$
$$E(\chi) = E_\chi$$
$$\tau_{prr} = 2r_\eta \frac{R_\chi^{'}}{R_\chi^3}$$
$$\tau_{pzz} = -2\tau_{prr}$$

(7.9.2)

The asymptotic scaling can be stated as [153]:

$$R(z) \propto z^{-1/4}$$

(7.9.3)

Just above the deposit point ($z=\chi$), asymptotic thinning conditions applied. R drops towards zero and E approaches E_∞. The electric field is not equal to E_∞, so we assumed a slightly larger value, E_χ.

TABLE 7.7 Constants Which Were Used in Electrospinning Simulation

Constant	Quantity ($\times 10^{-3}$)
Re	2.5
We	0.1
Fr	0.1
Pe	0.1
De	10
E	1
β	40
χ	20
E0	0.7
Eχ	0.5
r_η	0.9

The momentum, electric field and stress equations could be rewritten into a set of four coupled first order ordinary differential equations (ODE's) with the above mentioned boundary conditions. The numerical relaxation method is chosen to solve the generated boundary value problem.

The results of these systems of equations are presented in Figs. 7.25 and 7.26 that matched quite well with the other studies that have been published [60, 78, 153, 154, 159].

The variation prediction of R, R,' ER^2, ER^2' and E versus axial position (z) are shown in Fig. 7.25. Physically, the amount of counductable charges reduces with decreasing jet radius. Therefore, to maintain the same jet current, more surface charges should be carried by the convection. Moreover, in the considered simulation region, the density of surface charge gradually increased. As the jet gets thinner and faster, electric conduction gradually transfers to convection. The electric field is mainly induced by the axial gradient of surface charge, thus it is insensitive to the thinning of the electrospun jet:

$$\frac{d(\sigma R)}{dz} \approx -\left(2R\frac{dR}{dz}\right)/Pe \qquad (7.9.4)$$

Therefore, the variation of E versus z can be written as:

$$\frac{d(E)}{dz} \approx \ln \chi \left(\frac{d^2 R^2}{dz^2}\right)/Pe \qquad (7.9.5)$$

Downstream of the origin, E shoots up to a peak and then relaxes due to the decrease of electrostatic pulling force in consequence of the reduction of surface charge density, if the current was held at a constant value. However, in reality, the increase of the strength of the electric field also increases the jet current, which is relatively linear [78, 153]. As the jet becomes thinner downstream, the increase of jet speed reduces the surface charge density and thus E, so the electric force exerted on the jet and thus R' become smaller. The rates of R and R' are maximum at $z=0$, and then relaxes smoothly downstream toward zero [153, 159]. According to the relation between R, E and z, ER^2 and ER^2' vary in accord with parts (c) and (d) in Fig. 7.25.

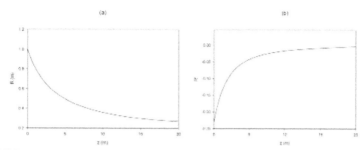

FIGURE 7.25 Solutions given by the electrospinning model for (a) R; (b) R'; (c) ER2; (d) ER2' and (e) E.

Figure 7.26 shows the changes of axial, radial shear stress and the difference between them, the tensile force (T) versus z. The polymer tensile force is much larger in viscoelastic polymers because of the strain hardening. T also has an initial rise, because the effect of strain hardening is so strong that it overcomes the shrinking radius of the jet. After the maximum value of T, it reduces during the jet thinning. As expected, the axial polymer stress rises, because the fiber is stretched in axial direction, and the radial polymer stress declines. The variation of T along the jet can be nonmonotonic, however, meaning the viscous normal stress may promote or resist stretching in a different part of the jet and under different conditions [153, 159].

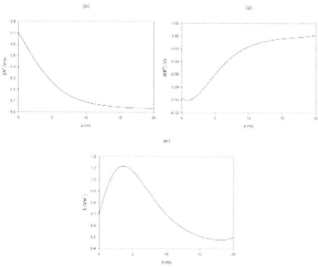

FIGURE 7.26 Solutions given by the electrospinning model for (a) τ_{prr}; (b) τ_{pzz} and (c) T.

7.11   APPLIED NUMERICAL METHODS FOR ELECTROSPINNING

Mathematics is indeed the language of science and the more proficient one is in the language the better. There are three main steps in the computational modeling of any physical process: (i) problem definition, (ii) mathematical model, and (iii) computer simulation [237–239].

(i) The first natural step is to define an idealization of our problem of interest in terms of a set of affiliate quantities, which it would be wanted to measure. In defining this idealization we expect to obtain a well-posed problem, this is one that has a unique solution for a given set of parameters. It might not always be possible to insure the integrity of the realization since, in some instances; the physical process is not entirely conceived.

(ii) The second step of the modeling process is to represent the idealization of the physical reality by a mathematical model: the governing equations of the problem. These are available for many physical phenomena. For example, in fluid dynamics the Navier–Stokes equations are considered to be an accurate representation of the fluid motion. Analogously, the equations of elasticity in structural mechanics govern the deformation of a solid object due to applied external forces. These are complex general equations that are very difficult to solve both analytically and computationally. Therefore, it needs to introduce simplifying assumptions to reduce the complexity of the mathematical model and make it amenable to either exact or numerical solution.

(iii) After the selection of an appropriate mathematical model, together with suitable boundary and initial conditions, it could be proceed to its solution. In this part, it will be considered the numerical solution of mathematical problems, which are described by partial differential equations (PDEs).

The sequential steps of analyzing physical systems can be summarized as presents in Fig. 7.27.

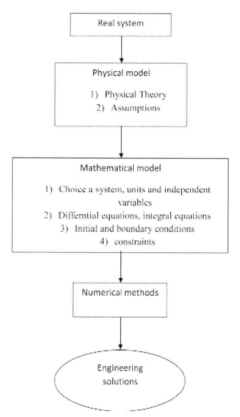

FIGURE 7.27   The diagram of sequential steps of analyzing physical systems.

Numerical analysis is the study of algorithms, which uses numerical approximations for the problems of mathematical analysis. The overall goal of the field of numerical analysis is the design and analysis of techniques to give approximate but accurate solutions to hard problems. Numerical analysis naturally finds applications in all fields of engineering and the physical science. Numerical analysis is also concerned with computing (in an approximate way) the solution of differential equations, both ordinary differential equations and partial differential equations. Ordinary differential equations appear in celestial mechanics (planets, stars and galaxies), numerical linear algebra is important for data analysis, stochastic differential equations and Markov chains are essential in simulating living cells for medicine and biology [240, 241]. Partial differential equations (PDEs) are used to describe most of the existing physical phenomena. The most general and practical way to produce solutions to the PDEs is to use numerical methods and computers. It is essential that the solutions produced by the numerical scheme are accurate and reliable. Partial differential equations are solved by first discretizing the equation, bringing it into a finite-dimensional subspace. This can be done by several methods such as a finite element method, a finite difference method, or (particularly in engineering) a finite volume method. The theoretical justification of these methods often involves theorems from functional analysis. This reduces the problem to the solution of an algebraic equation [242, 243]. In the table, the advantages of simulation illustrate by comparison between doing experiments and simulations (Table 7.8).

TABLE 7.8   The Advantages of Simulation Versus Doing Experiments

Experiments	Simulations
Expensive	Cheap (er)
Slow	Fast (er)
Sequential	Parallel
Single-purpose	Multiple-purpose

Now, some numeric methods which applied by researchers in the investigation on electrospinning models and simulations are over looked.

7.11.1   KUTTA-MERSON METHOD

One hundred years ago, C. Runge was completing his famous paper. This work, published in 1895, extended the approximation method of Euler to a more elaborate scheme, which was able to have a greater accuracy by using Taylor series expansion. The idea of Euler was to bring up the solution of an initial value problem forward by a sequence of small time-steps. In each step, the rate of change of the solution is treated as constant and is found from the formula for the derivative evaluated at the beginning of the step [244, 245].

The system of ordinary differential equations (ODEs) arising from the application of a spatial discretization of a system of PDEs can be very large, especially in three-dimensional simulations. Consequently, the constraints on the methods used for integrating these systems are somewhat different from those, which have taken much of the advance of numerical methods for initial value problems. On their high accuracy and modest memory requirements, the Runge–Kutta methods have become popular for simulations of physical phenomena. The classical fourth-order Runge–Kutta method requires three memory locations per dependent variable but low-storage methods requiring only two memory locations per dependent variable can be derived. This feature is easily achieved by a third-order Runge–Kutta method but an additional stage is required for a fourth-order method. Since the primary cost of the integration is in the evaluation of the derivative function, and each stage requires a function evaluation, the additional stage represents a significant increase in expense. For the same reason, error checking is generally not performed when solving very large systems of ODEs arising from the discretization of PDEs [245–247].

In 2004 and in 2006, Reznik and et al. experimentally and theoretically studied on the shape evolution of small compound droplets in the normal form and at the exit of a core-shell system in the presence of a sufficiently strong electric field respectively [248, 249].

In the first study they considered an axisymmetric droplet of an incompressible conducting viscous liquid on an infinite conducting plate. It was neglected in the gravity effects so the stationary shape of the droplet is spherical. The droplet shaped as Fig. 7.28 a spherical segment rests on the plate with a static contact angle of the liquid/air/solid system.

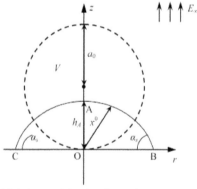

FIGURE 7.28   The initial shape of the droplet at the moment when the electric field is to be applied.

They estimated the relevant characteristic dimensionless parameters of the problem and used the Stokes equations for the liquid motion within the droplet then the equations solved numerically using the Kutta–Merson method [249].

In the second work, a core-shell nozzle (Fig. 7.29) consists of a central cylindrical pipe and a concentric annular pipe surrounding it. When the process took place without an applied electric field, the outer surface of the droplet and the interface between its components acquire near-spherical equilibrium shapes owing to the action of the surface and interfacial tension, respectively. If after the establishment of the equilibrium shape an electric field is applied to the compound nozzle and droplet attached to an electrode immersed in it, with a counter electrode, say a metal plate, located at some distance from the droplet tip, the latter undergoes stretching under the action of Maxwell stresses of electrical origin. At first, the problem was defined then the droplet evolution determined by time stepping using the Kutta-Merson method.

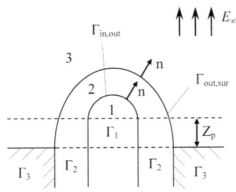

FIGURE 7.29   Core-shell droplet at the exit of a core-shell nozzle.

Both the core and the shell fluids are considered as leaky dielectrics, whose electric relative permittivities and conductivities are denoted. The electric boundary conditions describe the jump in the normal component of the electric field and electric induction at the boundaries. The governing equations rendered dimensionless and subject to the boundary conditions were solved with the aid of equivalent boundary integral formulations in which a set of corresponding integral equations was solved [248].

In 2010, Holzmeister and et al considered a simple two-dimensional model can be used for describing the formation of barb electrospun polymer nanowires with a perturbed swollen cross-section and the electric charges "frozen" into the jet surface. This model is integrated numerically using the Kutta-Merson method with the adoptable time step. The result of this modeling is explained theoretically as a result of relatively slow charge relaxation compared to the development of the secondary electrically driven instabilities which deform jet surface locally. When the disparity of the slow charge relaxation compared to the rate of growth of the secondary electrically driven instabilities becomes even more pronounced, the barbs transform in full scale long branches. The competition between charge relaxation and rate of growth of capillary and electrically driven secondary localized perturbations of the jet surface is affected not only by the electric conductivity of polymer solutions but also by their viscoelasticity. Moreover, a nonlinear theoretical model was able to resemble the main morphological trends recorded in the experiments [232].

7.11.2 FINITE ELEMENT METHOD

The finite element method is a numerical analysis technique for obtaining approximate solutions to a vast variety of engineering problems. All finite element methods involve dividing the physical systems, such as structures, solid or fluid continua, into small subregions or elements. Each element is an essentially simple unit, the behavior of which can be readily analyzed. The complexities of the overall systems are accommodated by using large numbers of elements, rather than by resorting to the sophisticated mathematics required by many analytical solutions [250, 251].

A typical finite element analysis on a software system requires the following information [252]:

1. Nodal point spatial locations (geometry)
2. Elements connecting the nodal points
3. Mass properties
4. Boundary conditions or restraints
5. Loading or forcing function details
6. Analysis options

And the FEM Solution Process [253]:

1. Divide structure into pieces (elements with nodes) (discretization/ meshing)
2. Connect (assemble) the elements at the nodes to form an approximate system of equations for the whole structure (forming element matrices)
3. Solve the system of equations involving unknown quantities at the nodes (e.g., displacements)
4. Calculate desired quantities (e.g., strains and stresses) at selected elements

One of the main attractions of finite element method is the ease with which they can be applied to problems involving geometrically complicated systems. The price that must be paid for flexibility and simplicity of individual elements is in the amount of numerical computation required. Very large sets of simultaneous algebraic equations have to be solved, and this can only be done economically with the aid of digital computers [254]. Therefore, the advantage of this method is that for a smooth problem where the derivatives of the solution are well behaved, the computational cost increases algebraically while the error decreases exponentially fast and the disadvantage of it is that the method leads to nonsingular systems of equations that can easily solve by standard methods of solution. This is not the case for time-dependent problems where numerical errors may grow unbounded for some discretization [238].

The finite element method is used for spatial discretization. Special numerical methods are used and developed for viscoelastic fluid flow, including the DEVSS and log-conformation techniques. For describing the moving sharp interface between the rigid particle and the fluid both ALE (Arbitrary Lagrangian Euler) and XFEM (extended finite element) techniques are being developed and employed [255]. In the development of numerical algorithms for the stable and accurate solution of viscoelastic flow problems, like electrospinning process, applying the finite element method to solve constitutive equations of the differential type is useful [256].

In 2011, Chitral and et al analyzed the electrospinning process based on an existing electrospinning model for viscoelastic liquids using a finite element method. Four steady-state equations concluding Coulomb force, an electric force imposed by the external electric field, a viscoelastic force, a surface tension force, a gravitational force, and an air drag force were solved as a set of equations with this method [257].

7.11.3 BOUNDARY INTEGRAL METHOD (BOUNDARY ELEMENT METHOD)

Boundary integral equations are a classic tool for the analysis of boundary value problems for partial differential equations. The term "boundary element method" (BEM) (Fig. 7.30) denotes any method for the approximate numerical solution of these boundary integral equations. The approximate solution of the boundary value problem obtained by BEM has the distinguishing feature that it is an exact solution of the differential equation in the domain and is parameterized by a finite set of parameters living on the boundary [258].

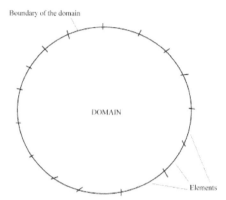

FIGURE 7.30   The idea of boundary element method.

The BEM has some advantages over other numerical methods like finite element methods
(FEM) or finite differences [258–260]:

1. Only the boundary of the domain needs to be discretized. Especially in two dimensions where the boundary is just a curve this allows very simple data input and storage methods.

2. Exterior problems with unbounded domains but bounded boundaries are handled as easily as interior problems.

3. In some applications, the physically relevant data are given not by the solution in the interior of the domain but rather by the boundary values of the solution or its derivatives. These data can be obtained directly from the solution of boundary integral equations; whereas boundary values obtained from FEM solutions are in general not very accurate.

4. The solution in the interior of the domain is approximated with a rather high convergence rate and moreover, the same rate of convergence holds for all derivatives of any order of the solution in the domain. There are difficulties, however, if the solution has to be evaluated close to, but not on the boundary.

Some main difficulties with BEM are the following [258, 260, 261]:

1. Boundary integral equations require the explicit knowledge of a fundamental solution of the differential equation. This is available only for linear partial differential equations with constant or some specifically variable coefficients. Problems with inhomogeneities or nonlinear differential equations are in general not accessible by pure BEM.

Sometimes, however, a coupling of FEM and BEM proves to be useful.

2. For a given boundary value problem there exist different boundary integral equations and for each of them several numerical approximation methods. Thus, every BEM application requires that several choices be made. To evaluate the different possibilities, one needs a lot of mathematical analysis. Although the analysis of BEM has been a field of active research in the past decade, it is by no means complete. Thus there exist no error estimates for several methods that are widely used. From a mathematical point of view, these methods, which include very popular ones for which computer codes are available, are in an experimental state, and there might exist problems of reliability.

3. The reason for the difficulty of the mathematical analysis is that boundary integral equations frequently are not ordinary Fredholm integral equations of the second kind. The classical theory of integral equations and their numerical solution concentrates on the second kind integral equations with regular kernel, however. Boundary integral equations may be of the first kind, and the kernels are in general singular. If the singularities are not integrable, one has to regularize the integrals, which are then defined in a distributional sense. The theoretical framework for such integral equations is the theory of pseudodifferential operators. This theory was developed 20 years ago and is now a classic part of Mathematical Analysis, but it is still not very popular within Applied Mathematics.

4. If the boundary is not smooth but has corners and edges, then the solution of the boundary value problem has singularities at the boundary. This happens also if the boundary conditions are discontinuous, e.g., in mixed boundary value problems. BEM clearly has to treat these singularities more directly than FEM. Because the precise shape of the singularities frequently contains important information, for example, stress intensity factors in fracture mechanics, this is a positive aspect of BEM. But besides practical problems with the numerical treatment of these singularities, nonsmooth domains also present theoretical difficulties. These have so far been satisfactorily resolved only for two-dimensional problems. The analysis of BEM for three-dimensional domains with corners and edges is still in a rather incomplete stage.

The paper of Kowalewski which published in 2009, can be effectively addressed by use of a boundary element method (BEM). In essence, the boundary element method is a statement of the electrostatic problem (Poisson equation) in terms of boundary integrals, as such, it involves only the discretization of boundary surfaces, which in our case would be the electrode surfaces and the outer shell of the fiber. The model was, essentially, a time-dependent three-dimensional generalization of known slender models with the following differences:

(i)   The electric field induced by the generator and by the charges on the fiber is explicitly resolved, instead of being approximated from local parameters.

(ii) Electrical conductivity is neglected. Indeed, the convection of surface charges is believed to strongly overcomes bulk conduction at locations distant from the Taylor cone by a few fiber radii, since we are mostly interested in the description of the bending instability, this assumption appears reasonable [262].

7.11.4   (INTEGRAL) CONTROL-VOLUME FORMULATION

All the laws of mechanics are written for a system, which is defined as an arbitrary quantity of mass of fixed identity. Everything external to this system is denoted by the term surroundings, and the system is separated from its surroundings by its boundaries. The laws of mechanics then state what happens when there is an interaction between the system and its surroundings [263].

Typically, to understand how a given physical law applies to the system under consideration, one first begins by considering how it applies to a small, control volume, or "representative volume." There is nothing special about a particular control volume, it simply represents a small part of the system to which physical laws can be easily applied. This gives rise to what is termed a volumetric, or volume-wise formulation of the mathematical model [264, 265].

In fluid mechanics and thermodynamics, a control volume is a mathematical abstraction employed in the process of creating mathematical models of physical processes. In an inertial frame of reference, it is a volume fixed in space or moving with constant velocity through which the fluid (gas or liquid) flows. The surface enclosing the control volume is referred to as the control surface [266, 267].

Control-volume analysis is "more equal," being the single most valuable tool to the engineer for flow analysis. It gives "engineering" answers, sometimes gross and crude but always useful [263].

The advantage of this method, over that of the finite-element method, is the following. In the finite-element method, it is necessary to construct a grid over the flaw as well as the entire region surrounding the flaw, and to solve for the fields at all points on the grid. In contrast, in the volume-integral method, it is only necessary to construct a grid over the flaw and solve for the currents in the flaw; the Green's function takes care of all regions outside the flaw. This removes the complicated gridding requirements of the finite-element method, and reduces the size of the problem tremendously. That is the reason which this method can obtain much more accurate probe responses, and in much less time than a finite-element code, while running on a small personal computer or workstation. And without the complicated gridding, problems can be set up much more quickly and easily [268].

The famous researcher in electrospinning process, Feng, considered the steady stretching process is important in that it not only contributes to the thinning directly for Newtonian flows. The jet is governed by four steady-state equations representing the conservation of mass and electric charges, the linear momentum balance, and Coulomb's law for the electric field, which used control-volume balance for analyzing them [269].

7.11.5   RELAXATION METHOD

Relaxation method, an alternative to the Newton iteration method, is a method of solving simultaneous equations by guessing a solution and then reducing the errors that result by successive approximations until all the errors are less than some specified amount. Relaxation methods were developed for solving nonlinear systems and large sparse linear systems, which arose as finite-difference discretizations of differential equations [270, 271]. These iterative methods of relaxation should not be confused with "relaxations" in mathematical optimization, which approximate a difficult problem by a simpler problem, whose "relaxed" solution provides information about the solution of the original problem [272]. In solving PDEs problem with this method, it's necessary to turn them to the ODEs equations. Then ODEs have to be replaced by approximate finite difference equations. The relaxation method determines the solution by starting with a guess and improving it, iteratively. During the iterations the solution is improved and the result relaxes towards the true solution. Notice that the number of mesh points may be important for the properties of the numerical procedure. In Relax setting, increase the number of mesh points if the convergence towards steady state seems awkward [273, 274].

The advantage of this method is a relative freedom in its implementation. It can be used for smooth problem. This is useful in particular when an analytical solution to the model is not available and can handle models, which exhibit saddle-point stability. Therefore, a "large stepping" in the direction of the defect is possible, while the termination point is defined by the condition that the vector field becomes orthogonal. The relaxation algorithm can easily cope with a large number of problems, which arise frequently in the context of multidimensional, infinite-time horizon optimal control problems [275, 276].

A researcher in a thesis named "A Model for Electrospinning Viscoelastic Fluids" studied the electrospinning process of rewriting the momentum, electric field and stress equations as a set of six coupled first order ordinary differential equations. It made a boundary value problem, which can best be solved by a numerical relaxation method. Within relaxation methods the differential equations are replaced by finite difference equations on a certain mesh of points covering the range of integration. During iteration (relaxation) all the values on the mesh are adjusted to bring them into closer agreement with the finite difference equations and the boundary conditions [277].

7.11.6 LATTICE BOLTZMANN METHOD WITH FINITE DIFFERENCE METHOD

Lattice Boltzmann methods (LBM) (or Thermal Lattice Boltzmann methods (TLBM)) are a class of computational fluid dynamics (CFD) methods for fluid simulation. Instead of solving the Navier–Stokes equations, the discrete Boltzmann equation is solved to simulate the flow of a Newtonian fluid with collision models such as Bhatnagar-Gross-Krook (BGK). LBM is a relatively new simulation technique for complex fluid systems and has attracted interest from researchers in computational physics. Unlike the traditional CFD methods, which solve the conservation equations of macroscopic properties (i.e., mass, momentum, and energy) numerically, LBM models the fluid consisting of fictive particles, and such particles perform consecutive propagation and collision processes over a discrete lattice mesh. Due to its particulate nature and local dynamics, LBM has several advantages over other conventional CFD methods, especially in dealing with complex boundaries, incorporating of microscopic interactions, and parallelization of the algorithm. A different interpretation of the Lattice Boltzmann equation is that of a discrete-velocity Boltzmann equation. By simulating streaming and collision processes across a limited number of particles, the intrinsic particle interactions evince a microcosm of viscous flow behavior applicable across the greater mass. The LBM is especially useful for modeling complicated boundary conditions and multi phase interfaces [278–280]. The idea that LBE is a discrete scheme of the continuous Boltzmann equation also provides a way to improve the computational efficiency and accuracy of LBM. From this idea, the discretization of the phase space and the configuration space can be done independently. Once the phase space is discretized, any standard numerical technique can serve the purpose of solving the discrete velocity Boltzmann equation. It is not surprising that the finite difference, finite volume, and finite element methods have been introduced into LBM in order to increase computational efficiency and accuracy by using nonuniform grids [281]. The Lattice Boltzmann method has a more detailed microscopic description than a classical finite difference scheme because the LBM approach includes a minimal set of molecular velocities of the particles. In addition, important physical quantities, such as the stress tensor, or particle current, is directly obtained from the local information. However, the LBM scheme may require more memory than the corresponding finite difference scheme. Another motivation is that the boundary conditions are more or less naturally imposed for a given numerical scheme [282]. In the ordinary LBM, this discrete BGK equation is discretized into a special finite difference form in which the convection term does not include the numerical error. But considering the collision term, this scheme is of second-order accuracy. On the other hand, the discrete BGK equation can be discretized in some other finite-difference schemes or in the finite-volume method and other computational techniques for the partial differential equations, and these techniques are considered as a natural extension of numerical calculations [283–285].

There is a method which couples Lattice Boltzmann method for the fluid with a molecular dynamics model for polymer chain. The Lattice Boltzmann equation using single relaxation time approximations (Fig. 7.31).

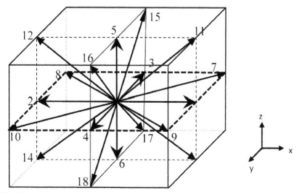

FIGURE 7.31   Lattice scheme.

In a thesis named "Modeling Electrospinning Process and a Numerical Scheme Using Lattice Boltzmann Method to Simulate Viscoelastic Fluid Flows," the researcher wrote the conservation of mass and momentum based on the lattice scheme which presented above. It was assumed that the fluid is incompressible and the continuum/macroscopic governing equations were written then forward step finite difference method in time applied for solving obtained equations [286].

7.11.7 SEMI-INVERSE METHOD

From the mathematical point of view, for the problems which the analytical solution may be very hard to attain, even in the simplest boundary value problem, because the set of equation forms a nontrivial system of nonlinear, partial differential equations generating often non unique solutions. To solve the resulting boundary-value problems, inverse techniques can be used to provide simple solutions and to suggest experimental programs for the determination of response functions. Two powerful methods for inverse investigations are the so-called inverse method and the semi-inverse method. They have been used in elasticity theory as well as in all fields of the mechanics of continua [287–289].

In the inverse method, a solution is found in priori such that it satisfies the governing equation and boundary condition. We can obtain the solution through this way for a luck case, but it is not a logical way. For the semi-inverse method, certain assumptions about the components of displacement strain are made at the beginning. Then, the solution is confined by satisfying the equations of equilibrium and the boundary conditions [289].

In the semi-inverse method is systematically studied and many examples are given to show how to establish a variational formulation for a nonlinear equation. From the given examples, we found that it is difficult to find a variational principle for nonlinear evolution equations with nonlinear terms of any orders [290].

In the framework of the theory of Continuum Mechanics, exact solutions play a fundamental role for several reasons. They allow investigating in a direct way the physics of various constitutive models to understand in depth the qualitative characteristics of the differential equations under investigation and they provide benchmark solutions of complex problems.

The mathematical method used to determine these solutions is usually called the semi-inverse method. This is essentially a heuristic method that consists in formulating a priori a special ansatz on the geometric and/or kinematical fields of interest, and then introducing this ansatz into the field equations. Luck permitting, these field equations reduce to a simple set of equations and then some special boundary value problems may be solved. Another important aspect in the use of the semi-inverse method is associated with fluid dynamics with the emergence of secondary flows and in solid mechanics with latent deformations. It is clear that "Navier-Stokes fluid" and an "isotropic incompressible hyperelastic material" are intellectual constructions [287, 291].

The most interesting features of this method are its extreme simplicity and concise forms of variational functionals for a wide range of nonlinear problems [292, 293]. An advantage of the semi-inverse method is that it can provide a powerful mathematical tool in the search for variational formulations for a rather wide class of physic problems without using the well-known Lagrange multipliers, which can result in variational crisis (the constrains can not eliminated after the identification of the multiplier or the multipliers become zero), furthermore, to use the Lagrange multipliers, we must have a known variational principle at hand, a situation which not always occurs in continuum physics [294, 295]. Although, using this method has not any special difficulties, there are primary difficulties with ill-posed problems [296, 297].

In 2004, Wan and et al studied in a model of steady state jet, which introduced by Spivak but they considered the couple effects of thermal, electricity, and hydrodynamics. Therefore, the model consists of modified Maxwell's equations governing electrical field in a moving fluid, the modified Navier-Stokes equations governing heat and fluid flow under the influence of electric field, and constitutive equations describing behavior of the fluid. The set of conservation laws can constitute a closed system when it is supplemented by appropriate constitutive equations for the field variables such as polarization. The most general theory of constitutive equations determining the polarization, electric conduction current, heat flux, and Cauchy stress tensor has been developed by Eringen and Maugin. By the semi-inverse method, it could be obtained various variational principles for electrospinning. After obtaining a set of equations, they were solved by the semi - inverse method [298].

7.12 CONCLUDING REMARKS OF ELECTROSPINNING MODELING

Comprehensive investigation on the nanoscience and the related technologies is an important topic these days. Due to the rising interest in nanoscale material properties, studying on the effective methods of producing nanomaterial such as the electrospinning process plays an important role in the new technologies progress. Since the electrospinning process is dependent on a lot of different parameters, changing them will lead to significant variations in the process. In this section, we have attempted to analyze the each part of process in detail and investigate some of the most important relationships from ongoing research into the fundamental physics that govern the electrospinning process. The idea has been to provide a few guiding principles for those who would use electrospinning to fabricate materials with the initial scrutiny. We have outlined the flow procedure involved in the electric field and those associated with the jetting process. The relevant processes are the steady thinning jet, whose behavior can be understood quantitatively using continuum equations of electrohydrodynamics, and the ensuing fluid dynamical instabilities that give rise to whipping of the jet. The current carried by the jet is of critical importance to the process and some scaling aspects concerning the total measured current have been discussed. The basics of electrospinning modeling involve mass conservation, electric charge conservation, momentum balance, coulomb's law and constitutive equations (Ostwald-de Waele power law, Giesekus equation and Maxwell equation), which are discussed in detail. These relations play an important role in setting final features. The dominant balance of forces controlling the dynamics of the electrospinning process depends on the relative magnitudes of each underlying physical effect entering the set of governing equations. Due to the application of the obtained nano web and material for different usages, surveying on microscopic and macroscopic properties of the product is necessary. Therefore, for achieving to this purpose, the relationships are pursued. Also for providing better understanding of electrospinning simulation, a model has been implemented for a straight jet in a relaxation method and simulated for the special viscoelastic polymers. The model capability to predict the behavior of the process parameters was demonstrated using simulation in the last part. The plots obviously showed the changes of each parameter versus axial position. During the jet thinning, the electric field shoots up to a peak and then relaxes. The tensile force also has an initial rise and then reduces. All the plots show similar behavior with the results of other researchers. At the end of this study, various numerical methods, which applied in the different electrospinning simulations, were revised.

7.13   NUMERICAL STUDY OF COAXIAL ELECTROSPINNING

Electrospinning provides a simple and highly versatile method for the large-scale fabrication of nanofibers. The technique has attracted significant attention in the last decade, since it allows for a straightforward, relatively easy and cheap method of manufacturing polymer nanofibers. Recently, advances in the technique of electrospinning have allowed this method to be used to directly fabricate core shell, core-sheath and hollow nanofibers of composites, ceramics, and polymers with diameters ranging from 20 nm–1 mm. By replacing the single capillary with a coaxial spinneret in electrosinning set up, it is possible to generate core-sheath and hollow nanofibers made of polymers, ceramics or composites [299–302]. These nanofibers, as a kind of one dimensional nanostructure, have attracted special attention in several research groups, because these structures could further enhance material property profiles and these unique core-sheath structures offer potential in a number of applications including nanoelectronics, microfluidics, photonics, and energy storage [300, 303, 304]. This process leads to encapsulation of core material (not necessarily polymeric), which is beneficial for storage and drug delivery of bioactive agents. The shell material which is polymeric substance can be carbonized or calcinated. Coaxial electrospinning is of particular interest for materials that will not easily form fibers via electrospinning on their own [305].

Similarly to electrospinning, (Fig. 7.32) coelectrospinning employs electric forces acting on polymer solutions in dc electric fields and resulting in significant stretching of polymer jets due to a direct pulling and growth of the electrically driven bending perturbations [306]. In this method, a plastic syringe with two compartments containing different polymer solutions or a polymer solution (shell) and a nonpolymeric Newtonian liquid or even a powder (core) is used to initiate a core-shell jet. At the exit of the core-shell needle attached to the syringe appears a core-shell droplet, which acquires a shape similar to the Taylor cone due to the pulling action of the electric Maxwell stresses acting on liquid. Liquid in the cone, being subjected to sufficiently strong (supercritical) electric field, issues a compound jet, which undergoes the electrically driven bending instability characteristic of the ordinary electrospinning process. Strong jet stretching resulting from the bending instability is accompanied by enormous jet thinning and fast solvent evaporation. As a result, the core-shell jet solidifies and core-shell fibers are depositing on a counter electrode (Fig. 7.32) [299, 306–308].

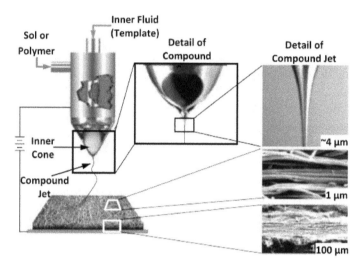

FIGURE 7.32   Experimental setup for coelectrospinning to produce core-shell nanofibers [309].

7.13.1 LITERATURE REVIEW OF COAXIAL ELECTROSPINNING

Coaxial electrospinning or coelectrospinning of core-shell micro and nanofibers was born about 10 years ago as a branch of nanotechnology which bifurcated from a previously known electrospinning. Through electrospinning, coelectrospinning inherited roots in polymer science and electrohydrodynamics, while some additional genes from textile science and optical fiber technology were spliced in addition. It also engulfed emulsion electrospinning. Co-electrospinning rapidly became widely popular and its applications proliferated into such fields as biotechnology, drug delivery and nanofluidics. It also triggered significant theoretical and experimental efforts directed at a better understanding and control of the process. The literature dealing with coelectrospinning already consists of hundreds of papers, while its simulation and modeling have not been investigated sufficiently.

Greiner's group [301] have demonstrated that core-sheath nanofibers could be fabricated by coelectrospinning two different polymer solutions through a spinneret comprising two coaxial capillaries. They manufactured nano and micro fibers in two stages process, which started with ordinary electrospinning of core polymer (stage one), and followed by the coating deposition of the shell polymer (stage two) [310]. Li and Xia prepared TiO_2/polymer composites and ceramic hollow nanofibers by electrospinning two immiscible liquids through a coaxial, two-capillary spinneret, followed by selective re-moval of the cores [302]. Using a similar setup, Loscertales and co-workers have demonstrated that polymer-free, inorganic nanotubes could also be fab-ricated by co-electrospinning an aged inorganic sol and an immiscible (or poorly miscible) liquid such as olive oil or glycerin, followed by selective removal of the inner liquid [309].

Recently, both coaxial electrospray and coaxial electrospinning of two im-miscible liquids became interesting research fields besides single-liquid elec-trospray and electrospinning. An electrified coaxial liquid jet can break into compound droplets in which one substance is coated with the other, or can make core-shell fibers on the submicrometer scale under different conditions [311–314]. In the case of electrospraying, the jets should be rapidly atomized into tiny core-shell droplets, with no viscoelasticity or jet bending involved, whereas in the case of coelectospinning, polymer jet should stay intact to make nanofibers, and viscoelasticity and jet bending are the dominant phe-nomena. Loscertales et al. [309] initially proposed that a coaxial spinneret could be used to produce a core sheath jet. However, in their experimental studies, instability led to the breakup of their jet and the formation of core-sheath nanoparticles. McCann et al. [299] solved the instability problem by cospinning two immiscible solutions, followed by gelation (or cross-linking) and stabilization of the sheath.

7.13.2 COAXIAL ELECTROSPINNING MODELING

As discussed above, when two immiscible liquids are injected at appropriate flow rates through two electrified capillary needles, one of them inside the other, the menisci of both liquids adopt conical shapes with an outer menis-cus surrounding the inner one. A liquid thread is issued from each one of the vertex of the two menisci in such a way that a compound, coaxial jet of two coflowing liquids is formed downstream [315, 316].

Several theoretical works dealt with different aspects of coelectrospinning with the goal of better understanding its underlying physics and enhancing its efficiency. Using core-shell nozzles does not necessarily imply formation of core-shell jets. The core material may be not entrained into the shell jet, which results in monolithic instead of core-shell jets. The detailed numerical simu-lations of flows developing in a core-shell droplet at the exit of a core-shell nozzle under the action of the pulling electric Maxwell stresses showed that the electric charges very rapidly escape to the outer surface of the forming jet

[317]. As a result, core entrainment is possible only due to viscous tractions. The core entrainment was predicted to be facilitated by a core nozzle protruding from a coaxial shell nozzle, which was demonstrated experimentally [317, 318].

A complete knowledge of the governed equations of the electrified co-axial jets generated via EHD as a function of the flow rates and the physical properties of the two liquids would be highly desirable; mainly, the current transported by them, the diameters of the inner and outer jets, and the size of the droplets resulting from their break up. However, the task of finding the above dependences is undoubtedly extremely complex, much more complex for coaxial electrospinning than for ordinary one since the number of parameters and unknowns is larger in the former problem than in the later one [315].

Because of its complexity, even the knowledge of simple cone jet electrospinning process is still incomplete, in spite of the fact that the equations and boundary conditions that govern its electro-hydrodynamic behavior are well known. The existence of very disparate scales in the cone–jet problem, a free interface whose position must be consistently calculated from the solution of the problem, and the time-dependent break up of the jet are different aspects of the problem which greatly contribute to its numerical complexity.

Competing models on the emitted current and jet diameter in single electrospinning, based on different simplifying hypothesis, are found in the literature. Researchers predict different expressions for the emitted current except for a nondimensional constant, which either depends on the dielectric constant of the liquid [148, 319] or it is independent of it, [313]. Even more unsatisfactory is the prediction of the jet diameter for which the existing models give different dependences on the flow rate and on the physical properties of the liquid. Unfortunately, the existing experimental measurements do not suffice to either completely support or reject some of these models since, for example, the jet diameter predicted by the different models are comparable to the accuracy with which it can be experimentally sized [320]. Therefore, more precise knowledge of the electrospinning must come from either numerical simulations or more refined experimental measurements. A numerical simulation of the transition zone between the cone and the jet, which can throw light on this complex problem, has been carried out by [321].

Lopez-Herrera et al. [315] applied an experimental investigation on the electrified coaxial jets of two immiscible liquids issuing from a structured Taylor cone. The effect of the flow rates of both liquids on the current transported by these coaxial jets and on the size of the compound droplets and final coaxial jet diameter was investigated. Their suggested scaling laws fitted well the scaling law reported by [319] and [148], but also exhibit a good agreement with the other competing scaling law jet size derived analytically by [311]. A draw back of this model is that the small differences between the jet diameters predicted by the two models are comparable to the accuracy with which can be experimentally sized. Therefore, their measurements did not provide a definitive proof for ruling out the wrong model [315].

Artana [322] carried out a temporal linear instability analysis of a cylindrical electrified jet flowing inside a cylindrical coaxial electrode. Hohman [60, 61] performed a complete instability analysis of a charged jet using an asymptotic expansion of radial direction to obtain a mechanism of electrospinning. Feng [153] introduced non-Newtonian rheology into a theoretical model for electrospinning. Loscertales [316] first presented an electrospray method to generate the compound droplets by using two immiscible coaxial liquid jets. The electrified coaxial jet can be divided into outer-driving and inner-driving states. The outer-driving state means that outer fluid has higher conductivity and free charges are located on the outer interface, and the inner driving state means inner fluid has higher conductivity and free charges are located on the inner interface.

López-Herrera [315] carried out an experiment on coaxial jets generated from an electrified Taylor cone and studied the scaling law between electric current and drop size under the influence of flow rate and other liquid properties. Chen [312] found experimentally a series of different modes in the coaxial jet electrospray with outer driving liquid. Sun et al. [301] Li and Xia, [302] and Yu et al. [323] investigated the mechanism of coaxial electrospinning and successfully produced composite or hollow nanofibers of different materials. However, little has been done so far in the theoretical analysis of the instability of the electrified coaxial jet. Li et al. [324] performed temporal linear instability analysis of an electrified coaxial jet inside a coaxial electrode and the analytic dispersion equation is derived in their model.

Boubaker presented a model to core-shell structured polymer nanofibers deposited via coaxial electrospinning. His investigations were based on a modified Jacobi-Gauss collocation spectral method, proposed along with the Boubaker Polynomials Expansion Scheme (BPES), for providing solution to a nonlinear Lane-Emden-type equation [325].

7.13.3 ANALYTICAL AND NUMERICAL MODELS

Coaxial electrospray and electrospinning are important branches of electrohydrodynamics that concern the fluid dynamics with the electric force effects

[223, 315, 316, 326, 327]. To create and maintain a steady cone–jet mode in coaxial process is a complex process that requires appropriate equilibrium of different forces. Since coaxial electrospinning has many potential advantages over other micro encapsulation/nanoencapsulation processes, it is of great interest to study the multiphysical mechanism and derive the analytical and numerical models for this process. Some of the most important equations, which govern the process, are mentioned in following sections.

7.13.3.1　FORMULATION

To better understand the governing mechanism of coaxial electrospray, it is important to establish a theoretical framework in advance. The complete set of governing equations can be given by the fluid dynamic equations (i.e., the Navier-Stokes equations) and the electrical equations (i.e., the Maxwell's equations). For Newtonian fluids of uniform constitution, the governing equations for each phase can be expressed as:

$$\frac{1}{\rho}\frac{d\rho}{dt} + \nabla u = 0 \qquad (7.13.1)$$

$$\rho\frac{du}{dt} = -\nabla p + \mu\nabla^2 u + \rho g + f \qquad (7.13.2)$$

$$\nabla D = q, \nabla B = 0, \nabla \times E = -\frac{\partial B}{\partial t}, \nabla \times H = J + \frac{\partial D}{\partial t} \qquad (7.13.3)$$

where the notation $d/dt = \partial/\partial t + u\nabla$ is the material derivative and the quantities ρ, u, p, μ, g, f_c, D, q, B, E, H and J stand for the density, velocity vector, pressure, dynamic viscosity, gravitational acceleration, electric force, electric displacement vector, free charge density, magnetic induction, electric intensity field, magnetic intensity field and conduction current density, respectively. Equation (7.13.1) represents the conservation of mass; Eq. (7.13.2) expresses the momentum equation and Eq. (7.13.3) show the well-known Maxwell's equations.

The above equations can be simplified with reasonable assumptions. In most studies, the liquids and the gas can be considered as incompressible fluids. Thus, Eq. (7.13.1) is simplified as $\nabla u=0$. For simplicity, the linearly constitutive relations $D=\mu E$, $B=\cap H$ and $J=KE$ are introduced, with μ, $\cap$ and K standing for permittivity, permeability and electrical conductivity, respectively. Since the magnetic field is very weak, the Maxwell's equations can be simplified as the following: $\nabla D = q$, $\nabla E = 0$, $dp/dt + \nabla j = 0$. The detailed derivation of the governing equations can be found in [328, 329].

Artana and his co-workers considered a liquid jet flowing vertically downwards out from an injector and into a gas at room pressure. Depending on the velocity of the liquid, one obtains different kinds of jets. The one of interest here corresponds to the second wind regime or to the regime of atomization that is to say for very high velocity (about 100 m/s). For these regimes the aspect of the jet looks like a pulverization shaped as a cone composed of sparse droplets in most of its volume, except in the region of its revolution axis where the density of the droplets is very high. This jet flows through one coaxial cylindrical electrode brought to a certain potential that we will consider positive. The potential of the electrode V_0 is maintained constant and the injector is earthed (see Fig. 7.33). The analysis of stability is done with an infinite jet inside an infinite electrode.

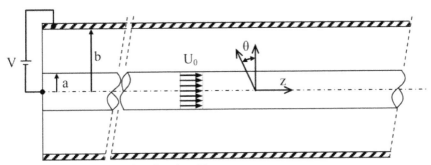

FIGURE 7.33   Schema of analyzed problem.

They assumed the two fluids to be incompressible and the motion to be irrotational. They neglect the effects of gravity, magnetic fields, viscosity and mass transfer at the interface. Liquid and gas are considered as isothermal and incompressible and their electrical properties are those of an ohmic conductor for the liquid phase and of a perfect dielectric for the gaseous phase, both having uniform conductivity and dielectric constant. The electric charge on the jet was at the jet surface and there was no free charge source in the bulk of the liquid or of the gaseous phase. The velocity profile in a typical stream wise station was invariant with the axial coordinate. To solve the problem they considered three regions: the regions of the liquid, gas and the interface. they used Gibbs's model [330] for the interface and we formulate the conservation equations as Slattery [331].

7.13.3.2   FLUID MECHANICS EQUATIONS

The mass conservation leads to each phase [322]:

$$\nabla \cdot U_i = 0 \qquad (7.13.4)$$

With Ui the velocity of the phase $i=1$ for the liquid and $i=2$ for the gas. At the interface:

$$\nabla_\zeta \cdot U_\zeta = 0, \qquad (7.13.5)$$

where U_ζ is the velocity vector of the interface and ∇_ζ the surface divergence.

As the motion being irrotational, Artana et al. [322] considered a potential function for the velocity φi and they obtained the Laplace equation $\Delta\varphi_i=0$.

As there is *no mass transfer* between phases

$$\left(U_i - U_\zeta\right) \cdot n = 0, \qquad (7.13.6)$$

or

$$\left(\nabla_1 - U_2\right) . n = 0, \qquad (7.13.7)$$

n being normal to the interface. As for *momentum conservation*, it leads to in each phase:

$$\rho_i \frac{dU_i}{dt} - \nabla^\circ T_i = 0 \qquad (7.13.8)$$

ρi and Ti are the mass density and Maxwell's constraint tensor of the phase i and

$$\frac{d}{dt} = \frac{\partial}{\partial_t} + U_i \cdot \nabla \qquad (7.13.9)$$

is the material derivative.
at the interface:

$$-\nabla_\zeta^\circ \left(T_\zeta\right) - \left(T_1 - T_2\right)^\circ n = 0, \qquad (7.13.10)$$

where $\mathbf{T}_f$ is the constraint tensor in this region.

The expression of the Maxwell constraint tensor is:

$$T_i = -p_i I + T_i^{el} \qquad (7.13.11)$$

where pi is the static pressure and $\mathbf{I}$ the identity tensor. The components of $\mathbf{T}^{el}_i$, using Einstein's notation, are:

$$\left(T^{el}_i\right)_{jk} = \varepsilon_i\left(E_i\right)_k\left(E_i\right)_j - \left(\frac{\varepsilon_i}{2}\right)\delta_{kj}\left(E_i\right)_m\left(E_i\right)_m, \qquad (7.13.12)$$

with εi and $(Ei)j$ the permittivity and the electric field in the jth direction, of the phase i. The expression of the surface constraint tensor is given by:

$$T_\zeta = \gamma_{i-j}\left(I - n\otimes n\right) \qquad (7.13.13)$$

According to [322]:

$$\nabla_\zeta{}^\circ T_\xi = \gamma_{i-j}\langle v_n\rangle n + grad\,\zeta\lambda_{i-j}, \qquad (7.13.14)$$

$<v_n>$ being the mean curvature and $\gamma i j$ the surface tension. As in our problem $\gamma i j$ is constant, grad ζ the surface gradient of this magnitude vanishes.

7.13.3.3 ELECTRICAL EQUATIONS

The Maxwell's equations can be simplified with some assumptions. In the bulk, it can be written that [322]:

$$\nabla\cdot D_i = 0 \qquad (7.13.15)$$

$$\nabla\times E_i = 0, \qquad (7.13.16)$$

here Ei and Di are the electric Field strength and the dielectric displacement. By using an electrical potential function V, it can be expressed:

$$E_i = -\nabla V_i \qquad (7.13.17)$$

That leads to Laplace equation $\Delta V=0$.

At the interface, the continuity of the tangential electric field, Gauss law at the surface and conservation of charge considering the rate of change of a surface element can be expressed as [332]:

$$n\times\|E_i\| = 0, \qquad (7.13.18)$$

$$\|\varepsilon_i E_i\|\cdot n = q_\zeta, \qquad (7.13.19)$$

$$\frac{dq_\zeta}{dt} + q_\zeta\left(U_\zeta\cdot n\right)\nabla_n + \nabla_\zeta J_\zeta + n\|J_i\| - \left(U_\zeta\cdot n\right)\|q_i\| = 0 \qquad (7.13.20)$$

with q_ζ the surface charge, J_ζ the surface current density, Ji the volume current density, qi the volume charge density and we notice a jump in a magnitude from a phase to another with the symbol $\| \ \|$.

7.13.3.4   THE ELECTRIC FIELD OF THE NONPERTURBED JET

In the nonperturbed case negative electrical charges are uniformly placed on the surface of the jet, which is consequently isopotential. In the liquid media this leads immediately to

$$V_1\left(r,\theta,z\right)=0 \text{ and } E_1\left(r,\theta,z\right)=-\nabla V_1=0 \qquad (7.13.21)$$

In the gas, the resolution of Poisson equation

$$\Delta V_2 = 0 \text{ with } V_2\left(a,\theta,z\right)=0 \text{ and } V_2\left(b,\theta,z\right)=V_0, \qquad (7.13.22)$$

which gives

$$V_2\left(r,\theta,z\right)=V_0\frac{\ln\left(\frac{r}{a}\right)}{\ln\left(\frac{b}{a}\right)} \qquad (7.13.23)$$

The electric field is then

$$E_2\left(r,\theta,z\right)=-\frac{V_0}{\ln\left(\frac{b}{a}\right)}\frac{i_r}{r} \qquad (7.13.24)$$

Artana et al. [322] studied the linear instability analysis of an electrified co-axial jet. In fact, the mentioned nonperturbed state is totally ideal as it never occurs because of the high instability character of the flow and the many possible sources of perturbation like, for instance, the roughness of the orifice of the injector which induces modifications on the ideal velocity and potential.

They simulated the real phenomenon, by artificially perturbing the considered solution. Hence, they perturbed the interface between the liquid and the gas and determined the modifications induced on the velocity, the electrical potential and the pressure. The perturbation, which was arbitrary, was decomposed into a Fourier series and was considered as a set of small elementary waves propagating on the surface of the jet. As they considered a linear analysis, they accepted that there is no interaction between the different modes (or waves) and so the analysis can be undertaken for each individual mode. The amplitude of some of them was diminish and for others it grew until it became large enough to give rise to droplets. Therefore, the study of the creation of the droplets came down to the one of the stability of the jet submitted to a perturbation, whatever its origin [322, 333]. Artana et al. [322] supposed that the flow was subjected to a modification at its interface. The coordinates of each point at the interface:

$$OM_\zeta = (a, \theta, z, t)$$

(7.13.25)

Became

$$OM_\zeta = (r_s(\theta, z, t), \theta, z)$$

(7.13.26)

after modification. Where:

$$r_s = a + \eta, and\, \eta$$

(7.13.27)

is the perturbation that depends on the space variables and on time t. Because they analyzed the initial stages of growth or decay, they studied directly a single mode and η was written as:

$$\eta = \eta_0 \exp\left[\left(\omega t + i(kz + n\theta)\right)\right]$$

(7.13.28)

At the interface, the normal n to the interface at any point is

$$n = \left(1, -\left(in/a\right)\eta, -ik\eta\right)$$

(7.13.29)

and the mean curvature $<\upsilon_n>$ is:

$$\left\langle \upsilon_n \right\rangle = \nabla \cdot n = \frac{1}{r_s} - \frac{1}{r_s^2}\frac{\partial^2 \eta}{\partial \theta^2} - \frac{\partial^2 \eta}{\partial z^2}$$

(7.13.30)

The perturbation of the interface led to solutions for the velocity and for the electric field, which were different from the ones obtained in the nonperturbed case. The governing equations of the fluid mechanics problem still were $\Delta\varphi_i=0$ and $\Delta v_i=0$. After that, they solved them with a moving interface (boundaries changing with time). Since they confined their investigations to the linear stability analysis, they neglected all the terms proportional to η^2 or to higher exponents.

7.13.3.5   THE PERTURBED STATE

7.13.3.6   VELOCITY FIELD

Using Slattery's formulation [331] and considering that $U_2(b, \theta, z)=0$ and that the velocity is finite (A) M, they arrived to:

$$U_1 = C_1 \nabla \left(I_n(kr)\eta \right) \text{ and } U_2 = U_0 + C_2 \nabla \left(\left(\frac{I_n(kr)}{C_3} + \frac{K_n(kr)}{C_4} \right) \eta \right) \quad (7.13.31)$$

With

$$C_1 = \frac{\omega}{kI'_n(ka)}, C_2 = \frac{\omega - iU_0 k}{k}, C_1 = I'_n(ka)(1-\lambda) \text{ and } C_4 = K'_n(ka)\left(1 - \frac{1}{\lambda}\right) \quad (7.13.32)$$

With

$$\lambda = \frac{I'_n(kb) K'_n(ka)}{K'_n(kb) I'_n(ka)} \cdot E_1 = 0 \quad (7.13.33)$$

In and *Kn* are the modified Bessel's functions of first and second kind, *I'n* and *K'n* being their derivatives with respect to the variable r.

7.13.3.7   ELECTRIC FIELD

The perturbed solution depends on the electrical state of the jet, generally. Artana and his co-workers perturbed a cylindrical column of liquid flowing through a cylindrical coaxial electrode and that, in the nonperturbed state, the electrical charges were in electrical equilibrium on the surface of the liquid. The electrical state of the jet after perturbation depends on the way the electrical charges can move when the surface is perturbed. They considered two cases [322]:

In the first one the electrical relaxation time of charges was small enough compared to a certain characteristic time of the deformation (in their case, the period of the perturbation). "Relaxation is quicker than deformation." The jet remains isopotential at any time.

In the second case, which was the opposite extreme of the first one, the charges were in a way "linked" to the fluid particles and follow the dilatation and stretching of the surface of the jet which consequently does not remain isopotential. This was the case of jet with charges on its surface supporting a perturbation with a large growth rate (very high-velocity jet). They considered these two cases separately, it being understood that the reality of the electrical state of the surface was somewhere between them.

(a) Isopotential Case

If the surface of the liquid remains in electrical equilibrium despite the motion induced by the perturbation, the boundary conditions needed for the determination of the electric field in the gas are $V_2(b)=V_0$ at the electrode and $V_2(a+\eta)=0$ at the interface.

Gauss law and the continuity of the tangential component of the electric Field at the interface immediately gave the solution in the liquid. They obtained

$$E_1 = 0 \text{ and } E_2 = C_5 \nabla \left(\ln \left(\frac{r}{a} \right) - \left(\frac{I_n(kr)}{C_6} + \frac{K_n(kr)}{C_7} \right) \eta \right) \quad (7.13.34)$$

With

$$C_5 = \frac{V_0}{\ln \left(\frac{b}{a} \right)}, C_6 = aI_n(ka)(1-\beta) \text{ and } C_7 = aK_n(ka)\left(1 - \frac{1}{\beta}\right), (7.13.35)$$

Being

$$\beta = \frac{K_n(ka)I_n(kb)}{I_n(ka)K_n(kb)} \quad (7.13.36)$$

(b) Nonisopotential Case

As it was mentioned above, in this case it is supposed that the electrical charges, uniformly distributed on the surface of the jet in the nonperturbed state, move by following the fluid particles of the surface. Now the distribution of the electrical charges on the surface of the jet is not uniform any longer. From a physical point of view, the extremely rapid motion of the jet surface inhibits the rearrangement of the charges at the surface motivated by the electric forces. This assumption is equivalent to considering a problem in which the mobilities of the charges in the bulk and the interface are null.

So in the equation of charge conservation, any contribution of the current density in the rate of change of the surface charge density can be disregarded. The contribution of the surface current density is reduced to the convection term $\nabla_\zeta(q_\zeta U_\zeta)$. However, as the linear terms are kept only, this convection term has no contribution either. The linearized equation can finally be written as

$$\frac{\partial q_\zeta}{\partial t} + q_\zeta \left(U_\zeta \cdot n \right) \nabla_n = 0 \qquad (7.13.37)$$

The integration in time of this equation then gives the following solution

$$q_\zeta (\theta, z, t) = q_0 \frac{a}{r_s}, \qquad (7.13.38)$$

where q_0 is the surface density of the electrical charges of the nonperturbed state (t=0). The boundary conditions for the solution of Laplace equation that determines the value of the electric field are deduced from the following considerations:

- The potential at any point of the liquid (especially for r=0) is finite.
- In the gas we have V_2 (b) $=V_0$.
- At the interface we use the traditional results deduced from Maxwell's equations continuity of tangential component of the electric field and discontinuity of the normal component of the displacement vector.

Considering the electrical potential functions

$$V_i = A_i \ln \left(\frac{r}{a} \right) + R_i \left(r \right) \exp \left(i \left(kz + n\theta \right) + \omega t \right) \text{ with } R_i \left(r \right) = B_{i1} I_n \left(kr \right) + B_{i2} K_n \left(kr \right) \quad (7.13.39)$$

and with the cited boundary conditions, we have

$$E_1 = C_5 \nabla \left(\frac{I_n \left(kr \right)}{C_8} \eta \right) \text{ and } E_2 = -C_5 \nabla \left(\ln \left(\frac{r}{a} \right) - \frac{K_n \left(kr \right)}{C_9} \eta \right) \qquad (7.13.40)$$

with

$$C_8 = a I_n \left(ka \right) \left(1 - \frac{\varepsilon_1}{\varepsilon_2} \frac{I_n' \left(ka \right) K_n \left(ka \right)}{I_n \left(ka \right) K_n' \left(ka \right)} \right), \qquad (7.13.41)$$

$$C_9 = a K_n \left(ka \right) \left(\frac{\varepsilon_2}{\varepsilon_1} \frac{K_n' \left(ka \right) I_n \left(ka \right)}{K_n \left(ka \right) I_n' \left(ka \right)} - 1 \right), \qquad (7.13.42)$$

ε_1 and ε_2 being the permittivities of the two media. In the expressions of the electric field in the gaseous phase Artana et al. assume that the product kb is large enough to consider the quotient $K_n(kb)/I_n(kb) \approx 0$, condition usually verified.

7.13.3.8 PRESSURE FIELD

Knowing the velocity and electric fields with the equation of conservation of linear momentum the pressure in both media can be determined. The term dU_i/dt can be written as

$$\frac{dU_i}{dt} = \frac{\partial U_i}{\partial t} + \left(U_i^\circ \nabla\right)U_i,$$ (7.13.43)

Where

$$\left(U_i^\circ \nabla\right)U_i = \left(\nabla \times U_i\right) \times U_i + \frac{1}{2}\nabla\left(U_i^2\right)$$ (7.13.44)

The motion supposed to be irrotational and the linearized Navier-Stokes equation then becomes

$$\rho_i \frac{dU_i}{dt} - \nabla^\circ \left(p_i I + T_{iel}\right)$$ (7.13.45)

The solution of Navier-Stokes equation in terms of pressure is

$$p_1 = C_{10} I_n\left(kr\right)\eta \text{ and } p_2 = C_{11} + C_{12}\left(\frac{I_n\left(kr\right)}{C_{13}} + \frac{K_n\left(kr\right)}{C_{14}}\right)\eta$$ (7.13.46)

With

$$C_{10} = -\rho_1 \frac{\omega^2}{k I_n'\left(ka\right)}, C_{11} = -\rho_2 \frac{U_0^2}{2}, C_{12} = -\rho_2 \frac{\left(\omega - iU_0 k\right)^2}{k}$$ (7.13.47)

$$C_{13} = I_n'\left(ka\right)\left(1-\lambda\right) \text{ and } C_{14} = K_n'\left(ka\right)\left(1-\frac{1}{\lambda}\right)$$ (7.13.48)

7.13.3.9 THE DISPERSION EQUATION

The dispersion equation is obtained by substitution of all terms in the momentum conservation equation at the interface. This equation can be expressed as:

$$\frac{\gamma_{i-j}}{a^2}k\left(1-n^2-\left(ak\right)^2\right) + \rho_1 \frac{I_n\left(ka\right)}{I_n'\left(ka\right)}\omega^2 - \rho_2 a\omega^2\left(\omega - ikU_0\right)^2 - \frac{\varepsilon_0 V^2}{a^3 \ln^2\left(b/a\right)}k\left(1+ak\varsigma\right) = 0$$ (7.13.49)

Considering a temporal analysis, it can be written:

$$\omega_r^2 = \frac{k}{a^2\left(1-\chi\right)\rho_2\alpha}\left[-\gamma_{i,j}\left(1-n^2-\left(ak\right)^2\right) + \varepsilon_0 \frac{V^2}{a\ln^2\left(b/a\right)}\left(1+ak\varsigma\right)\right] - \left(\frac{U_0 k}{\left(1-\chi\right)}\right)^2 \chi$$ (7.13.50)

$$\omega_i = \frac{U_0 k}{1-\chi} \tag{7.13.51}$$

where

$$\chi = \frac{\rho_1 I_n(ak)}{\rho_2 I_n'(ak)\alpha} \text{ and } \alpha = \frac{I_n(ka)}{I_n'(ka)(1-\lambda)} + \frac{K_n(ka)}{K_n'(ka)\left(1-\frac{1}{\lambda}\right)} \tag{7.13.52}$$

and where ω_i and ω_r are, respectively, the real part and the imaginary part of the growth rate u. In the equipotential case,

$$\varsigma = \frac{I_n'(ka)}{I_n(ka)(1-\beta)} + \frac{K_n'(ka)}{K_n(ka)\left(1-\frac{1}{\beta}\right)} \tag{7.13.53}$$

and in the nonequipotential one,

$$\varsigma = -\left(\frac{K_n(ka)}{K_n'(ka)} - \frac{\varepsilon_2 I_n(ka)}{\varepsilon_1 I_n'(ka)}\right)^{-1} \tag{7.13.54}$$

These equations, taken in the same conditions as the ones of the articles of, respectively, Levich [334], Taylor [335], Melcher [336], Bailey [337] or Lin and Kang [338], give the same equation that each of these authors has found.

Reznik et al. [339] introduced a numerical model which can predict velocity profile of flow at the central pipe which assume the axisymmetric Poiseuille velocity profile and the corresponding flow in the annular pipe, which were imposed according to the analogous boundary conditions:

$$u_z = \frac{2Q_1}{\pi r_{in}^2}\left(1 - \left(\frac{r}{r_{in}}\right)^2\right) \quad 0 < r < r_{in} \tag{7.13.55}$$

$$u_z = \frac{2Q_2}{\pi r_{out}^2 \omega}\left[1 - \left(\frac{r}{r_{out}}\right)^2 - \left(1 - \left(\frac{r_{in}}{r_{out}}\right)^2\right)\frac{\ln\left(\frac{r}{r_{out}}\right)}{\ln\left(\frac{r_{in}}{r_{out}}\right)}\right] \quad r_{in} < r < r_{out} \tag{7.13.56}$$

$$\omega = 1 - \left(\frac{r_{in}}{r_{out}}\right)^4 + \frac{\left(1 - \left(\frac{r_{in}}{r_{out}}\right)^2\right)^2}{\ln\left(\frac{r_{in}}{r_{out}}\right)} \tag{7.13.57}$$

where r_{in} and r_{out} are the radii of the core and shell nozzles and Q_1 and Q_2 are the corresponding volumetric flow rates. In their model, the fluid motion is governed by the Stokes equations, the electric tractions are related to the Maxwell stresses equations, and the continuity of the tangential component of the electric field and the balance of the free charge at the surfaces were considered as governing equations. Using dimentionless parameters and applying the Kutta-Merson method as a numerical method, velocity profiles were achieved [317].

In electrohydrodynamics problems, the electric force affects the movement of fluids, which changes the distribution of charges within the fluids and on the interfaces. Therefore, the electric force is coupled with surface/interface tension, viscous force, inertia force and so on. Considering that there are two interfaces in a coaxial electrospray problem, the kinematic, dynamic and electrical boundary conditions are required for each interface. The kinematic interface condition for each interface can be written as:

$$\frac{\partial F}{\partial t} + u\nabla F = 0 \qquad (7.13.58)$$

where F is the interface function. The dynamic boundary condition for each interface can be expressed as:

$$\left\| T^m + T^e \right\| n = \gamma \nabla n + (\delta - nn)\nabla\gamma \qquad (7.13.59)$$

where $\|$ and $\|$ indicated the jump of corresponding quantity across the interface. T^m, T^e, n, γ and δ represent the hydrodynamic stress tensor, electrical Maxwell tensor, normal unit vector, surface tension and identity matrix, respectively. The electrical boundary conditions for each interface can be given by:

$$n\|D\| = q_s \qquad (7.13.60)$$

$$n\|E\| = 0 \qquad (7.13.61)$$

Equation (7.13.60) expresses the Gauss law in which the surface charge density q_s, satisfies the surface charge conservation law. Equation (7.13.61) represents the continuity of the tangential component of the electric field. For viscous fluids, the tangential component of the velocity should be continuous on the interface:

$$C_k = 1 + 2E_k \qquad (7.13.62)$$

In addition to the above boundary conditions, there are other solution-dependent boundary conditions such as the finiteness of velocity and electric intensity field at the symmetric axis. These governing equations and boundary conditions are applicable in most cases no matter whether the fluids are liquids or gasses. It must be pointed out that solving the above problem is very difficult because of the following challenges:

The two interfaces of the problem are unknown, the length scales of the capillary needles and the jets are very disparate, and the breakup of jets is time dependent. Therefore, modeling a coaxial electrospray process is a complicated procedure involving a larger number of unknowns and parameters. Further simplifications and assumptions are necessary in order to study the coaxial electrified jet under either the outer-driving or the inner-driving flow conditions.

Lopez-Herrera et al. [315] introduced two additional governing equations inorder to explore the influence on the spraying process of the properties of the liquids: i.e., the electrical conductivity K, dielectric constant β, interfacial tension of the liquid couple γ, viscosity μ, etc.

The first equation was the normal balance across the outer interfaces:

$$\frac{\gamma_{01}}{R+\delta} \approx P_1 - P_0 + \frac{1}{2}\varepsilon_0\left[E_{n0}^2 - \beta E_{n1}^2\right]$$

(7.13.63)

And the second one was the normal balance across the inner interfaces:

$$\frac{\gamma_{12}}{R} \approx P_2 - P_1 + \frac{1}{2}\varepsilon_0\beta E_{n1}^2$$

(7.13.64)

Where δ is the thickness of the layer oil, R is the radius of the inner interface, p is the pressure, E_n is the normal component of the electric field, and subscripts 0, 1, and 2 refer to air, oil and driving liquid, respectively. In Eqs. (7.13.63) and (7.13.64), it was assumed that E_{n2} is much smaller than E_{n0} and E_{n1}. By adding these two equations and neglecting δ as compared to R, one arrives at [315]:

$$\frac{\gamma_{12} + \gamma_{01}}{R} \approx P_2 - P_0 + \frac{1}{2}\varepsilon_0 E_{n0}^2$$

(7.13.65)

This is different from a single-axial electrospray process where analytical and numerical models can be obtained and the key process parameters can be analyzed systemically.

7.13.4   TWO-PHASE FLOW PATTERNS MODEL IN COAXIAL ELECTROSPINNING

By comparison, electrospinning composite nanofibers with distinct layers give rise to extra problems of fluid dynamics. Among them, flow patterns of two-phase flows within capillary tubes are often studied first of all as a basis for advanced research. Figure 7.34 illustrates three flow patterns—bubbly, slug, and annular—generally observed in liquid–liquid (e.g., water–oil) capillary pipe flows or gas–liquid experiments conducted in reduced gravity environments (or conditions of similarity) to understand gas–liquid behavior in outer space. Sometimes two transitional flow patterns are particularly labeled: bubbly slug and slug-annular [340].

Flow pattern maps have been drawn from experiments, but a widely accepted theory is not accessible yet. A flow pattern model based on the Weber number, however, is regarded as relatively effective in distinguishing annular flows from slug flows. For gas-liquid two-phase flows, slug flows change to slug-annular transitional flows at unity of the gas Weber number and annular flows occur if the gas Weber number is greater than 20 [340, 341].

Hu et al. [340] presented a mathematic physical model to study the flow patterns of two-phase flows in coaxially electrospinning composite nanofibers, where shear-thinning non-Newtonian properties and strong electric fields were two characteristics that are of particular concern. They applied numerical simulations using commercial computational fluid dynamics (CFD) software as a preprocessor. By means of these, they tried to study the mechanism of coaxial electrospinning and provide useful and least heuristic information for the development of advanced composite nanofibers.

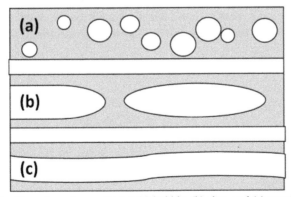

FIGURE 7.34   Two-phase flow patterns: (a) bubbly, (b) slug, and (c) annular.

According to this two phase flow pattern, Fig. 7.35 illustrates a two-tube system ready to coaxially electrospin: The regions I and II are, respectively, filled with core and shell liquids. A coaxial two-phase flow may form if an electric field is activated in the x direction. Only steady flows are concerned to avoid theoretical obstacles at the start up of electrospinning, for instance, strong singularity at the outer tube exit [340].

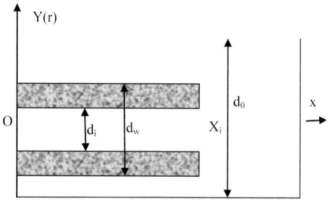

FIGURE 7.35 Shematic arrangement of the two-tube system ready to coaxially electrospin in both rectangular coordinates (x, y, z) and the cylindrical polar coordinates (r, θ, x).

7.13.4.1 TWO PHASE PATTERN MODEL FOR LAMINAR FLOW

For incompressible liquids of constant density flowing in an electric field, the continuum and momentum equations can read)neglecting gravity .(where V, ρ, and p are velocity vector, liquid density, and pressure, in the order given; $\overleftrightarrow{T}$ and $\overleftrightarrow{\sigma}^M$ are stress tensors due to viscosity and the electric field, respectively [342].

$$\nabla \cdot \vec{V} = 0 \tag{7.13.66}$$

$$\rho \frac{D\vec{V}}{Dt} = -\nabla p + \nabla \cdot \overleftrightarrow{T} + \nabla \cdot \overleftrightarrow{\sigma}^M \tag{7.13.67}$$

Since most liquids suitable for electrospinning nanofibers are shear-thinning non-Newtonian fluids [343], the viscous stress tensor T_J in Eq. (7.13.68) is modeled according to the power law

$$\ddot{T} = \alpha \dot{\gamma}^{n-1} \ddot{D} \qquad (7.13.68)$$

in which the deformation rate tensor $D^{\leftrightarrow}$ has components:

$$D_{ij} = \frac{\partial u_i}{\partial x_j} + \frac{\partial u_j}{\partial x_i} \qquad (7.13.69)$$

the shear rate

$$\dot{\gamma} = \sqrt{\ddot{D} : \ddot{D}} \qquad (7.13.70)$$

α is the measure of the average viscosity of a liquid; n is the measure of the deviation of a liquid from Newtonian fluids.

The electric stress tensor $\sigma^{\leftrightarrow M}$ in Eq. (7.13.71 (is related to the density of free charges ρ^e, the gradient of an isotropic electric contribution, and dielectric permeability ε in the electric field E) ε_0 taken as absolute permeability (.

$$\nabla \cdot \ddot{\sigma}^M = \rho^e \vec{E} + \nabla \left[\frac{1}{2} \varepsilon_0 \rho \left(\frac{\partial \varepsilon}{\partial \rho} \right)_T \vec{E} \cdot \vec{E} \right] - \frac{1}{2} \varepsilon_0 \vec{E} \cdot \vec{E} \nabla \varepsilon \qquad (7.13.71)$$

In Eq. (7.13.71(, the second term at the right-hand side,

$$\nabla \left[\frac{1}{2} \varepsilon_0 \rho \left(\frac{\partial \varepsilon}{\partial \rho} \right)_T \vec{E} \cdot \vec{E} \right] \qquad (7.13.72)$$

vanishes in incompressible flows with constant density. To simplify the last term,

$$-\frac{1}{2} \varepsilon_0 \vec{E} \cdot \vec{E} \nabla \varepsilon \qquad (7.13.73)$$

where, permeability is assumed to be homogenous in each liquid but its gradient exists on the core-shell interface. By introducing a phase function c to indicate the two liquids, Δ_ε is related to $\varepsilon \Delta c$. Because the permeability gradient usually resists liquids to flow, $-\Delta_\varepsilon = \varepsilon \Delta c$ is reasonable for $c=1$ as the core and $c=0$ as the shell. Hence, a more feasible form of the last term in Eq. (7.13.71) becomes

$$-\frac{1}{2} \varepsilon_0 \vec{E} \cdot \vec{E} \nabla \varepsilon = \frac{1}{2} \varepsilon \varepsilon_0 \vec{E} \cdot \vec{E} \nabla c \qquad (7.13.74)$$

The phase indicator c, an unknown function tracking interface movement, is considered to satisfy the convective equation

$$\frac{\partial c}{\partial t} + \vec{V} \cdot \nabla c = 0 \tag{7.13.75}$$

Solved by the volume of fluid method, the phase indicator c acts as a fractional volume function in Eq. (7.13.75) the effect of surface tension is addressed by adding a source term into the momentum.

$$\vec{F} = \frac{2\bar{\rho}}{\rho_{core} + \rho_{shell}} \sigma \kappa \nabla c \tag{7.13.76}$$

Where σ is the surface tension coefficient, k is the interface curvature, and

$$\bar{\rho} = c\rho_{core} + (1-c)\rho_{shell} = c\rho_{core} + (1-c)\rho_{shell} \tag{7.13.77}$$

Can represent the liquid density of the core – ρ_{core}, the shell – ρ_{shell}, or the mixture of the two, depending on the c value. With these treatments above, the final momentum equation for the laminar two-phase flow in the process of coaxially electrospinning composite nanofibers yields [340]:

$$\bar{\rho}\frac{D\vec{V}}{Dt} = -\nabla p + \nabla \cdot \left(\alpha\dot{\gamma}^{n-1}\vec{D}\right) + \rho^e\vec{E} + \frac{1}{2}\varepsilon\varepsilon_0\vec{E}\cdot\vec{E}\nabla c + \frac{2\bar{\rho}}{\rho_{core} + \rho_{shell}}\sigma\kappa\nabla c \tag{7.13.78}$$

In Eq. (7.13.78) $\rho^e E$ acts as a body force to drive liquids to move, whereas all other forces, whereas all other forces, $\nabla \cdot \left(\alpha\dot{\gamma}^{n-1}\vec{D}\right)$ (caused by viscosity), $\frac{1}{2}\varepsilon\varepsilon_0\vec{E}\cdot\vec{E}\nabla c$ (induced by electric charges on interfaces), and $\frac{2\bar{\rho}}{(\rho_{core} + \rho_{shell})}\sigma\kappa\nabla c$ (due to surface tension), resist liquids to flow.

For study limited to steady flows, time derivatives in Eqs. (7.13.77) and (7.13.78) are necessary for numerical computational techniques instead of physical description. Zero velocity and hydrostatic pressure are assigned as initial conditions. Boundary conditions are set at tube walls, inlet at $x=0$ and outlet at $x=x_0$.

Walls: $\vec{V} = 0$,

$$x = 0 : \partial\vec{V}/\partial x = 0, \partial p/\partial x = 0, c = 1 \text{ (core) and 0 (shell)} \tag{7.13.79}$$

$$x = x_0 : \partial\vec{V}/\partial x = 0, \partial p/\partial x = 0, \partial c/\partial x = 0$$

Thus a two-phase laminar flow problem is posed by the governing equations (the Eq. (7.13.72) and (7.13.78)), the interface convective equation (the Eq. (7.13.77)), and boundary conditions (the Eq. (7.13.79)).

7.13.5 SCALING LAW

Since the cone and the jet are in disparate scales, they are usually studied separately. As for the cone, Marin et al. [344] studied the coaxial electrospray within a bath containing a dielectric liquid and observed that a sharp tip in the inner dielectric meniscus would be formed without mass emission. They presented an analytical model of the flow in the inner and the outer menisci based on different simplifying hypotheses. The fluid dynamic equations were simplified in a low Reynolds number limit and under the assumption that the electrical effects inside the liquid bulk were negligible. The electrical equations were also reduced into the Laplace equation of the electric field. After the boundary conditions on the two interfaces were applied and the assumptions of self-similarity and very thin conductive layer were made, the velocity, the pressure, fields and the electrical shear stress at the outer surface were finally calculated. As for the jet and the resulting droplets, Lopez-Herrera et al., derived the scaling laws of the diameter of the coaxial electrified jet and the current transported throughout the jet by experimenting with different liquids, such as water, sunflower oil, ethylene-glycol and Somos [315]. The dimensionless parameters were defined based on the reference characteristic values of the flow rate Q_0, the current I_0, and the diameter d_0, as given by:

$$Q_0 = \frac{r_{eff} s_0}{\rho \sigma}, I_0 = \left(\frac{r_{eff}^2 s_0}{\rho}\right)^2, d_0 = \left(\frac{Q_0 s_0}{\sigma}\right)^{1/3} \qquad (7.13.80)$$

where y_{eff} denotes the effective value of the surface tension. The results indicated that the current I/I_0 on the driving flow rate Q/Q_0 closely followed a power law of $(Q/Q_0)^{1/2}$, similar to that in single-axial electrospinning. It was also found that the mean diameter of the droplets resulted from the breakup of the coaxial jets scaled linearly with both inner and outer flow rates in the case of outer driving; whereas that diameter was closely dependent of the ratio of inner and outer flow rates in the case of inner driving. Marín et al. obtained the diameter of the coaxial jets d as a function of the flow rate Q [344]. They found that the experimental results fitted in the $Q^{1/2}$ law as below:

$$\frac{d}{d_0} = 1.25 \left(Q/Q_0\right)^{1/2} \qquad (7.13.81)$$

Mei et al. found the particle encapsulation conditions relevant with the flow rates and the material properties in the case of the inner driving flow [327]. Let r* be the charge relaxation length and R* be the inertial length:

$$r^* = \left(Q \varepsilon \varepsilon_0 / \sigma \right)^{1/3}, R^* = \left(\frac{\rho Q^2}{\gamma} \right)^{1/3} \tag{7.13.82}$$

The particle encapsulation conditions were therefore expressed as:

$$\frac{r^*}{r_1^*} < 500, \frac{R_0^*}{R_1^*} < 0.015 \tag{7.13.83}$$

where the subscripts O and I indicate the outer liquid and the inner liquid, respectively. Furthermore, the flow rates of the inner needle and outer needle may affect the range of the stable cone–jet, and thus affect the jet size and the particle size. Chen et al. found that the working range for the stable cone–jet could be expanded by increasing the inner liquid flow rate and by decreasing the outer liquid flow rate in the case of outer driving [345]. It has been shown previously that the particle size decreases as the applied voltage increases in a stable cone–jet mode. Similar reduction of the particle size can also be achieved by reducing the flow rate, which can be explained by easier overtaking of the electrical force over the hydrodynamic forces in reduction of flowing materials [316]. In practice, stable cone–jet mode should be adjusted at the higher applied voltage and lower flow rates in order to get the smaller particle size.

7.13.6　INSTABILITY ANALYSIS

Coaxial electrospraying is a new effective technique to form micro/nano capsules that are monodisperse and controllable. It has many applications in the drug industry, food additives, paper manufacture, painting and coating processes. In experiments, when two immiscible liquids are emitted from two homocentric capillary tubes, respectively, under appropriate flow and electric field conditions, a stationary Taylor cone is formed. At the tip of the Taylor cone, a steady axisymmetric coaxial jet with nearly uniform diameter arises. The coaxial jet is intrinsically unstable, and breaks up into micro compound droplets at some distance downstream. This is called the cone-jet coaxial electrospraying mode [316].

Recently, many experiments have been carried out to investigate the mechanism of coaxial electrospraying and the scaling laws between important quantities, for example, Loscertales et al. [316], Lopez-Herrera et al. [315], Chen et al. [312] and Marin et al. [344]. Theoretical and numerical work by Li, Yin and Yin [346, 347] has analyzed the linear instability of an inviscid coaxial jet having a conducting annular liquid under a radial electric field, where both the equipotential and nonequipotential cases were studied; Higuera [348] performed a brief but valuable numerical simulation of a stationary electrified coaxial jet in the framework of the leaky dielectric model and quasi-unidirectional approximation.

The breakup process of a liquid jet under an electric field is closely related to the growth and propagation of unstable disturbance waves at the interface between fluids. Therefore, instability analysis is useful in predicting the breakup modes of liquid jets, and also in predicting the intact jet length and droplet size [349–351]. In the instability analysis of electrically charged single liquid jets, two kinds of electric field, that is, radial and axial, are usually encountered. Also, the electrical properties of liquids may be of a perfect conductor, perfect dielectric or leaky dielectric (dielectric with finite conductivity). Thus, many cases can arise due to the variation of the electric field imposed on the jet and electrical properties of the liquid. For instance, Turnbull [352], analyzed the temporal linear stability of conducting and insulating liquid jets in the presence of both radial and axial electric fields; Saville [225] and Mestel [353] studied the stability of a charged leaky dielectric liquid jet under a tangential electric field, paying particular attention to the effect of liquid viscosity; Lopez-Herrera, Riesco-Chueca and Ganan-Calvo [148, 311, 313] researched the stability of a viscous leaky dielectric liquid jet under a radial electric field, taking into account the effect of ambient air flow. Garcıa et al. and Gonzalez, Garcia and Castellanos [354] investigated the effect of AC radial electric fields on the instability of liquid jets. Huebner and Chu [355] and Son and Ohba [356], respectively, explored the jet instability under axisymmetric and nonaxisymmetric disturbances. More recently, nonlinear effects have been specially studied in order to explain the experimental phenomena, such as the formation of satellite droplets (e.g., Lopez-Herrera [313, 315], Gañán-Calvo and Perez-Saborid [313]; Lopez-Herrera and Gañán-Calvo [357]; Zakaria [358]; Elhefnawy, Agoor and Elcoot [359]; Elhefnawy, Moatimid and Elcoot [360]; Moatimid [361]).

In most of the published reports, the liquid jet is usually assumed to be either perfectly conducting or perfectly dielectric. In practice, most liquids used in experiments are leaky dielectric. Unlike prefect conductors or dielectrics, for leaky dielectrics free charge may occur in the fluid bulk and therefore electromechanical coupling occurs not only at interface but probably also in the bulk. Furthermore, electric stresses on an interface are no longer perpendicular to it, because free charge accumulated on the interface may modify the electric field. From this point of view, electric stresses tangential to the interface are inevitable, and must be balanced by viscous stresses. For perfect conductors or perfect dielectrics, the tangential component of electric stress vanishes, because free charge is reset instantaneously on the interface to keep the interface equipotential for the former, and it is absent for the latter [225].

In Li et al. studies [346, 347], the assumption that the inner and outer liquids are inviscid was made in the instability analysis of the electrified coaxial jet. However, with such an assumption the tangential component of electric stress on the interface cannot be balanced due to the absence of liquid viscosity. Moreover, liquid viscosity may play an important role in the jet instability, because the diameter of the coaxial jet generated in electrospraying experiments is very small, usually of the order of several tens of micrometers. Therefore, from this physical point of view, it is incorrect to neglect the viscosity of liquid.

To gain further insight into the outer-driving coaxial electrospraying and to study the axisymmetric instability behavior of the charged coaxial jet under a radial electric field, Li et al. [346, 347] proposed a viscous leaky dielectric model, based on the theory of the Taylor– Melcher leaky dielectric model [225, 336].

Accordingly the outer liquid was assumed to be a leaky dielectric, acting as the driving liquid. Furthermore, the conductivity of the outer liquid was assumed to be large enough so that free charge is relaxed to the interface instantaneously. The inner liquid in the outer-driving coaxial electrospraying, for a generic case, should be considered as a leaky dielectric too. Though the conductivity of the inner fluid is much smaller than that of the outer liquid, it is sufficient to have charge relaxed and transferred to the interface, where the distribution of charge density is determined by the surface charge conservation equation. In such a leaky dielectric case, both the conductivity ratio of inner to outer liquid and the electrical relaxation time of the inner liquid are taken into account. The initial steady state may be solved following the method provided by Higuera [348]. In the outer-driving electrospraying experiments the electrical conductivity of the inner liquid is at least two or three orders of magnitude smaller than that of the outer liquid [315]. Also, in the numerical simulation of Higuera [348], it was found that the charge density at the inner interface is about two orders smaller than that at the outer interface. Therefore, the conductivity of the inner liquid is negligible, and free charge can be supposed to lie approximately only on the outer air–liquid interface in theoretical studies. In Li et al. [346, 347] model, the inner liquid was approximated as a perfect dielectric and there was no free charge at the inner liquid–liquid interface. Such an approximation was deduced from the generic leaky dielectric model, assuming that the outer-to-inner conductivity ratio approaches infinite.

It is well known that the breakup of liquid jets is closely associated with the jet instability [362–364]. Therefore, the hydrodynamic instability theory can be used for coaxial electrospray analysis and has successfully predicted the experimental observations [365, 366]. The instability theory deals with the mathematical analysis of the response of disturbances with small amplitudes superposed on a laminar basic flow. If the flow returns to its original laminar state, it is recognized as stable. However, if the disturbance grows and the flow changes into a different state, it is recognized as unstable. When analyzing the instability problems, the governing equations and the boundary conditions described above are used and the classical method of expansion of normal modes is usually implemented. This method analyzes the development of perturbations in space only, in time only, or in both space and time. The analysis results may provide theoretical insight and practical guidance for coaxial electrospray process control.

For instability analysis in a cylindrical coordinate (z, r, θ), the arbitrary and independent perturbations are typically decomposed into Fourier series like exp $\omega t + i$ ($kz + n\theta$), where θ, k and n stand for the frequency, the axial wave number and the azimuthal wave number, respectively. In the case of coaxial electrospray, a local temporal method is used for instability analysis.

This method assumes a real axial wave number and pursues a complex frequency since its linear dispersion relation is relatively easy to solve. To the best of the authors' knowledge, this is the most commonly used method so far for instability analysis of coaxial electrospray. Other methods are waiting for development in the future. Li et al. studied the instability of an inviscid coaxial jet under an axial electric field [366]. They also studied the axisymmetric and nonaxisymmetric instabilities of a viscous coaxial jet under a radial electric field [324]. These studies solved the governing equations and boundary conditions based on a number of assumptions and simplifications, such as: the inner and the outer liquids were assumed to be perfect conductors, perfect dielectrics or leaky dielectrics; the free charges were relaxed to the interface instantaneously, and the effects of gravitational acceleration and temperature were ignored. Figure 7.2 sketches a simplified physical model for coaxial electrospray. It consists of a cylindrical inner liquid (1) of radius R_1, an annular outer liquid (2) of outer radius R_2, and Fig. 7.36 an ambient gas (3) that is stationary air in an unperturbed state. The basic flows should be assumed uniform or with specific shapes. Si et al. used the uniform velocities U_i with i = 1, 2, 3 for the inner liquid, the outer liquid and the ambient gas (in this case $U_3 = 0$) and the uniform axial electric field E_0 [367]. The corresponding dispersion relations were derived and written in an explicit analytical form, and the Eigen values were computed by numerical methods. In general, the dispersion relation could be expressed in the form of:

$$f(\omega, k, n, U_i, E_0, K, ...) = 0 \qquad (7.13.84)$$

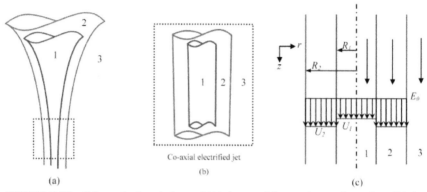

(a) Co-axial electrified jet (b) (c)

FIGURE 7.36 Schematic descriptions of (A) the coaxial cone–jet configuration; (B) the coaxial electrified jet; (C) the simplified theoretical model. 1, the inner liquid; 2, the outer liquid; 3, the ambient gas.

The involved dimensionless parameters include: dimensionless wave number $\alpha = kR_2$, dimensionless frequency $\beta = \omega R_2/U_2$, Weber number $We = \rho_2 U_2^2 R_2/\gamma_2$ dimensionless electrostatic force $E = \varepsilon E_0^2/\rho_2 U_2$, density ratios $S = \rho_1/\rho_2$ and $Q = \rho_3/\rho_2$, velocity ratio $U = U_1/U_2$, diameter ratio $R = R_1/R_2$, electric permittivity ratios $\varepsilon_{1p} = \varepsilon_1/\varepsilon_3$ and $\varepsilon_{2p} = \varepsilon_2/\varepsilon_3$, conductivity ratio $K = K_1/K_2$, and interfacial tension coefficient ratio $\gamma = \gamma_1/\gamma_2$.

The instability analysis yielded the following three unstable modes: paravaricose mode, parasinuous mode and transitional mode. The paravaricose mode occurs when the phase difference of initial perturbations at the inner and the outer interfaces was about 180°. The parasinuous mode occurs at a phase difference of approximately 0°. The transitional mode occurs when the initial perturbation is changed from in phase to out of phase. In particular, the maximal growth rate of dimensionless frequency β_{rmax} dominates the jet breakup because the perturbation for β_{rmax} grows most quickly (i.e., the perturbation grows exponentially in the dimensional form of exp $\{\omega_r t\}$). The corresponding axial wave number plays an important role in fabricating MPs and NPs because the wave number is closely associated with the wavelength of perturbations (i.e., $\alpha = 2\pi R_2/\lambda$, where λ stands for the wavelength). The larger max is, the smaller the size of resulting MPs and NPs becomes. As a result, allows us to study the effects of the electric field, the electrical conductivity, the electrical permittivity and the other important hydrodynamic parameters on the instability of the coaxial jet. It also allows us to predict the different flow modes and the corresponding transitions.

7.13.7 INSTABILITY MODEL FOR A VISCOUS COFLOWING JET

Consider an infinitely long coflowing jet with two immiscible liquids surrounded by the ambient air, as sketched in Fig. 7.37. The inner liquid cylinder has a radius R_1, and the outer liquid annulus has an exterior radius R_2. An earthed annular electrode of radius R_3 is positioned surrounding the jet, and a voltage V_0 is imposed on the jet surface. The electrical property of the outer liquid is assumed to be leaky dielectric; the inner liquid and air are perfect dielectrics. A basic radial electric field of magnitude $-V_0/[r \ln (R_2/R_3)]$ is thus formed in the air. Free charge is assumed to be relaxed on the interface between the air and outer liquid instantaneously, owing to the conductivity of the outer liquid [346–347]. The density of surface charge on the unperturbed air–liquid interface is $-\varepsilon_3 V_0/[R_2 \ln (R_2/R_3)]$.

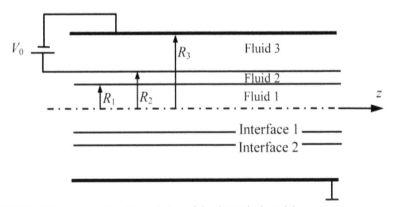

FIGURE 7.37   Schematic of description of the theoretical model.

The liquids and air are considered to be incompressible and Newtonian and the flow assumes to be axisymmetric. Effects of gravitational acceleration, magnetic field and temperature are ignored [346, 347]. The governing equations of the flow are:

$$\nabla \cdot u_i = 0, \, i = 1, 2, 3, \tag{7.13.85}$$

$$\rho_i \left(\frac{\partial u_i}{\partial t} + u_i \cdot \nabla u_i \right) = -\nabla p_i + \mu_i \nabla^2 u_i, \, i = 1, 2, 3, \tag{7.13.86}$$

where **u**, ρ, p and μ are the velocity, density, pressure and dynamic viscosity, respectively. The subscripts 1, 2 and 3 denote the inner liquid, the outer liquid and the ambient air, respectively.

When the jet is perturbed by an arbitrary disturbance, both the inner liquid–liquid interface and the outer air–liquid interface depart from their original equilibrium positions. For infinitesimal disturbances, their new positions can be expressed as $r = R_j + \eta_j$, $j = 1, 2$, where η_j is the displacement of the interface from R_j, and the subscripts 1 and 2 denote the inner and outer interfaces, respectively [346, 347]. Considering only the axisymmetric instability, we have:

$$\eta_j = \eta_j(z, t) \tag{7.13.87}$$

The boundary conditions include the no-slip condition at the electrode, that is,

$$u_3 = 0 \text{ at } r = R_3; \tag{7.13.88}$$

the continuity of the velocity at the inner and outer interfaces, that is,

$$u_2 = u_3 \text{ at } r = R_2 + \eta_2; \tag{7.13.89}$$

$$u_1 = u_2 \text{ at } r = R_1 + \eta_1; \tag{7.13.90}$$

the finiteness of the velocity at the symmetric axis, that is,

$$u_1 < \infty \text{ at } r = 0; \tag{7.13.91}$$

the kinematic boundary conditions at the interfaces, that is,

$$u_{1,2} = \left(\frac{\partial}{\partial t} + u_{1,2} \cdot \nabla \right) \eta_1 \text{ at } r = R_1 + \eta_1, \tag{7.13.92}$$

$$u_{2,3} = \left(\frac{\partial}{\partial t} + u_{2,3} \cdot \nabla \right) \eta_2 \text{ at } r = R_2 + \eta_2, \tag{7.13.93}$$

where u is the radial velocity component; and the dynamic boundary conditions at the interfaces, that is,

$$(T_2 - T_1) \cdot n_1 - \gamma_1 (\nabla \cdot n_1) n_1 = 0 \text{ at } r = R_1 + \eta_1, \tag{7.13.94}$$

$$\left(T_3 - T_2\right) \cdot n_2 - \gamma_2 \left(\nabla \cdot n_2\right) n_2 = 0 \text{ at } r = R_2 + \eta_2, \qquad (7.13.95)$$

where T is the stress tensor, γ is the surface tension, n is the normal unit vector and $\nabla \cdot n$ is the curvature. For the axisymmetric case,

$$n_j = \frac{\left(1, -\eta_{jz}\right)}{\sqrt{1+\eta_{jz}^2}} \text{ and } \nabla \cdot n_j = \frac{1}{\sqrt{1+\eta_{jz}^2}} \left(\frac{1}{R_j + \eta_j} - \frac{\eta_{jzz}}{1+\eta_{jz}^2}\right) \quad j = 1, 2, \qquad (7.13.96)$$

where η_z and η_{zz} are the first- and second-order partial derivatives of η with respect to z, respectively. In the presence of an electric field, the stress tensor T includes not only the hydrodynamic stress tensor but also the electrical Maxwell tensor, that is,

$$T = T^h + T^e, \text{ with } T^h = -p\delta + \mu\left[\nabla u + \left(\nabla u\right)^T\right], \qquad (7.13.97)$$

where δ is the identity matrix and E is the electric field intensity.

The governing equations and boundary conditions related to the electric field are needed to close the problem. As free charge is absent in the bulk, the Maxwell equations in the liquids and air reduce to [346, 347]:

$$\nabla \cdot E_i = 0 \text{ and } \nabla \times E_i = 0, \, i = 1, 2, 3, \qquad (7.13.98)$$

Introduce an electrical potential function ψi, satisfying the Laplace equation:

$$\nabla^2 \psi_i = 0, \, i = 1, 2, 3, \qquad (7.13.99)$$

And the electric field intensity $Ei = -\nabla \psi i$.

The electrical boundary conditions are:

(a) zero electrical potential at the annular electrode, that is,

$$\psi_3 = 0 \text{ at } r = R_3; \qquad (7.13.100)$$

(b) The finiteness of the electric field at the symmetric axis, that is,

$$E_1 = 0 \text{ at } r = 0; \qquad (7.13.101)$$

(c) Continuity of the tangential component of the electric field at the inner and outer interfaces, that is,

$$n_j \times \left[E\right] = 0 \text{ at } r = R_j + \eta_j, \, j = 1, 2, \qquad (7.13.102)$$

where the symbol [,] indicates the jump of the corresponding quantity across the interface; (d) continuity of the normal component of the electric displacement at the inner interface, that is,

$$n_1 \cdot \left(\varepsilon_2 E_2 - \varepsilon_1 E_1 \right) = 0 \text{ at } r = R_1 + \eta_1 ; \qquad (7.13.103)$$

where the surface charge density qs satisfies the surface charge conservation equation

$$\frac{\partial q_s}{\partial t} + u \cdot \nabla q_s - q_s n \cdot (n \cdot \nabla) u + [\sigma E] \cdot n = 0 \qquad (7.13.104)$$

The four terms on the left-hand side of Eq. (7.13.104) represent the contributions of charge accumulation, surface convection, surface dilation and bulk conduction, respectively [225].

Before the instability analysis, the basic velocity profile of the jet in the unperturbed state should be obtained. As the jet is perfectly cylindrical and the flow is axisymmetric, the basic velocity field is unidirectional, that is, $u=W(r)$ ez, where W is the axial velocity component and ez is the unit vector in the axial direction. Therefore, the momentum Eq. (7.13.86) reduces to

$$\left(\frac{d^2 W_i}{dr^2} + \frac{1}{r} \frac{dW_i}{dr} \right) = -\frac{G_i}{\mu_i}, i = 1, 2, 3, \qquad (7.13.105)$$

where $Gi = -\partial pi/\partial z$ is the negative of the stream wise pressure gradient [351]. According to the balance of forces on the interfaces, the pressure gradients in the liquids and air should be equal, that is, $Gi =G$. Integrating (the Eq. (7.13.105)) and using the continuity conditions of the velocity and shear force on the interfaces, the solutions of Wi are obtained. Choosing μ_2, R_2, and U_2 (the velocity of the jet at the outer interface, $=G (R^2_3 - R^2_2)/ (4\mu 3)$) as the scales of dynamic viscosity, length and velocity, respectively, the solutions Wi in the following dimensionless form become:

$$W_3 = \frac{b^2 - r^2}{b^2 - 1}, \qquad (7.13.106)$$

$$W_2 = 1 + \mu_{r3} \frac{1 - r^2}{b^2 - 1}, \qquad (7.13.107)$$

$$W_1 = 1 + \mu_{r3} \frac{1 - a^2}{b^2 - 1} + \frac{\mu_{r3}}{\mu_{r1}} \frac{a^2 - r^2}{b^2 - 1}, \qquad (7.13.108)$$

where two radius ratios are $a =R_1/R_2$, $b=R_3/R_2$, and two viscosity ratios are μr_1 $=\mu_1/\mu_2$, $\mu r_3 =\mu_3/\mu_2$. Note that the logarithm function ln r included in general solution of Eq. (7.13.105) by Chen [351], vanishes owing to the absence of the gravitational acceleration. It can be seen from Eqs. (7.13.106)–(7.13.108) that the basic velocity profile is closely associated with the relative viscosity of the inner liquid and outer liquid and that of air and the outer liquid. First, suppose the ambient air is almost inviscid, that is, $\mu r_3 \leq 1$. According to Eq. (7.13.107) the basic axial velocity of the outer liquid is nearly uniform, that is, $W_2 \approx 1$. For the inner liquid, there are two cases: if its viscosity is of the order of the air viscosity (i.e., $\mu r_1 \sim O(\mu r_3)$), such as a dense gas (without lose of generality, the inner liquid can be gas), it can be seen from Eq. (2.19c) that the corresponding basic velocity profile is still parabolic; conversely, if its viscosity is much larger than that of the air (i.e., $\mu r_1 \geq \mu r_3$), such as various polymers and oils, the basic velocity profile is also approximately uniform with the same magnitude as the outer liquid. Therefore, in general, according to the magnitude of the inner liquid viscosity, two cases are involved [346].

Case I: $\mu r_1 \sim O(\mu r_3)$, the basic axial velocity profile is

$$W_1 = 1 + \mu_{r3} \frac{1-a^2}{b^2-1} + \frac{\mu_{r3}}{\mu_{r1}} \frac{a^2-r^2}{b^2-1}, \qquad (7.13.108)$$

where the relative velocity ratio $A = U_0/U_2 - 1$ (U_0 is the velocity of the jet at the symmetric axis $r = 0$, $\approx 1 + (\mu r_3/\mu r_1)(a^2/(b^2-1))$). Case II: $\mu r_1 \geq \mu r_3$, the basic velocity profile is

$$W_3 = 1 + \frac{1-r^2}{b^2-1}, W_2 \cong 1, W_1 \cong 1 \qquad (7.13.109)$$

Figure 7.38 shows two typical basic velocity profiles; the axial velocity profile of the inner liquid is apparently parabolic for $\mu r_1 = 0.018$, corresponding to case I, and appears to be uniform for $\mu r_1 = 43$, corresponding to case II. In the following instability analysis we assume that the outer liquid is viscous and the air is inviscid in both cases and that the inner liquid is inviscid in case I and is viscous in case II. The advantage of this is that an analytical dispersion relation can be obtained. Moreover, the continuity of tangential force seems to be satisfied inherently for the simplified velocity profiles in both cases. Numerical studies of the instability of a viscous coaxial jet with leaky dielectric liquids in a radial electric field have to our knowledge not be reported before [346].

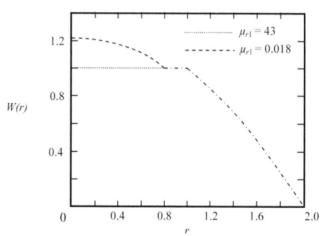

FIGURE 7.38 The influence of the relative viscosity of the inner liquid on the basic velocity profile for $\mu_r s=0.018$, $a=0.8$ and $b=2$.

If an axisymmetric jet of a viscous liquid is subjected to an axial electric field, its basic axial velocity profile is essentially parabolic, owing to the action of electrical shear stress [353]. However, as is well known, if both the viscosity and the nonuniform velocity profile are considered, the dispersion relation in an analytical form is beyond reach. In such a case either the velocity profile is assumed to be uniform [225, 315], or the viscosity of the liquid is assumed to be low or high [353], or the wavelength of the disturbance is assumed to be long [352]. But in this paper a radial electric field not an axial electric field is studied. In the presence of a radial electric field the electrical shear stress in the unperturbed state is absent as discussed above, and so the basic velocity profile for viscous liquids can reasonably be considered to be uniform.

The Froude number Fr $=U^2/gL$ (U is the characteristic velocity; g is the gravitational acceleration; L is the characteristic length) measures the relative effect of gravity. If the characteristic length of the coaxial jet is not so small, Fr is finite. In such a case, gravity may be as important as the axial electric field. Both gravity and the axial electric field induce nonuniformity of the basic axial velocity, but their effects may be opposite: if the axial electric field is in the same direction as gravity (which is in accordance with most experimental situations), the electrical shear force makes the velocity at the interface larger than in the liquid bulk, and with the spatial evolution of the jet its effect is diffused into the bulk owing to liquid viscosity; on the other hand, gravity, acting as a pressure gradient, induces a larger velocity in the liquid center. Consequently, the action of the axial electric field and gravity may reach an equilibrium state as the jet evolves spatially and the well-developed jet has an axial uniform velocity. In such a case, the uniform velocity profile approximation is appropriate [347].

For infinitesimal axisymmetric disturbances, the perturbation of the interface (and also of the other physical quantities), is decomposed into the form of a Fourier exponential, that is, ηj $(z, t) = \hat{\eta}j \exp(\omega t + ikz)$, $j = 1, 2$, where $\hat{\eta}j$ is the initial amplitude of the perturbation at the interface, ω is the complex wave frequency, the real and imaginary parts of which are the temporal growth rate and frequency, respectively, k is the real axial wavenumber related to the wavelength by $k = 2\pi/\lambda$, and the imaginary unit $i = \sqrt{-1}$. Substituting the perturbation expression into the governing equations and boundary conditions the dispersion relation between ω and k is obtained. The derivation details are given in Appendix A.

Choosing ρ_2, γ_2, $\rho_2 U_2^2$, ε_3 and $-V_0/[R_2 \ln(R_2/R_3)]$ as the characteristic scales of density, surface tension, pressure, electrical permittivity and electric field intensity, respectively, the dispersion relation is written in the following dimensionless form:

$$D(k,\omega) = \left(\frac{\hat{\eta}_2}{\hat{\eta}_1}\right)_1 \left(\frac{\hat{\eta}_1}{\hat{\eta}_2}\right)_2 - 1 = 0, \qquad (7.13.110)$$

For case I, where $(\hat{\eta}_2/\hat{\eta}_1)_1$ and $(\hat{\eta}_1/\hat{\eta}_2)_2$ are the amplitude ratio of the initial disturbances at the interfaces, respectively, expressed as

$$\left(\frac{\hat{\eta}_2}{\hat{\eta}_1}\right)_1 = \left[\frac{l^2+k^2}{a}\Delta_4\left(\frac{Eu\,Re\varsigma}{\omega}+1+\frac{l^2}{k^2}\right)-\frac{2k^2}{a}\Delta_3\left(\frac{Eu\,Re\varsigma}{\omega}+2\right)\right]^{-1}$$
$$\times\left[\mathrm{Re}^2 H_1\Delta_3\Delta_4 - 4lk^2\Delta_3\left(\Delta_6-\frac{1}{la}\Delta_4\right)+\Delta_4\left((l^2+k^2)\Delta_1-\frac{2k}{a}\Delta_3\right)\frac{l^2+k^2}{k}+\left(\frac{2k^2}{a}\Delta_3-\frac{l^2+k^2}{a}\Delta_4\right)\frac{Eu\,Re\varsigma}{\omega}\right] \quad (7.13.111)$$

and

$$\left(\frac{\hat{\eta}_1}{\hat{\eta}_2}\right)_2 = \left[4k^2\Delta_3-\frac{(l^2+k^2)^2}{k^2}\Delta_4+k\left((l^2+k^2)\Delta_2\Delta_4-2lk\Delta_3\Delta_5\right)\frac{Eu\,Re\varsigma}{\omega}-\mathrm{Re}^2\,\Delta_3\Delta_4 Euk\varsigma J\right]^{-1}$$
$$\times\left[\mathrm{Re}^2 H_2\Delta_3\Delta_4 - k\Delta_4\left((l^2+k^2)\Delta_2+2k\Delta_3\right)\left(\frac{Eu\,Re\varsigma}{\omega}+1+\frac{l^2}{k^2}\right)+2lk^2\Delta_3\left(\Delta_5-\frac{1}{l}\Delta_4\right)\left(\frac{Eu\,Re\varsigma}{\omega}+2\right)\right] \quad (7.13.112)$$

The amplitude ratio $(\hat{\eta}_2/\hat{\eta}_1)_1$ comes mainly from the dynamic balance at the inner interface and $(\hat{\eta}_1/\hat{\eta}_2)_2$ from the dynamic balance at the outer interface. The symbols appearing in the dispersion equation are

$$H_1 = \frac{iS\omega}{k}\left(i\omega \frac{I_0(ka)}{I_1(ka)} - \frac{2\Lambda}{a} \right) + \frac{\Gamma}{Wea^2}\left(1 - (ka)^2\right) \qquad (7.13.113)$$

$$H_2 = \frac{iQ\omega}{k}\left(i\omega L - \frac{2}{b^2 - 1} \right) + Eu\left(1 + k(\varsigma + 1)J\right) - \frac{1}{We}\left(1 - k^2\right) \qquad (7.13.114)$$

$$J = \frac{I_1(k)K_0(kb) + K_1(k)I_0(kb)}{I_0(k)K_0(kb) - K_0(k)I_0(kb)} \qquad (7.13.115)$$

$$\varsigma = \frac{-k\left(1 + \dfrac{l^2}{k^2}\right)\dfrac{\Delta_2}{\Delta_3} + 2l\dfrac{\Delta_5}{\Delta_4} + \dfrac{\text{Re}\,\omega J}{k}}{\dfrac{\text{Re}}{k}\left(\varepsilon_{r2}(\omega + \tau)\kappa - \omega J\right) + \dfrac{Eu\,\text{Re}}{\omega}\left(k\dfrac{\Delta_2}{\Delta_3} - l\dfrac{\Delta_5}{\Delta_4}\right)} \qquad (7.13.116)$$

$$\xi = \frac{\left(1 + \dfrac{l^2}{k^2}\right)\dfrac{1}{\Delta_3} - \dfrac{2}{\Delta_4}}{\dfrac{\text{Re}}{k}\left(\varepsilon_{r2}(\omega + \tau)\kappa - \omega J\right) + \dfrac{Eu\,\text{Re}}{\omega}\left(k\dfrac{\Delta_2}{\Delta_3} - l\dfrac{\Delta_5}{\Delta_4}\right)} \qquad (7.13.117)$$

$$L = \frac{I_0(k)K_1(kb) + K_0(k)I_1(kb)}{I_1(k)K_1(kb) - K_1(k)I_1(kb)}, \qquad \kappa = -\frac{I_1(ka)\Delta_1 - \dfrac{\varepsilon_{r2}}{\varepsilon_{r1}}I_0(ka)\Delta_3}{I_1(ka)\Delta_0 - \dfrac{\varepsilon_{r2}}{\varepsilon_{r1}}I_0(ka)\Delta_2} \qquad (7.13.118)$$

And $\Delta_0 - \Delta_6$ are

$$\Delta_0 = I_0(ka)K_0(k) - K_0(ka)I_0(k), \quad \Delta_1 = I_0(ka)K_1(k) + K_0(ka)I_1(k), \qquad (7.13.119)$$

$$\Delta_2 = I_1(ka)K_0(k) + K_1(ka)I_0(k), \Delta_3 = I_1(ka)K_1(k) - K_1(ka)I_1(k) \qquad (7.13.120)$$

$$\Delta_4 = I_1(la)K_1(l) - K_1(la)I_1(l) \quad \Delta_5 = I_1(la)K_0(l) + K_1(la)I_0(l), \qquad (7.13.121)$$

$$\Delta_6 = I_0(la)K_1(l) + K_0(la)I_1(l), \qquad (7.13.122)$$

where $In(x)$ and $Kn(x)$ $(n = 0, 1)$ are the nth-order modified Bessel functions of the first and second kinds. Here the wavenumber k and complex frequency

ω have been normalized by $1/R_2$ and U_2/R_2, respectively. In the following numerical section, we mainly take case I as an example to illuminate the unstable modes and behaviors of a viscous coflowing jet under a radial electric field. Case II is calculated when studying the effect of the inner liquid viscosity and in part of the thin layer approximation. The dispersion relation for case II is given in Appendix B.

The dimensionless parameters involved in the dispersion relation include:

- the density ratios $S = \rho_1/\rho_2$ and $Q = \rho_3/\rho_2$,
- the interfacial tension coefficient ratio $\Gamma = \gamma_1/\gamma_2$,
- the electrical permittivity ratios $er_1 = \varepsilon_1/\varepsilon_2$ and $er_2 = \varepsilon_2/\varepsilon_3$,
- the relative electrical relaxation time $\tau = R_2\sigma_2/U_2\varepsilon_2$,
- the Reynolds number Re $= \rho_2 U_2 R_2/\mu_2$,
- the Weber number We$= \rho_2 U_2^2 R_2/\gamma_2$,
- the electrical Euler number $Eu = \varepsilon 3\ V_0^2/\rho_2 U_2^2 R_2^2\ \ln^2(R_2/R_3)$.

The last three dimensionless parameters represent the relative magnitudes of the viscous force, surface tension and electrical force to the inertia force, respectively. In most experiments, R_3 is much larger than R_2, that is, the radius ratio $b \geq 1$. In the dispersion relation (the Eq. (7.13.110)), the parameter b appears only in J and L, for which we have the limits:

$$J|_{b \to \infty} = \frac{-K_1(k)}{K_0(k)} \text{ and } L|_{b \to \infty} = \frac{-K_0(k)}{K_1(k)} \qquad (7.13.123)$$

For large Reynolds numbers (Re ≥ 1), the dispersion relation (7.13.110) and the amplitude ratios (7.13.111) and (7.13.112) can be simplified dramatically to:

$$D(k,\omega) = ak^2 \left(T_1\Delta_3 - \Delta_1\right)\left(T_2\Delta_3 + \Delta_2\right) + 1 = 0, \qquad (7.13.124)$$

and

$$\left(\frac{\hat{\eta}_2}{\hat{\eta}_1}\right)_1 = -ak\left(T_1\Delta_3 - \Delta_1\right), \left(\frac{\hat{\eta}_1}{\hat{\eta}_2}\right)_2 = k\left(T_2\Delta_3 + \Delta_2\right) \qquad (7.13.125)$$

with

$$T_1 = -\omega^{-2}\left[\frac{iS\omega}{k}\left(i\omega\frac{I_0(ka)}{I_1(ka)} - \frac{2\Lambda}{a}\right) + \frac{\Gamma}{Wea^2}\left(1 - (ka)^2\right)\right] \qquad (7.13.126)$$

$$T_2 = -\omega^{-2}\left[iQ\omega\left(-i\omega L + \frac{2}{b^2 - 1}\right) - Euk\left(1 + kJ\right) + \frac{k}{We}\left(1 - k^2\right)\right] \qquad (7.13.127)$$

If the relative velocity ratio $\Lambda=0$, the dispersion relation is consistent with that for the electrified coaxial jet in the equipotential case where Λ is fixed to 1 [346].

In addition, the dispersion relation (the Eq. (7.13.110)) can be reduced to that for a single-liquid jet in a simple way. If the radius ratio a approaches zero, the jet consists only of the outer leaky dielectric liquid. The inner interface vanishes, and the numerator of (7.13.112) is zero, yielding

$$\mathrm{Re}^2\, H_2 - k\left(\left(l^2+k^2\right)\frac{\Delta_2}{\Delta_3}+2k\right)\left(\frac{Eu\,\mathrm{Re}\,\varsigma}{\omega}+1+\frac{l^2}{k^2}\right)+2lk^2\left(\frac{\Delta_5}{\Delta_4}+\frac{1}{l}\right)\left(\frac{Eu\,\mathrm{Re}\,\varsigma}{\omega}+2\right)=0 \quad (7.13.128)$$

where $\dfrac{\Delta_2}{\Delta_3}=\dfrac{-I_0(k)}{I_1(k)}$, $\dfrac{\Delta_5}{\Delta_4}=\dfrac{-I_0(l)}{I_1(l)}$ and

$$\varsigma=\frac{k\dfrac{I_0(k)}{I_1(k)}\left(1+\dfrac{l^2}{k^2}\right)-2l\dfrac{I_0(l)}{I_1(l)}\dfrac{\Delta_5}{\Delta_4}+\dfrac{\mathrm{Re}\,\omega J}{k}}{\dfrac{\mathrm{Re}}{k}\left(\varepsilon_{r2}\left(\omega+\tau\right)\kappa-\omega J\right)+\dfrac{Eu\,\mathrm{Re}}{\omega}\left(-k\dfrac{I_0(k)}{I_1(k)}+l\dfrac{I_0(l)}{I_1(l)}\right)} \quad (7.13.129)$$

After some algebra, the above equation is written in a clearer form:

$$\varsigma\omega^2+\frac{2\omega}{\mathrm{Re}}\left(2k^2\varsigma-1\right)+\frac{4k^2}{\mathrm{Re}^2}\left(k^2\varsigma-l^2\varsigma_v\right)+T+\left(1+\frac{Euk\varsigma}{EJ\omega^2}\left(l^2\varsigma_v-k^2\varsigma\right)\right)^{-1}$$

$$\left[\frac{2Eu\varsigma k^2}{E\,\mathrm{Re}\,\omega}\left(2+\frac{1}{kJ}\right)\left(k^2\varsigma-l^2\varsigma_v\right)+\frac{Eu\varsigma}{E}\left(2k^2\varsigma+\frac{k\varsigma l^2\varsigma_v}{J}+kJ\right)\right]=0 \quad (7.13.130)$$

Where

$$E=\frac{\varepsilon_{r2}}{kJ}\left(1+\frac{\tau}{\omega}\right)-\varsigma$$

$$\varsigma_v(l)=\frac{I_0(l)}{lI_1(l)},\; T=\frac{iQ\omega}{k}\left(i\omega L-\frac{2}{b^2-1}\right)+Eu\left(1+kJ\right)-\frac{1}{We}\left(1-k^2\right) \quad (7.13.131)$$

Equation (7.13.130) is the dispersion relation for a viscous jet with a leaky dielectric liquid, which exactly corresponds with Lopez-Herrera et al. [315, 368].

7.13.7.1   NUMERICAL RESULTS OF VISCOUS COFLOWING JET MODEL

The dimensionless dispersion relation (the Eq. (7.13.110)) is a quartic equation for the complex frequency ω. Given an axial wavenumber k, there are generally four eigenvalues corresponding to four different modes, but only two of the modes are unstable in the Rayleigh regime [351]; these are usually called the para-sinuous mode and the para-varicose mode. Suppose the interface perturbation amplitude ratio $\hat{\eta}_1/\hat{\eta}_2 = |\hat{\eta}_1/\hat{\eta}_2|\exp(i(\theta_1 - \theta_2))$, where $|\hat{\eta}_1/\hat{\eta}_2|$ is the relative magnitude of the amplitudes, and $\theta = \theta_1 - \theta_2$ is the corresponding phase difference. The para-sinuous mode means that the inner liquid–liquid interface and the outer air–liquid interface are perturbed almost in phase, that is, θ approaches 0°; and the para-varicose mode means that the two interfaces are perturbed nearly out of phase, that is, θ approaches 180°. Under most experimental situations [315], the para-sinuous mode is less stable than the para-varicose one and is dominant in the jet instability, promoting the formation of compound droplets. However, as long as the unstable para-varicose mode exists, coaxial electrospraying is negatively influenced by it. On the other hand, under a sufficiently intense electric field and sufficiently large flow rate, those nonaxisymmetric modes may become comparable to the axisymmetric ones, and even become dominant [346, 347, 356].

The theoretical model for the axisymmetric instability and maintaining the axisymmetric modes dominant through controlling the values of the dimensionless parameters was studied by Li et al. [347]. The non axisymmetric instability deserves special study. They solved (7.13.110) numerically in order to study the behavior of the coaxial jet under the radial electric field, mainly taking case I as an example. For convenience of calculation and comparison, a set of dimensionless parameters was chosen as a reference set. In case I, they took water and air as the outer and inner liquids, respectively, because water is the most common leaky dielectric and air is a common perfect dielectric, corresponding to the case of the outer-driving coaxial electrospraying [347]. Their physical properties can be found in [315]. The reference dimensionless parameters are $Q = 0.001$, $S = 0.001$, $a = 0.8$, $b = 10$, $\Lambda = 0.2$, Re = 10, We = 10, $\Gamma = 1$, Eu = 0.15, $\varepsilon r_1 = 1$, $\varepsilon r_2 = 80$ and $\tau = 1$. In the calculation, the dimensionless parameters were fixed to the reference values unless stated otherwise.

7.13.7.2   EFFECT OF LIQUID VISCOSITY AND COMPARISON WITH THE INVISCID MODEL

In order to study the effect of the viscosity of the outer liquid, Li et al. [347] first established an inviscid model. In the inviscid model, both the inner liquid and the outer liquid are assumed to be inviscid, and the velocities of the inner and outer liquids in the basic state are assumed to be uniform with a discontinuity at the inner and outer interfaces. Denoting the base axial velocities of the inner and outer liquids by U_1 and U_2, respectively, a new dimensionless parameter $\Lambda\dagger = U_1/U_2 - 1$ is obtained. For leaky dielectrics, the dimensionless dispersion relation of this inviscid model is also expressed in the form of Eq. (7.13.110), but with different perturbation amplitude ratios:

$$\left(\frac{\hat{\eta}_2}{\hat{\eta}_1}\right)_1 = \frac{k^2 a H_1 \Delta_3 + k a \omega^2 \Delta_1}{\omega^2} \tag{7.13.132}$$

$$\left(\frac{\hat{\eta}_1}{\hat{\eta}_2}\right)_2 = -\frac{k^2 H_2 \Delta_3 - k \omega^2 \Delta_2}{\omega^2 + E u k^3 J \xi \Delta_3} \tag{7.13.133}$$

where

$$H_1 = -\frac{S(\omega + i\Delta^1 k)\,^2 I_0(ka)}{k I_1(ka)} + \frac{T}{Wea^2}(1 - (ka)^2) \tag{7.13.134}$$

$$H_2 = -\frac{Q\omega^2}{k}L + Eu(1 + k(\xi + 1)J) - \frac{1}{We}(1 - k^2) \tag{7.13.135}$$

$$\zeta = \frac{k\omega(J - \Delta_2/\Delta_3)}{k\varepsilon_{r2}(\omega + r)k - k\omega J} \tag{7.13.136}$$

$$\xi = \frac{\omega/\Delta_3}{k\varepsilon_{r2}(\omega + \tau)k - k\omega J} \tag{7.13.137}$$

The other symbols are the same as in the viscous model. The coordinate system in this inviscid model is still moving with velocity U_2.

Note that the dispersion relation for the inviscid model is a little different from for the case of large Reynolds numbers, because the tangential dynamic continuity condition at the outer interface is treated differently in these two cases. In the large-Re case the tangential component of the electrical stress at the outer interface is very small, or even vanishes. In such a case the jet surface can be regarded to be approximately equipotential. However, in the inviscid leaky dielectric case, the tangential dynamic condition is not satisfied because the viscous shear is not taken into account, and the jet may be non-equipotential. The dispersion relation given by Eqs. (7.13.132) and (7.13.133) reduces to that for the equipotential case only if the relative electrical relaxation time τ approaches infinity [346]. In addition, if τ approaches zero it is reduced to the dispersion relation for the nonequipotential case [347].

The influence of the viscosity of the outer liquid on the unstable modes is shown by Li and his co-workers studies [342, 347] in Fig. 7.39, where the relative velocity ratio $\Lambda = 0$. For comparison, the curves for the large Reynolds number limit (marked with '∞') and the inviscid leaky dielectric model with the relative velocity ratio $\Lambda\dagger=0$ (dashed) are also plotted. It is clear that the growth rates of both the para-varicose mode in Fig. 39(a) and the para-sinuous mode in Fig. 39(b) are greatly enhanced as the Reynolds number increases, indicating that the viscosity of the outer liquid has a remarkable stabilizing effect on the jet instability.

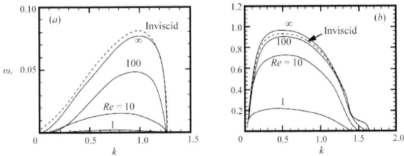

FIGURE 7.39   The influence of the Reynolds number on the growth rates of (a) the para varicose mode and (b) the para sinuous mode, for 0. The dashed curves are for the inviscid leaky dielectric model.

For the para-sinuous mode, the large-Re limit, that is, the equipotential case, is the most unstable. However, for the para-varicose mode, the inviscid leaky dielectric model possesses the maximum growth rate. In general these two special cases are very close. The reason may be that the reference value of the relative electrical permittivity of the outer liquid is so high ($\varepsilon r_2 =80$) that the denominators of Eqs. (7.13.136) and (7.13.137) are large enough to make the effects of ζ and ξ negligible. On the other hand, it is shown in the figure that the viscosity of the outer liquid has no apparent effect on the range of the unstable axial wavenumbers. The cut-off wavenumber kc is approximately 1.25 for the para-varicose mode and 1.6 for the para-sinuous mode.

In case I, the basic velocity profile of the inner liquid is parabolic, where a dimensionless parameter Λ is involved. Apparently, this parameter represents the effect of the shear in the axial velocity on the jet instability. As the density ratio S (=0.001) in the reference state is so small that the effect of

Λ cannot be shown clearly, in the calculation the value of S is chosen to be 0.1. Although the density ratio S is increased, the para-varicose mode is little influenced, the para-sinuous mode is stabilized slightly as Λ increases. If the density of the inner liquid is comparable to that of the outer liquid, the effect of Λ becomes more significant. The calculation shows that the growth rate of the para-varicose mode is also reduced as Λ increases. It is well known that the para-sinuous and para-varicose modes are associated primarily with the inner and outer interfaces [346, 347, 351], respectively. Apparently, the parameter Λ influences the mode closely connected with the inner interface more strongly. In general, Li and his co-workers result are in accordance with the fact that the shear in the axial velocity has a stabilizing effect on the jet instability [353]. Similar to the liquid viscosity, the shear in the axial velocity has a negligible effect on the unstable wavenumber range. The critical wavenumber is approximately the same as shown in Fig. 7.39.

In coaxial electrospraying experiments with two liquids, the inner one is usually highly viscous [312, 368]. However, in case I the viscosity of the inner liquid is neglected. Therefore, it is necessary to study the jet instability in case II, in which the inner liquid viscosity is allowed to be comparable to or larger than that of the outer liquid. We take sunflower oil as the inner dielectric liquid. Its physical properties can be found in Lopez-Herrera et al. [368].

7.13.7.3 *DISCUSSION ON THE ELECTRICAL RELAXATION TIME AND TWO LIMITING CASES*

For EHD leaky dielectrics, there are two important characteristic times, that is, the electrical relaxation time and the hydrodynamic time. The electrical relaxation time $\tau e \sim \varepsilon/\sigma$ measures the speed of charge relaxation, that is, the contribution of conduction to charge transportation. The hydrodynamic time can have several definitions, such as the capillary time $\tau c \sim (\rho L^3/\gamma)^{1/2}$, the viscous diffusion time $\tau v \sim \rho L^2/\mu$, and the convective flow time $\tau F \sim L/U$. In the most models, the convective flow time $\tau F \sim R_2/U_2$ is chosen as the characteristic hydrodynamic time, which measures the contribution of convection to charge transportation. As a result, a dimensionless parameter $\tau = (R_2\sigma_2)/(U_2\varepsilon_2)$ measuring the relative magnitude of the hydrodynamic time and the electrical relaxation time, that is, the relative importance of conduction and convection, is involved in the dispersion relation (the Eq. (7.13.110)). In this section, two limiting cases are derived according to the relative magnitude of the electrical relaxation time and the hydrodynamic time in a more generic sense. Selecting an appropriate hydrodynamic time τh, the surface charge conservation Eq. (7.13.104) is nondimensionalized as follows:

$$\frac{\partial q_s}{\partial t} + u.\nabla q_s - q_s n.(n.\nabla)u + \frac{\tau_h}{\tau_e}[\sigma E].n = 0. \tag{7.13.138}$$

For well-conducting liquids, the electrical relaxation time τe is usually several orders of magnitude smaller than the hydrodynamic time τh, that is, $\tau h/\tau e$, indicating that charge is transported mainly by conduction and the effect of convection is negligible. This case is called the small electrical relaxation time limit (SERT). In this limit the surface charge conservation Eq. (7.13.138) reduces to

$$[\sigma E].n = 0. \tag{7.13.139}$$

Note that the surface charge density is absent in Eq. (7.13.139). It can be obtained through the boundary conditions. In discussed model, the Eq. (7.13.139), which is obeyed at the outer air–liquid interface, implies that the outer liquid is equipotential (i.e., the equipotential case). Conversely, for relatively imperfectly conducting liquids with a relatively high velocity, the electric relaxation time τe may be much larger than the hydrodynamic time τh, that is, $\tau h/\tau e$. In such a case charge convection becomes significant and the effect of conduction is negligible. It is called the large electrical relaxation time limit (LERT). In this limit Eq. (7.13.138) reduces to

$$\frac{\partial q_s}{\partial t} + u_s.\nabla q_s - q_s n_s.(n_s.\nabla)u_s = 0. \tag{7.13.140}$$

where only the bulk conduction disappears, corresponding to the nonequipotential case. In case LERT, charge at the interface cannot be reset instantaneously to maintain the interface equipotential. Equation (7.13.140) may serve as the surface charge conservation equation for this limit. In the study of the interfacial instability of a conducting liquid jet under a radial electric field [322, 347], both the equipotential (SERT) and nonequipotential (LERT) cases are considered. According to these two limits, Eqs. (7.13.111) and (7.13.112) are reduced, with the dispersion Eq. (7.13.110) unchanged in form. For SERT, the amplitude ratio of the initial disturbances is

$$\left(\frac{\hat{n}_2}{\hat{n}_1}\right)_1 = \frac{akRe^2 H_1\Delta_3\Delta_4 + a(l^2+k^2)^2\Delta_1\Delta_4 - 4alk^3\Delta_3\left(\Delta_6 - \frac{1}{la}\Delta_4\right) - 2k(l^2+k^2)\Delta_3\Delta_4}{(l^2+k^2)\Delta_4/k - 4k^3\Delta_3} \tag{7.13.141}$$

$$\left(\frac{\hat{n}_1}{\hat{n}_2}\right)_2 = \frac{kRe^2 H_2\Delta_3\Delta_4 - (l^2+k^2)^2\Delta_2\Delta_4 + 4lk^3\Delta_3\left(\Delta_5 + \frac{1}{l}\Delta_4\right) - 2k(l^2+k^2)\Delta_3\Delta_4}{4k^3\Delta_3 - (l^2+k^2)^2\Delta_4/k} \tag{7.13.142}$$

where H_1 is the same as Eq. (2.24a) and

$$H_2 = \frac{iQ\omega}{k}\left(i\omega L - \frac{2}{b^2 - 1}\right) + Eu(1 + kJ) - \frac{1}{We}(1 - k^2) \quad (7.13.143)$$

For LERT, the amplitude ratio of the initial disturbances has the same form as Eqs. (7.13.113) and (7.13.114), but with

$$\zeta = \frac{-k\left(1 + \frac{l^2}{k^2}\right)\frac{\Delta_2}{\Delta_3} + 2l\frac{\Delta_5}{\Delta_4} + \frac{Re\omega J}{k}}{\frac{Re\omega}{k}(\varepsilon_{r2}k - J) + \frac{EuRe}{\omega}\left(k\frac{\Delta_2}{\Delta_3} - l\frac{\Delta_5}{\Delta_4}\right)} \quad (7.13.144)$$

$$\xi = \frac{\left(1 + \frac{l^2}{k^2}\right)\frac{1}{\Delta_3} - \frac{2}{\Delta_4}}{\frac{Re\omega}{k}(\varepsilon_{r2}k - J) + \frac{EuRe}{\omega}\left(k\frac{\Delta_2}{\Delta_3} - l\frac{\Delta_5}{\Delta_4}\right)}. \quad (7.13.145)$$

Figure 7.40 illustrates the growth rates of the unstable modes for these two limit cases, solid curves for SERT (i.e., $\tau \to 0$) and dashed curves for LERT (i.e., $\tau \to \infty$) according to Li investigation, where several values of the electrical Euler number are considered. It can be seen from Fig. 7.40(a) that for the para-varicose mode only solid curves for SERT appear when the electric field exists. The difference between the two limit cases is remarkable, because for LERT the growth rate of the para-varicose mode decreases to zero when the electrical Euler number exceeds 0.01. However, for SERT the growth rate is enhanced by the electric field and the unstable region moves towards relatively short waves. On the other hand, for the para-sinuous mode as shown in Fig. 7.40(b), the difference between two limit cases is discernible but not significant for the range of Eu studied, indicating that the influence of the relative electrical relaxation time τ on the jet instability is small. This may be attributed to the large relative electrical permittivity of the outer liquid ($\varepsilon r_2 = 80$), which makes $k\varepsilon r_2(\omega + \tau)\kappa$ in the denominators of Eqs. (7.13.115) and (7.13.116) a large term, and consequently the influence of τ is weakened.

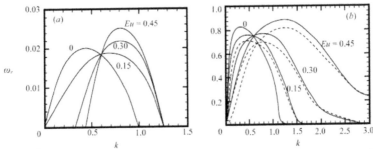

FIGURE 7.40   The growth rates of (a) the para varicose mode and (b) the para sinuous mode, for the limit case SERT (solid curve) and LERT (dashed curve) for different electrical Euler numbers.

7.13.7.4   EFFECT OF THE ELECTRIC FIELD TOGETHER WITH THE OTHER PARAMETERS ON THE JET INSTABILITY

For the inviscid coaxial jet under a radial electric field [347], in agreement with Li et al. [342, 346] studies, the radial electric field has a dual effect on both the para-varicose and para-sinuous modes, destabilizing them greatly when the axial wavenumber k exceeds a critical value, and stabilizing them if k is smaller than that value. Using viscous model, the dual effect of the electric field is persistent for both the para-varicose mode and para-sinuous mode, as shown in Fig. 7.41.

It is well known that when arbitrary disturbances are applied to the jet, the perturbation wave with maximum growth rate and those close to it grow faster than the others, which become dominant in the jet breakup process. Although two unstable modes occur in the jet instability process, the para-sinuous mode is much less stable than the para-varicose one in most situations [311, 312, 347]. Consequently, the most unstable wavenumber k_{max} comes from the para-sinuous mode. In Fig. 7.41(*b*), the value of k_{max} is amplified as the electrical Euler number increases, predicting that the most likely wavelength $\lambda = 2\pi R_{/}$ $kmax$ is diminished by the radial electric field.

It is necessary to study the effect of several important dimensionless parameters, such as the Weber number and Reynolds number, on the dominant wavenumber k_{max} and corresponding maximum growth rate ω_{max}. Figure 7.41(*a, b*) illustrates the effect of the Weber number and electrical Euler number on k_{max} and *ωmax*, respectively. The selected values of the Weber number are relatively small, since the coaxial jet is very thin and the surface tension is generally large in the experiments. It is found that the electric field influences k_{max} and ωmax slightly when the Weber number is relatively small (We<5). However, at relatively large Weber numbers (We>10), the electric field enhances $kmax$ and ωmax distinctly. The behavior of k_{max} and ω_{max} indicates that at small Weber numbers the jet instability is dominated primarily by the capillary force, while with the increase of We, the jet instability is dominated primarily by the electrical force [312, 347]. On the other hand, both the electrical Euler number and Weber number change the cut-off wavenumber kc significantly, as shown in Fig. 7.41(c). Obviously, kc is enlarged as We increases, especially at large Euler numbers. So the instability region may be extended into the first wind-induced regime, which reduces the formation of monodisperse droplets in coaxial electrospraying.

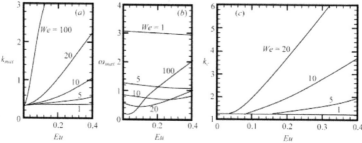

FIGURE 7.41 (a) The dominant axial wavenumber k_{max}, (b) the maximum growth rate ω_{max}, and (c) the cut off axial wave number k_c versus the electrical Euler number for different weber numbers.

7.13.7.5 DISCUSSION ON THE THIN LAYER APPROXIMATION A □ 1

For coflowing jets, one extreme case is that the thickness of the outer annular liquid layer is thin. In this section, simple dispersion relations were sought relatively applicable to the thin layer coating case $a \rightarrow 1$ for case I and case II. Some details are given in Appendix C; where case I is taken as an example. The thin layer approximation ultimately results in the dispersion relation [347]:

$$\omega^2 \left(QL - S\frac{I_0(k)}{I_1(k)}\right) + 2i\omega\left(\frac{Q}{b^2-1} - sA\right) + \frac{1+T}{We}k(1-k^2)$$

$$-Eu(1+kJ)\left(1 - \frac{\omega(1+kJ)}{\varepsilon_{r1}(\omega+\tau)\frac{I_1(k)}{I_0(k)} - \omega J}\right) = 0$$

(7.13.146)

for case I, and

$$\omega^2 \frac{S\, I_0(k)}{k\, I_1(k)} + \frac{2\mu_r \omega}{Re}\left(2k\frac{I_0(k)}{I_1(k)} - 1\right) + \frac{4u_r^2 k^2}{SRe^2}\left(k\frac{I_0(k)}{I_1(k)} - l_1\frac{I_0(l_1)}{I_1(l_1)}\right)$$

$$+\left|\frac{iQ\omega}{k}\left(i\omega L - \frac{2}{b^2 - 1}\right) + Eu(1 + kJ) - \frac{1+T}{We}(1 - k^2)\right. \tag{7.13.147}$$

$$\frac{SEu\left[k\omega\left(J\frac{I_0(k)}{I_1(k)}\right) + \frac{2\mu_r k^2}{S\, Re}\left(k\frac{I_0(k)}{I_1(k)} - l_1\frac{I_0(l_1)}{I_1(l_1)}\right)\right]^2}{Sk\omega\left(\omega J - \varepsilon_{r1}(\omega + \tau)\frac{I_1(k)}{I_0(k)}\right) + k^2 Eu\left(k\frac{I_0(k)}{I_1(k)} - l_1\frac{I_0(l_1)}{I_1(l_1)}\right)} = 0$$

for case II. The above expressions appear much simpler than for the coaxial jet. However, their validity needs to be evaluated. Note that under the thin layer approximation the viscosity and dynamic force of the outer liquid, as well as the electrical permittivity of the outer liquid, have no influence. Only the electrical permittivity of the inner liquid and the electrical relaxation time of the outer liquid layer play a role [346, 347]. Especially in the limit case $\tau \to \infty$ Eqs. (7.13.146) and (7.13.147), respectively, reduce to

$$\omega^2\left(QL - S\frac{I_0(k)}{I_1(k)}\right) + 2i\omega\left(\frac{Q}{b^2 - 1}SA\right) + \frac{1+T}{We}k(1 - k^2) - Eu(1 + kJ) = 0 \tag{7.13.148}$$

and

$$\omega^2 \frac{S\, I_0(k)}{k\, I_0(k)} + \frac{2\mu_r \omega}{Re}\left(2k\frac{I_0(k)}{I_1(k)} - 1\right) + \frac{4\mu_1^2 k^2}{S\, Re^2}\left(k\frac{I_0(k)}{I_1(k)} - l_1\frac{I_0(l_1)}{I_1(l_1)}\right)$$

$$+\frac{iQ\omega}{k}\left(i\omega L - \frac{2}{b^2 - 1}\right) + Eu(1 + kJ) - \frac{1+T}{We}(1 - k^2) = 0. \tag{7.13.149}$$

The dispersion Eqs. (7.13.148) and (7.13.149) accord with those for the single-liquid jet in the equipotential case but with double surface tension, the former for the inviscid liquid and the latter for the viscous liquid.

Under the thin layer approximation, only one unstable mode, that is, the para-sinuous mode, exists. Compared with the exact solution of the leaky dielectric model, if the viscosity of the inner liquid is neglected the thin layer approximation is more accurate when the density of the inner liquid is comparable to that of the outer liquid, but if the inner liquid viscosity is taken into account the thin layer approximation is accurate only for a relatively small electric field [346, 347].

7.14 APPLICATION OF COAXIAL ELECTROSPUN NANOFIBERS

Due to their excellent physico-chemical properties and flexible characteristics, coaxial electrospun core-sheath or hollow fibers can be used in various fields. For example, core-sheath fibers can exhibit an enhanced mechanical property compared to simple fibers, such as silk fibroin-silk sericin core-sheath fibers with a breaking strength of 1.93 MPa and a breaking energy of 7.21 J kg^{-1}, which are 82% and 92.8% higher than those of single silk fibroin fibers. To illustrate more specifically, in this section, the applications of these fibers in the fields including lithium ion batteries, solar cells, luminescence, supercapacitors, photocatalytic environmental remediation and filtration are critically reviewed with examples and design principles.

7.14.1 LITHIUM ION BATTERIES

Right now, the whole society faces a serious energy challenge with traditional fossil energy being used up and other new energies not ready for large-scale deployment. The widely used internal combustion engine using fossil energy has faced a bottleneck in the energy field development due to its low energy conversion resulting from its low energy efficiency, normally less than 30%. And the combustion of fossil fuels can cause severe environmental pollution. Urged by all these problems, scientists turn to a high-grade sustainable clean energy, electricity. And among the energy storage devices, batteries have been in the spotlight attracting considerable attention. Among all the batteries, the lithium ion battery has become the primary candidate in many applications, such as communication, transportation and regenerated energy sectors due to its higher voltage (about 3.6 V, two times higher than that of aqueous batteries), gravimetric specific energy (about 240 W h kg^{-1}, six times higher than that of lead acid batteries), long duration (500–1000 cycles), wide temperature range (20 to 60°C) and minimum memory effect [117, 372]. Figure 7.42 shows the schematic diagram of a typical commercial lithium ion battery with lithium alloy compound and graphite as the cathode and the anode, respectively. After the circuit is connected, electrons will flow from the anode to the cathode through an external circuit forming current driven by the chemical potential difference between the electrode materials, at the same time, lithium ions are transported in the same direction through the electrolyte inside the battery. At the cathode, lithium ions react with cathode material and electrons, and deposit there. During the charging process, both electrons and lithium ions go back though the previous pathway driven by the applied potential difference. Through the charge–discharge process, the stored chemical energy is finally converted into electricity. The chemical reactions during the charging and discharging processes are listed [116].

$$Cathod : \frac{1}{2}Li^+ + \frac{1}{2}e^{-1} + Li_{0.5}CoO_2 \Leftrightarrow LiCoO_2$$

$$Anode : LiC_6 \Leftrightarrow Li^+ + e^- + C_6$$

During the charging process, the lithium ions will insert into the graphite layer and combine with carbon atoms.

(a) (b)

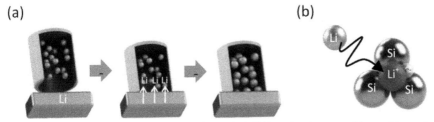

FIGURE 7.42 Schematic illustration of a typical lithium-ion battery with graphite and $LiCoO_2$ as anode and cathode materials, respectively.

This requires that the anode material should provide an extremely large surface area for a convenient combination. On the other hand, the insertion of lithium ions could cause electrode volume expansion, which affects the electric contact and battery capacity. And this expansion is a very serious problem in all lithium batteries. Therefore, different electrode materials have been studied aiming to eliminate the side-effect of the volume expansion and to support enough active sites. Hollow nanofibers are excellent candidates as anode material for lithium ion batteries. The hollow structure not only possesses very high specific surface area, but also buffers the volume expansion during the litigation process. And the hollow structure can integrate various components together to promote the performance of the lithium ion battery [116, 117, 372].

Graphite carbon is a very common commercial anode material for lithium ion batteries, but due to its low capacity and safety issues, other materials including transition metals, semiconductors and lithium alloys have been studied. Generally, carbon material is an excellent choice for an electrode in the battery system and it can be classified into graphitic (soft) and nongraphitic (hard) carbon. Specifically, soft carbon possesses a well-ordered lamellar structure, while hard carbon displays a relatively turbostatic arrangement. And significant property differences as the lithium ion battery anode result due to the structural diversity. Hard carbon possesses a high capacity (400–500 mA h g^{-1}), but poor capacity retention performance, which means that the high capacity will get attenuated very soon. Compared with hard carbon, soft carbon has a lower, but reversible capacity (200–300 mA h g^{-1}), however, it shows a very serious voltage hysteresis during the delithiation process, in which lithium ions are desorbed from the anode. A combination of the advantages of both materials can enhance the performance of lithium ion batteries and the coaxial electrospinning technique has been reported to achieve this goal. Other scientists considered adding other components to enhance the performance. With a high theoretical capacity of about 4000 mA h g^{-1}, nearly ten times that of the commercial graphite anode, silicon has been integrated into electrodes to improve the lithium ion battery performance. TiO$_2$ is another good choice due to its low cost, high working voltage, and structural stability during lithium insertion and extraction processes, although it has some drawbacks, it still attracts the attention. SnO$_2$ is also an excellent anode material due to its higher capacity (about 800 mA h g^{-1}) than that of graphite, high charge and discharge capacity, and fast electron transportation. All these materials could be fabricated into hollow composite nanofibers as anodes for lithium ion batteries to enhance their performance [116, 117, 373].

Lee et al. [374] have used coaxial electrospinning to fabricate hollow carbon nanofibers as anode materials and studied the carbonization temperature effect on the electrochemical performance. Both styrene-coacrylonitrile (SAN) and PAN are good choices as carbon precursors and their DMF solutions serve as the core and sheath solution, respectively. The as-spun fibers experienced a one-hour stabilization process at 270–300°C in an air atmosphere and a following one-hour carbonization at 800, 1000, 1200, 1600°C in a nitrogen atmosphere. During stabilization, the linear PAN molecules were converted to the ladder structure and got carbonized in the following process, at the same time, the core phase burned out resulting in the tubular structure.

Liu et al. [375] reported a core-sheath soft–hard carbon nanofiber web, which displayed an improved electrochemical performance as an anode in the lithium ion battery. A special terpolymer fibril (93 wt% acrylonitrile, 5.3 wt% methyl acrylate and 1.7 wt% itaconic acid) in DMF served as the sheath solution and mineral oil as the core solution. The as-spun fibers were stabilized for 6.5 h at 270°C in an air atmosphere and then carbonized for 1 h at 850°C under nitrogen protection. Finally, the soft–hard core-sheath carbon nanofibers were obtained with sheath PAN converted to hard carbon and core mineral oil decomposed to amorphous soft carbon. In this anode configuration, the hard sheath could prevent the deformation of soft core, which dominated the stable reverse capacity after a long service time. And an enhanced reversible capacity was obtained, 520, 450 and 390 mA h g^{-1} after 20 cycles at 25, 50 and 100 mA g^{-1}, respectively.

Lee et al. [374] have studied the electrochemical performance of the Si–C core-sheath fibers as the lithium ion battery anode. The fibers were prepared by following similar procedures63 and are briefly stated as follows. The Si nanoparticles with a diameter smaller than 100 nm were added to the core solution. The as-spun nanofibers were stabilized at 270–300 °C for one-hour in an air atmosphere and carbonized at 1000°C for one-hour in a nitrogen atmosphere with a heating speed of 10°C min. PAN converted to a carbon sheath and Si nanoparticles attached onto the inner wall of hollow fibers. Si–C hollow fibers are formed. Figure 7.43 illustrates the volume expansion mechanisms during the litigation process. Due to the large d-space between turbo stratic carbon layers, the electrode volume expansion could be caused by the combination of Si and lithium ions.

Hwang et al. [376] have fabricated core-sheath fibers containing different loadings of Si in the carbon core matrix as the anode for lithium ion batteries. Specifically, PAN was the sheath material, and poly(methyl methacrylate) (PMMA) and silicon nanoparticles were core materials. DMF served as a solvent for both phases. Proper quantity of acetone was also added to the core solution to prevent the mixing of core and sheath materials. PMMA worked as a stabilizer to encapsulate silicon nanoparticles and left enough space after burning out to buffer the volume expansion during the charging process. The asspun fibers were stabilized at 280°C for 1 h in an air atmosphere and further carbonized at 1000°C for 5 h in argon. The electrode was prepared as follows: about 70 wt% electrospun fibers, 15 wt% super P and 15 wt% poly (acrylic acid) (PAA) were mixed and added to 1-methyl-2-pyrrolidinone (NMP) to form slurry, which was then pasted onto a copper current collector. The prepared electrode was dried in a vacuum oven at 70°C for 6 h and then punched into circular discs. The more weight percent of silicon was added, the higher discharge capacity the samples had demonstrated. This is due to the superior capacity of silicon itself. For comparison, two more control materials have been prepared with the same method as electrodes. The first one was bare silicon nanoparticles with super P and the second one was carbon fibers decorated with Si nanoparticles. Compared with these two control electrodes, the desired electrode exhibited a high discharge capacity around 1250 mA h g^{-1} with a nearly 100% retention in the first 100 cycles at 0.242 A g^{-1} current rate. With further increase in the current rate up to 2.748 and 6.89 A g^{-1}, the electrode could still display a rather stable capacity in a larger cycle period with 99% retention after 300 cycles and 80.9% retention after 1500 cycles. This improved electrode performance is due to the extremely high theoretical specific capacity of silicon, nearly 4000 mA h g^{-1}, which is about ten times higher than those of commercial graphite anodes. And the performance is also determined by the stability of the solid electrolyte interphase (SEI) and the contact between Si nanoparticles and carbon matrix. Because the Si volume expansion at high current rate was not that considerable, the SEI layers were more stable and thus more reversible charge–discharge reaction occurred.

Han et al. [377] have investigated TiO_2 as the anode for lithium ion batteries. An improved electrochemical performance of the TiO_2 hollow fibers and nitrated TiO_2 hollow fibers was reported than that of solid ones. Ti $(OiPr)_4$ was used as a titanium precursor and PVP as a stabilizer in the sheath phase, and mineral oil was used as the core phase. After natural drying, extraction of mineral oil and calcination, hollow fibers were obtained. The nitridation step was processed through another annealing treatment under ammonia and the related schematic morphology changes are shown in Fig. 8a–c. The solid, hollow and nitrated hollow TiO_2 nanofibers exhibited initial columbic efficiencies of 75.8, 77.1 and 86.8%, and capacity retentions of 96.8, 98.8 and 100% after 100 cycles. And at different current rates, hollow fibers exhibited a higher capacity than solid fibers. This improved performance is due to the larger surface area (around 25%) for lithium ions to combine with active materials, and the decreased diffusion length of the lithium ions down to nearly 50%. The diffusion length is the maximum distance, through which lithium ions must diffuse to combine with anodic materials. Therefore, the longer the diffusion distance is, the more adverse it is for the performance of the lithium ion battery. And the high conductivity of the nitrated hollow nanofibers will also benefit its performance. The enhanced anode stability arises from the hollow structure, which provides space for volume expansion during litigation.

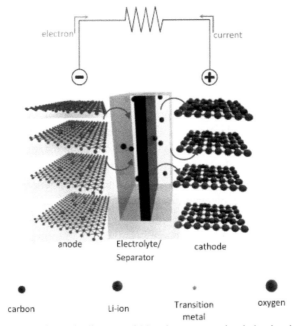

electron

current

$\ominus$ $\oplus$

anode   Electrolyte/   cathode
                Separator

carbon   Li-ion   Transition   oxygen
                                metal

FIGURE 7.43   The schematic diagram of (a) volume expansion behavior during contact-litigation and (b) ionization of Li atoms in the Si matrix.

7.14.2   SOLAR CELLS AND LUMINESCENCE

Other than serving as electrodes for lithium ion batteries, coaxial electrospun fibers have also been applied in other electrochemical fields, such as electrodes of solar cells and luminescence materials. Due to the nonrenewability of traditional fossil energy and its environmental impact, scientists have been searching and developing clean future energy resources, among which solar, wind, nuclear and even biomass energies have been used commercially. Particularly, solar energy has attracted extensive attention due to its environmentally friendly nature, stabilization and regeneration. Furthermore, the energy that the earth receives in just one hour can support the whole energy consumption of our world in one year. Thus solar energy has great potential to be developed for future deployments through conversion to electricity in solar cells. However, the charge recombination especially at the anode is a very serious problem that limits the efficiency [116].

Different strategies have been explored to suppress the charge recombination, for example, reduced recombination sites [378] and construction of an energy barrier. The core-sheath nanofibers are excellent candidates that can satisfy these two requirements. The tiny dimension could reduce the possibility that electrons recombine with the electrolyte and the sheath phase can naturally prevent serious combination as barriers. On the other hand, light is a very common, but very potential future energy. There are two ways to generate light, incandescence and luminescence. Incandescence describes that items such as tungsten filament can emit light when heated to a very high temperature. Compared with incandescence, luminescence refers to "cool light," such as the screens of electronic devices. Luminescence is caused by the jump of excited electrons back to the less excited or ground state. The energy difference is released in the form of "cool light" [379]. Luminescence can be classified into many different types according to input energy sources, such as chemiluminescence, electroluminescence, triboluminescence and bioluminescence. If the input energy is another kind of light, for example, ultraviolet radiation or X-ray, this luminescence is called fluorescence. Many materials have been found to be luminescent, including transition metal ions, rare earth metal complexes, heavy metals, quartz, feldspar and aluminum oxide with electron–hole pairs. Some organic materials are also luminescent. Luminescence makes materials detectable and scientists are trying to develop new signaling systems and devices with luminescent materials aiming to fabricate sensors with high efficiency and accuracy in detection, biological and medicine engineering. New luminescent materials have been prepared for light-emitting diodes (LEDs), which have longer service time, lower energy consumption and stronger brightness. The utilization of core-sheath nano-fibers could increase the contact area, which is beneficial for the excitation process. On the other hand, the fiber form can integrate optical and optoelectronic devices into textile [116, 117, 380].

7.14.3   FILTRATION

Nanofibers are a perfect choice for different types of separation, such as air purification, solution filtration and osmosis separation, due to their extremely high surface area to volume ratio, highly porous structure and chemical properties in some special cases, as well as the flexibility and low cost [52, 117]. Generally, the electrospun nanofiber mats can be used for filtration directly and some researchers have used fibers for waste water purification [381]. The unique structure of hollow nanofibers becomes the intrinsic advantage as the osmosis membrane material and the appropriate selection of sheath material could promise better performance in separation. Since the convenient fabrication of hollow fibers through the coaxial electrospinning process, it has opened a new door to this field [303]. Anka et al. [382] have reported the wonderful separation performance of the PAN hollow fiber membrane upon treatment with Indigo carmine dye and sodium chloride solutions. They used PAN–DMF solution as the sheath and mineral oil as the core. The as-spun fibers were sunk into hexane to remove the core phase. The fabrication and filtration processes are illustrated in Fig. 7.44 (a) and (b). The results showed that the dye and sodium salt were rejected 100 and 97.7%, respectively.

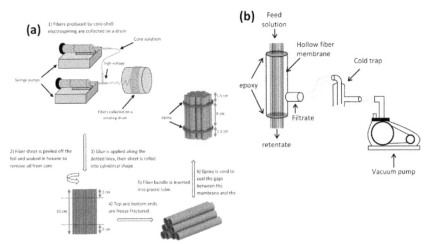

FIGURE 7.44   The schematic diagram of the hollow fiber membrane filtration setup. Reprinted with permission from American Chemical Society.

7.14.4   ELECTROSPUN CORE-SHELL NANOFIBERS AS DRUG-DELIVERY SYSTEMS

If drugs or bioactive agents are encapsulated by a shell polymer, core-shell electrospun nanofibers can be used for functional drug delivery. In this regard, coaxial electrospinning might be particularly suitable for making biomimetic scaffolds with drug delivery capability. The advantage is that it does not require the drug to be electrospinnable or for it to have good physico-chemical interaction with the carrier polymer [383–386]. In contrast, for the cases of drugs loaded by blend electrospinning, poor interaction between the drug and polymer [122, 123, 129] and drug nonelectrospinnability [387] both tremendously affect the drug distribution in the polymer matrix and consequently the release behavior.

The benefits of using core-shell nanofibers for such a purpose are quite obvious. Firstly, it will be able to preserve those labile biological agents such as DNA and growth factors from being deactivated or denatured even when the applying environment is aggressive. In fact, such protection begins as early as during the fabrication stage because, unlike blend electrospinning, the aqueous solution Secondly, core-shell nanofibers belong to reservoir type drug release device; therefore it will be possible to address the burst release problem noted in those electrospun fibers where drugs were usually incorporated through electrospinning a blend of the drug and polymer carrier [120, 123, 124, 129]. Furthermore, by manipulating the core-shell nano-/microstructure, desired and controlled releasing kinetics could be achieved.

7.14.4.1   DRUG RELEASE MODELING

Modeling loading and releasing drugs will provide a basis for further design and optimization of processing conditions to control the core-sheath nanostructure so as to achieve highly sustainable, controllable, and effective drug factor releases. In the context of tissue engineering applications, as delivery of growth factors is indispensable in the course of tissue regeneration, it is believed that coaxial electrospinning and the produced core-shell nanofibers will have great potential to locally regulate cellular process for a prolonged time through controlled release of these appropriate growth factors directly into the cell living microenvironment [305, 386].

Related to core-shell fibers, two controlling phenomena can be considered: diffusion through the polymer shell (barrier diffusion) or partition of the drug from the core to the shell. The diffusion through the shell polymer should not be too slow; otherwise this diffusion will be rate-limiting step. In this instance, the system behaves as monolith fibers and not core-shell fibers (reservoir system). Shell porosity must also be carefully controlled since the drug from the core will be released through water-filled channels rather than through the barrier/shell polymer. Composite fibers that contain drug vehicles such as microspheres and nanoparticles are also a type of reservoir system (double barrier system) in which the drug molecules have to diffuse through longer pathways: the polymer comprising the vehicle and the "shell" polymer.

The power law equation, which was developed considering that the main mechanism for drug release is drug diffusion through the polymer or solvent diffusion inside the polymer that produces polymer relaxation/chain rearrangement is the most widely used equation in works concerning drug release:

$$\frac{me}{m_{cot}} = a_0 + kt^n \qquad (7.14.1)$$

where m_t/m_{tot} is the fractional release of the drug at time t; a_0 is a constant, representing the percentage of burst release, k is the kinetic constant and n is the release exponent, indicating the mechanism of drug release (which can either be Fickian drug diffusion or polymer relaxation and an intermediate case combining the two).

Other models consider different phenomena that control the release such as desorption due to the fact that under the assumption of diffusion control, 100% release of the drug is expected, but this was not verified experimentally. In the desorption model, the release is not controlled by diffusion, but by desorption of the drug from fiber pores or from the fiber surface. Thus, only the drug on the fiber and pore surfaces can be released, whereas the drug from the bulk can only be released when the polymer starts to degrade. These assumptions are similar to the theory of mobile agent that can be released by diffusion and the immobilized agent, which can be released through degradation. The equation 14 is based on a pore model, in which the effective drug diffusion coefficient, Deff is considered and not the actual diffusion coefficient in water, D (with Deff/D $\leq$ 1) because desorption from the pore is the rate limiting step and not drug diffusion in water, which is relatively fast.

$$\frac{m_e}{m_{cot}} = a\left[1 - exp\left(-\frac{\pi^2}{8}\frac{t}{\tau_r}\right)\right] \qquad (7.14.2)$$

where the porosity factor a $a = m_{s0}/(m_{s0} + m_{b0}) < 1$; with m_{s0} and m_{b0} being the initial amount of drug at the fiber surface and the initial amount of drug in the fiber bulk, respectively; m_t is the drug amount released at time t, while the total initial amount of drug in the fiber is $m_{tot} = m_{s0} + m_{b0}$ and τ_r is the characteristic time of the release process [388].

Various release kinetics exist for drug release and the most desirable one is the zero-order kinetics in which the drug is released at constant rate, independent of. Usually, zero-order kinetics is achieved for reservoir systems such as core-shell fibers or composite fibers in which the drug is properly encapsulated in the core of the fiber or in other vehicles (micro/nanoparticles).

Burst stage in this kind of system is diminished (or nonexistent) because there is no drug deposited on the surface of the fibers. As the controlling release phenomena is drug partition from one phase to another and not diffusion, there is no decrease in release rate over time as expected in a diffusion-controlled system (the release rate depends on the concentration gradient and on the length of diffusion path; as release proceeds, the concentration gradient decreases and the diffusion length increases and both contribute to slowing down the release rate).

The change of drug concentration on the fiber surface can be represented by adsorption and desorption terms with constant rates k_{ads}, k_{des}, respectively:

$$\frac{\partial c_B}{\partial_t} = k_{ada} \times C_A \times (C_B^{max} - C_B) - K_{des} \times C_A \qquad (7.14.3)$$

where C_B is drug concentration expressed in relation to the weight of the polymer material (kg/kg), C_A is drug concentration in the fluid (kg/m³).

Spatiotemporal distribution of the drug in the fluid surrounding fibers is described by diffusion equation and additional mass source term describing drug released from the fibers:

$$\frac{\partial C_A}{\partial t} = D_A \nabla^2 C_A + p_p \times \frac{\partial C_B}{\partial t} \qquad (7.14.4)$$

where ρ_p is polymer density (kg/m³).

7.15   ADVANTAGES AND DISADVANTAGES OF CO-ELECTROSPINNING PROCESS

Co-electrospinning is of particular interest for those core materials that will not form fibers via electrospinning by themselves. Here the shell polymer can serve as a template for the core material leading to cable-type structures. Cores shell fibers of this type will certainly foster applications, for example, in the field of microelectronics, optics and medicine, which justifies intensive research in this direction. Co-electrospinning can also facilitate manufacturing of polymer nanotubes and it allows one to eliminate the vapor deposition stage characteristic of three-stages TUFT-process and to develop a novel two-stage process [301, 389].

This method cannot only be used to spin the unspinnable polymers (polyaramid, nylon, and polyaniline) into ultrafine fibers, but also ensures keeping functionalizing agents like antibacterial and biomolecules agents inside nanofibers. Inspite of all advantages of coaxial electrospinning process, controlling of coelectrospinning is more difficult than simple electrospinning because supervising the flow rates and the physical properties of the two liquids, the current transported by them, the diameters of the inner and outer jets, and the size of the droplets resulting from their break up are major concerns [325].

7.16   GENERAL ASSUMPTIONS IN COAXIAL ELECTROSPINNING PROCESS

In all offered models for coaxial electrospinning following assumptions are considered:

1. The gas and the liquids are incompressible Newtonian fluids;
2. The effects of gravity, magnetic field, and mass transfer at the interfaces are neglected;
3. The fluids are assumed inviscid, and the flow is irrotational;
4. In most cases, liquid 2 (the driving liquid) is a perfect conductor having an infinite conductivity. While liquid 1 is a perfect dielectric, which does not influence the flow because charges stay on the outer interface (jet boundary) according to electrostatics;
5. Charges are introduced from the anode and transported by means of convection and conduction of liquid 2;
6. All fluids are assumed having constant conductivity and dielectric properties;
7. Models have neglected the elastic effect due to the evaporation of the solvent.

7.17   CONCLUSIONS AND FUTURE PERSPECTIVES

Despite the intensive research in the field of nanofibers a number of unanswered questions still remain to act as a driving force for further studies. The largest challenge is a complete understanding of the electrospinning mechanism. In order to control the properties, orientation and mass production of the nanofibers, it is necessary to understand quantitatively how electrospinning transforms the fluid solution through a millimeter-sized needle into solid fibers having diameters that are four-to-five orders smaller. The next bottleneck in the electrospinning is the process efficiency and repeatability. Furthermore, the construction of a proper, three-dimensional scaffold remains a technological challenge, while from the point of view of drug delivery the drug loading has to be increased and the initial burst release has to be reduced in many cases.

Coaxial electrospinning is an emerging technique with only 10 years of history. Although the technique has multiple advantages over traditional microencapsulation/nanoencapsulation processes, further advancement is challenged by the complex physics of the process and a large number of design, material and process parameters contributing to the process outcome.

As an innovation of conventional electrospinning, coaxial electrospinning has attracted much attention and has been studied deeply. The parameters, including both the internal physical properties and the external operation factors have been analyzed theoretically and experimentally. In particular, the interaction between different phases due to the contact and the heterogeneousness, which is also the characteristic of coaxial electrospinning, is discussed in detail. The evaporation rate, polymer concentration, conductivity and flow rate have been systematically reviewed in achieving smooth and uniform morphology as well as fine dimension. Generally, a stronger and longer electrospinning process can result in more smooth fibers. All strategies to enhance the electrical force and elongate the electrospinning whipping stage will favor the final fibers. However, there will always be a suitable operation range for all parameters, only within which nonbeaded fibers can be obtained. The lower limitation is the critical value, beyond which the electrospinning process can take place. And the higher limitation can prevent the strong effect on the formation of core-sheath fibers. The interaction due to the property difference should promise the formation of the core-sheath Taylor cone and the smooth electrospinning process. Too huge a difference may lead to a significantly different behavior during electrospinning. Through additional treatment, hollow nano- fibers can be fabricated. The following extraction process or thermal treatment can remove the core phase, maintaining the complete sheath structure. In addition, the thermal treatment is much quicker and can stabilize and carbonize the polymer matrix at the same time, which can increase its efficiency.

Nearly all polymer solutions and their polymer composite matrices could be fabricated into core-sheath and hollow nanofibers through the coaxial electrospinning technique. Through appropriate post-treatments of the electrospun fibers, various inorganic nanofibers or nanotubes have been developed with much wider applications including photocatalysis, lithium ion batteries supercapacitors, solar cells, etc. The obtained nanofibers possess extremely high specific surface area and combined properties, which have shown superior performance over the traditional electrospun fibers. However, there are also some challenges, such as the dimension and order of the obtained nanofibers, the improved performance in their applications and the expanded applications of the core-sheath and hollow nanofibers. So far, the fine dimension of the core-sheath nanofibers is about 60 nm, which is much larger than that of the fibers through conventional electrospinning, about a few nanometers. The obtained fibers are in random order, not aligned. To obtain ultra-fine core-sheath nanofibers, the matrices should be selected properly for phases, as well as the operation parameters, collection method and post-treatment methods, such as thermal treatment. To develop the properties and performance of core-sheath and hollow nanofibers, new materials should be tested to integrate into both phases as desired, for example, biomaterials can be used as the matrix to increase the biocompatibility in applications of tissue engineering. Polymers can be integrated into a sheath phase to fabricate high strength materials with very low density. Different nanoparticles can be loaded to achieve suitable interactions during charging and discharging processes and catalysis. Core-sheath and hollow nanofibers are very promising potential materials in various applications due to their flexibility and excellent physico-chemical properties. Therefore, more amazing and meaningful applications of the materials are envisioned to be achieved in the future when deployed in other fields, such as the magnetic nanocomposites, semiconductors, carbon materials with high strength and conductivity, anticorrosion and fire-proof properties.

Current research efforts on this technique focus on concept approval and feasible study, with about 100 journal papers based on individually customized experimental setups, a limited number of material combinations and empirical process parameters. Numerical study of coaxial electrospinning is rare due to the complex physical nature of the phenomenon. The researchers working in this area can be categorized into two groups.

The first group is biomedical engineers and clinical researchers who are interested in modeling encapsulating biological agents in core-shell nanofibers and simulation drug loading and releasing. The second group includes experimentalists and theorists in the field of fluid mechanics who are interested in experimental and modeling works associated with coaxial electrospinning.

Discussed models for coaxial electrospinning have some assumptions inorder to simplify the process. These assumptions would restrict the accuracy of final results. So it is appropriate to reduce these approximations and make the virtual experimental environment more close to real one, which can be considered in future approaches.

The coaxial electrospinning model can be extended for situations where the inner liquid has high electric conductivity and the outer liquid is dielectric. Also, the effects of gravity, magnetic field, air drag, solvent evaporation and mass transfer at the interfaces can be considered. In all offered models diameter of both layers of core shell nano jets considered to be constant, while they reduce gradually in real. This fact can be taken into account too. Further investigation could also include the effect of the viscosity of the fluids on the instability of the coaxial jet.

APPENDIXES

APPENDIX A: DERIVATION OF THE DISPERSION RELATION FOR CASE I

In case I, the inner liquid and air are inviscid, with parabolic basic velocity profiles. In such a case, the radial component of the velocity satisfies the following Bessel equation [369]:

$$\frac{d^2 \hat{u}_{1.3}}{dr^2} + \frac{1}{r}\frac{d\hat{u}_{1.3}}{dr} - \left(k^2 + \frac{1}{r^2}\right)\hat{u}_{1.3} = 0. \tag{A1}$$

where the 'hats' stand for the initial perturbation amplitudes. The solutions of Eq. (A1) are the linear combination of the modified Bessel functions. Their coefficients are determined by using the boundary conditions (7.13.87) and (7.13.91)–(7.13.93). Suppose that the cylindrical coordinate system (r, θ, z) is moving with velocity U_2, then the solutions are

$$\hat{u}_1 = \omega \hat{n}_1 \frac{I_1(kr)}{I_1(ka)} \quad \hat{u}_3 = \omega \hat{n}_2 \frac{I_1(kr)k_1(kb) - k_1(kr)I_1(kb)}{I_1(k)K_1(kb) - K_1(k)I_1(kb)}. \tag{A2}$$

The axial velocity and pressure can be obtained using Eqs. (7.13.85) and (7.13.86). As the outer liquid is viscous, we decompose the velocity perturbation into two terms [315, 354], that is, $u_2 = up + uv$, where up and uv satisfy the following linearized equations:

$$\nabla \cdot u_p = 0. \qquad \frac{\partial u_p}{\partial t} = -\nabla p_2. \qquad \nabla \cdot u_v = 0. \qquad \frac{\partial u_v}{\partial t} = \frac{1}{Re}\nabla^2 u_v. \tag{A3}$$

For up, a potential function φ_2 ($up = \nabla \varphi_2$) is introduced, which satisfies the Laplace equation $\nabla^2 \varphi_2 = 0$. Therefore, the amplitude $\hat{\varphi}2$ satisfies the modified Bessel equation

$$\frac{d^2 \hat{\phi}_2}{dr^2} + \frac{1}{r}\frac{d\hat{\phi}_2}{dr} - k^2\hat{\phi}_2 = 0. \tag{A4}$$

The solution is $\hat{\varphi}_2 = A_1 I_0(kr) + A_2 K_0(kr)$, where A_1 and A_2 are coefficients to be determined by boundary conditions. The amplitudes of the radial and axial velocity components and the pressure can also be obtained [346–347].

For the viscous part, the radial momentum equation yields

$$\frac{d^2 \hat{u}_v}{dr^2} + \frac{1}{r}\frac{d\hat{u}_v}{dr} - \left(k^2 + Re\omega + \frac{1}{r^2}\right)\hat{u}_v = 0. \tag{A5}$$

The solution is $\hat{u}v = A_3 I_1(lr) + A_4 K_1(lr)$, where $A3$ and $A4$ are coefficients to be determined, and $l = \sqrt{k2 + Re\omega}$. Then the continuity equation gives the solution of the axial velocity component, $w\hat{v} = il(A_3 I_0(lr) - A_4 K_0(lr))/k$.

For the electric field, the perturbations of the electrical potentials, ψi, $i = 1, 2, 3$, also satisfy the Laplace equation. Further, their perturbation amplitudes $\hat{\psi} i$, also satisfy the modified Bessel Eq. (A4). Consequently their solutions are the linear combinations of two modified Bessel functions $I_0(kr)$ and $K_0(kr)$, which gives five coefficients $A_5 - A_9$ to be determined.

For the present problem, there are in all 12 unknown quantities: $A_1 - A_9$, $\hat{\eta}_1$, $\hat{\eta}_2$ and $\hat{q}s$ (the perturbation amplitude of the surface charge density). On the other hand, the boundary conditions (the Eqs. (7.13.92)–(7.13.93) and (7.13.97)–(7.13.102)) provide 12 equations to solve the unknown. These equations set up a homogeneous linear system. The system has nontrivial solutions only if the determinant of its coefficient matrix is null, which provides the dispersion relation we need. However, considering the size of the coefficient matrix, it is hard to obtain an explicit expression for the dispersion relation in such a way [346, 347].

Therefore, the equations were solved step by step as outlined in the following, aiming to obtain the dispersion relation in a more compact form.

For the outer liquid, according to the continuity equation,

$$\hat{\omega}_2 = \frac{i}{k}\left(\frac{d\hat{u}_2}{dr} + \frac{\hat{u}_2}{r}\right) \tag{A6}$$

Differentiating the above equation with respect to r and using the momentum equation in the radial direction, we have

$$\frac{d\hat{\omega}_2}{dr} = \frac{i}{k}\left(l^2\hat{u}_2 + Re\frac{d\hat{p}_2}{dr}\right) \tag{A7}$$

Then using the linearized kinematic boundary conditions at the interfaces, it can be expressed as:

$$\frac{d\hat{\omega}_2}{dr}\bigg|_{r=a+n_1} = \frac{i\omega}{k}[l^2\hat{n}_1 - Rek(A_1 I_1(ka) - A_2 K_1(ka)]. \tag{A8a}$$

$$\frac{d\hat{\omega}_2}{dr}\bigg|_{r=1+n_2} = \frac{i\omega}{k}[l^2\hat{n}_2 - Rek(A_1 I_1(k) - A_2 k_1(k)]. \tag{A8b}$$

Substituting the corresponding solutions into the linearized kinematic boundary conditions and the tangential dynamic boundary conditions, we obtain the expressions for $A_1 - A_4$. The process and the expressions are omitted for brevity. On the other hand, substituting the corresponding solutions into the electric field boundary conditions, $A_5 - A_9$ can be obtained. Finally, substituting the expressions for the corresponding quantities into the linearized normal dynamic boundary conditions, Eqs. (7.13.111) and (7.13.112) are obtained, together with the dispersion relation (the Eq. (7.13.110)).

APPENDIX B: DERIVATION OF THE DISPERSION RELATION FOR CASE II

In case II, the dispersion relation is also written in the form of Eq. (7.13.110), but with the amplitude ratio of the interface perturbation [346, 347]:

$$\left(\frac{\hat{n}_2}{\hat{n}_1}\right) = \left[\frac{\Delta_4}{a}\left(\frac{\pi}{H}+\omega\right)\left(\frac{EuRe\zeta}{\omega}+1+\frac{l_2^2}{k^2}\right) - \frac{\Delta_3}{a}\frac{\pi}{H}\left(\frac{EuRe\zeta}{\omega}+2\right)\right]^{-1}$$

$$\times\left[k\Delta_1\Delta_4\left(\frac{\pi}{H}+\omega\right)\left(1+\frac{l_2^2}{k^2}-\mu_r\left(1+\frac{l_1^2}{k^2}\right)\right)\right.$$

$$-l_2\Delta_3\Delta_6\frac{\pi}{H}\left(2-\mu_r\left(1+\frac{l_1^2}{k^2}\right)\right)$$

$$+\omega\Delta_3\Delta_4(2(1-\mu_r))\left(\theta-\frac{1}{a}\right)-\frac{Re\theta\pi}{k^2}\frac{\pi}{H})$$

$$+\frac{T}{Wea^2}(1-(ka)^2)Re\Delta_3\Delta_4 + \frac{EuRe\xi}{\omega}\frac{1}{a}\left(\frac{\pi}{H}\Delta_3 - (\frac{\pi}{H}+\omega)\Delta_4\right)].\qquad (B1a)$$

$$\left(\frac{\hat{n}_1}{\hat{n}_2}\right)_2 = \left[\Delta_4\left(1+\frac{l_2^2}{k^2}-\mu_r\left(1+\frac{l_1^2}{k^2}\right)\right)\left(-(l_2^2+k^2)+\frac{Sk^2E\Delta_1}{H\Delta_3\Delta_4}\right)\right.$$

$$+\Delta_3\left(2-\mu_r\left(1+\frac{l_1^2}{k^2}\right)\right)\left(2k^2 - \frac{Skl_2E\Delta_6}{H\Delta_3\Delta_4}\right)$$

$$-\frac{SRe\omega\theta E}{kH} - Re^2\Delta_3\Delta_4Euk\xi J + \frac{EuRe\xi}{\omega}\left(\frac{SkE}{aH}\left(\frac{1}{\Delta_4}-\frac{1}{\Delta_3}\right)\right.$$

$$+k^3\Delta_2\Delta_4\left(1+\frac{l_2^2}{k^2}\right)-2l_2k^2\Delta_3\Delta_5)]^{-1}\left[-\frac{SkE^2}{aH\Delta_3\Delta_4}+Re^2H_2\Delta_3\Delta_4\right.$$

$$-k\Delta_4((l_2^2+k^2)\Delta_2+2k\Delta_3)\left(\frac{EuRe\zeta}{\omega}+1+\frac{l_2^2}{k^2}\right)+2l_2k^2\Delta_3$$

$$\times\left(\Delta_5+\frac{1}{l_2}\Delta_4\right)-\frac{EuRe\zeta}{\omega}\frac{SkE}{aH}\left(\frac{1}{\Delta_4}-\frac{1}{\Delta_3}\right)].\qquad (B1b)$$

where

$$\zeta = \left[\varepsilon_{r2}(\omega + \tau)k - \omega J + \frac{Eu}{\omega}\left(\frac{S}{aH}\left(\frac{1}{\Delta_3} - \frac{1}{\Delta_4} \right) \right)^2 + k\left(k\frac{\Delta_2}{\Delta_3} - l_2\frac{\Delta_5}{\Delta_4} \right) \right]^{-1}$$

$$\times \left[\omega J - \frac{k^2\Delta_2}{Re\Delta_3}\left(1 + \frac{l_2^2}{k^2} \right) + \frac{2l_2 k\Delta_5}{Re\Delta_4} - \frac{S}{Re\,aH}\left(\frac{1}{\Delta_3} - \frac{1}{\Delta_4} \right)\left(\frac{1}{\Delta_3}\left(1 + \frac{l_2^2}{k^2} \right) - \frac{2}{\Delta_4} \right) \right]. \tag{B2}$$

$$\xi = \left[\varepsilon_{r2}(\omega + \tau)k - \omega J + \frac{lEu}{\omega}\left(\frac{S}{aH}\left(\frac{1}{\Delta_3} - \frac{1}{\Delta_4} \right) \right)^2 + k\left(k\frac{\Delta_2}{\Delta_3} - l_2\frac{\Delta_5}{\Delta_4} \right) \right]^{-1}$$

$$\times \left[\frac{k}{Re\Delta_3}\left(1 + \frac{l_2^2}{k^2} - \mu_r\left(1 + \frac{l_1^2}{k^2} \right) \right)\left(1 + \frac{S\Delta_1}{H}\left(\frac{1}{\Delta_3} - \frac{1}{\Delta_4} \right) \right) \right.$$

$$\left. - \frac{k}{Re\Delta_4}\left(2 - \mu_r\left(1 + \frac{l_1^2}{k^2} \right) \right)\left(1 + \frac{l_2 S\Delta_6}{kH}\left(\frac{1}{\Delta_3} - \frac{1}{\Delta_4} \right) \right) - \frac{S\omega\theta}{k^2 H}\left(\frac{1}{\Delta_3} - \frac{1}{\Delta_4} \right) \right] \tag{B3}$$

and the relative viscosity of the inner liquid $\mu r = \mu_1/\mu_2$. Note that κ, H_2 and the other symbols are the same as in case I. It is shown that the dispersion relation for case II reduces to that for case I as long as the viscosity of the inner liquid is neglected (i.e., $\mu r = 0$). The inner liquid viscosity makes the problem much more complicated [346, 347].

APPENDIX C: DERIVATION OF THE THIN LAYER APPROXIMATION $A \rightarrow 1$

In the instability analysis of an annular viscous liquid jet, the thin sheet approximation is usually derived [350, 370, 371]. If the inner and outer radii of the annular jet are supposed to approach infinity with the thickness of the liquid layer constant, a plane liquid sheet is obtained, as in Refs. [370] and [350]. Usually the approach of Ref. [371], is applied for expanding the dispersion relation under the thin layer limit. The derivation process is outlined below, taking case I as an example.

Define a quantity $\delta = 1 - a$, the dispersion Eq. (2.22) is expanded under the thin layer approximation $\delta \leq 1$. First, the expansions of $\Delta_0 - \Delta_6$ are

$$\Delta_0 = -\delta - \frac{\delta^2}{2} + o(\delta^3), \quad \Delta_1 = \frac{1}{k} + \frac{k\delta^2}{2} + o(\delta^3). \tag{C1a}$$

$$\Delta_2 = \frac{1}{k} + \frac{\delta}{k} + \frac{k\delta^2}{2}\left(1 + \frac{2}{k^2} \right) + o(\delta^3), \quad \Delta_3 = -\delta - \frac{\delta^2}{2} - \frac{k^2\delta^3}{6}\left(1 + \frac{3}{k^2} \right) + o(\delta^4), \tag{C1b}$$

$$\Delta_4 = -\delta - \frac{\delta^2}{2} - \frac{l^2\delta^3}{6}\left(1 + \frac{3}{l^2}\right) + o(\delta^4), \quad \Delta_5 = \frac{1}{l} + \frac{\delta}{l} + \frac{l\delta^2}{2}\left(1 + \frac{2}{l^2}\right) + o(\delta^3), \quad \text{(C1c)}$$

$$\Delta_6 = \frac{1}{l} + \frac{l\delta^2}{2} + o(\delta^3). \tag{C1d}$$

Then, the expansions of κ, ζ and ξ are

$$k = \frac{\varepsilon_{r1}}{\varepsilon_{r2}}\frac{I_1(k)}{I_0(k)} + k\left(1 - \frac{\varepsilon_{r1}}{\varepsilon_{r2}}\right)\left[1 + \frac{\varepsilon_{r1}}{\varepsilon_{r2}}\left(\frac{I_1(k)}{I_0(k)}\right)^2\right]\delta + o(\delta^2). \tag{C2a}$$

$$\zeta = \frac{\frac{k}{Re}\left(\frac{1}{\delta} + \frac{1}{2}\right)\left(\frac{l^2}{k^2} - 1\right) + \omega J + \frac{k}{Re}\left(\frac{l^2}{k^2} - 1\right)\left(\frac{1}{4} - \frac{k^2}{3}\right)\delta + o(\delta^2)}{\varepsilon_{r1}(\omega + \tau)\frac{I_1(k)}{I_0(k)} - \omega J + \left(\varepsilon_{r2}(\omega + \tau)ko_{(\delta)} - \frac{EuRek}{6}\right)\delta + o(\delta^2)} \tag{C2b}$$

$$\xi = \frac{\frac{k}{Re}\left(-\frac{1}{\delta} + \frac{1}{2}\right)\left(\frac{l^2}{k^2} - 1\right) + \frac{k}{Re}\left(\frac{l^2}{k^2} - 1\right)\left(\frac{1}{4} - \frac{k^2}{6}\right)\delta + o(\delta^2)}{\varepsilon_{r1}(\omega + \tau)\frac{I_1(K)}{I_0(k)} - \omega J + \left(\varepsilon_{r2}(\omega + \tau)ko_{(\delta)} - \frac{EuRek}{6}\right)\delta + o(\delta^2)} \tag{C2c}$$

where $\kappa O(\delta)$ represents a coefficient of $O(\delta)$ in the expansion of κ. Now we write the coefficients of the interface perturbation amplitudes $\hat{\eta}_1$ and $\hat{\eta}_2$ individually for each order, i.e., for $O(1)$,

$$(\hat{n}_1)_2: (k^2 - l^2)\frac{EuRe\xi o_{(1/\delta)}}{\omega}, \qquad (\hat{n}_2)_2: (l^2 - k^2)\frac{EuRe\xi o_{(1/\delta)}}{\omega}, \tag{C3}$$

$$(\hat{n}_1)_2: (k^2 - l^2)\frac{EuRe\xi o_{(1/\delta)}}{\omega}, \qquad (\hat{n}_2)_2: (l^2 - k^2)\frac{EuRe\xi o_{(1/\delta)}}{\omega}, \tag{C4}$$

for $O(\delta)$,

$$(\hat{n}_1)_1: Re^2EukJ\zeta o_{(1/\delta)} + (l^2 - k^2)\frac{EuRe\zeta o_{(1)}}{\omega} + \frac{(l^2+k^2)^2}{k^2}$$

$$-4k^2 + \frac{3}{2}(l^2 - k^2)\frac{EuRe\zeta o_{(1/\delta)}}{\omega} \tag{C5}$$

$$(\hat{n}_2)_1: -Re^2EukJ\xi o_{(1/\delta)} + (k^2 - l^2)\frac{EuRe\xi o_{(1)}}{\omega} + \frac{(l^2+k^2)^2}{k^2}$$

$$-4k^2 + \frac{3}{2}(k^2 - l^2)\frac{EuRe\xi o_{(1/\delta)}}{\omega} \tag{C6}$$

$$(\hat{n}_1)_2 : (k^2 - l^2)\frac{EuRe\zeta o_{(1)}}{\omega} - \frac{(l^2 + k^2)^2}{6} + 4k^2 + \frac{1}{2}(k^2 + l^2)\frac{EuRe\zeta o_{(1/\delta)}}{\omega}, \quad (C7)$$

$$(\hat{n}_2)_2 : (l^2 - k^2)\frac{EuRe\xi o_{(1)}}{\omega} - \frac{(l^2 + k^2)^2}{k^2} + 4k^2 + \frac{1}{2}(l^2 - k^2)\frac{EuRe\xi o_{(1/\delta)}}{\omega}, \quad (C8)$$

for $O(\delta^2)$,

$$(\hat{n}_1)_1 : Re^2(H_2)_{o_{(1)}} + \left(\frac{(l^2 - k^2)^2}{6} + 2(l^2 - k^2)\right)\frac{EuRe\zeta o_{(1/\delta)}}{\omega} + \frac{3}{2}(l^2 - k^2)\frac{EuRe\zeta o_{(1)}}{\omega}$$

$$+(l^2 - k^2)\frac{EuRe\zeta o_{(\delta)}}{\omega} + Re^2 EukJ\zeta o_{(1/\delta)} + \frac{3}{2}\frac{l^4 - k^4}{k^2} + l^2 - k^2, \quad (C9)$$

$$(\hat{n}_2)_1 : -Re^2 EukJ\left(\xi o_{\left(\frac{1}{\delta}\right)} + \zeta o_{(1)}\right) - \left(\frac{(l^2 - k^2)^2}{6} + 2(l^2 - k^2)\right)\frac{EuRe\xi o_{(1/\delta)}}{\omega}$$

$$+\frac{3}{2}(k^2 - l^2)\frac{EuRe\xi o_{(1)}}{\omega} + (k^2 - l^2)\frac{EuRe\xi o_{(\delta)}}{\omega} + \frac{1}{2}\left(\frac{(l^2 + k^2)^2}{k^2} - 4k^2\right) \quad (C10)$$

$$(\hat{n}_1)_2 : \left(-(l^2 + k^2)\left(\frac{l^2}{6} + \frac{1}{2}\right) + 2k^2\left(\frac{k^2}{6} + \frac{1}{2}\right)\right)\frac{EuRe\xi o_{(1/\delta)}}{\omega}$$

$$+\frac{k^2 - l^2}{2}\frac{EuRe\zeta O_{(1)}}{\omega} + (k^2 - l^2)\frac{EuRe\zeta O_{(\delta)}}{\omega} - \frac{(l^2 + k^2)^2}{2K^2} + 2k^2, \quad (C11)$$

$$(\hat{n}_2)_2 : Re^2(H_1)_{o_{(1)}} + \left((l^2 + k^2)\left(\frac{l^2}{6} + \frac{1}{2}\right) - 2k^2\left(\frac{k^2}{6} + \frac{1}{2}\right)\right)\frac{EuRe\xi o_{(1/\delta)}}{\omega}$$

$$+\frac{l^2 - k^2}{2}\frac{EuRe\xi o_{(1)}}{\omega} + (l^2 - k^2)\frac{EuRe\xi O_{(\delta)}}{\omega} - \frac{(l^2 + k^2)^2}{2k^2} + 2k^2, \quad (C12)$$

Substituting the above expansions into the dispersion relation, it is found that for $O(1)$ and $O(\delta)$ the dispersion relation is inherently satisfied. Therefore, the dispersion relation for the thin layer approximation is given by $O(\delta^2)$. The process of simplification is straightforward and is omitted for brevity. Ultimately the dispersion relation in the form of Eq. (7.13.145) is obtained.

APPENDIX D: TABLE OF APPLIED SYMBOLS AND THEIR DEFINITIONS.

Symbols	Definition	Units
R	Radius of jet	m
v	Jet velocity	m/s
Q	Flow rate	m³/s
I	Jet current	A
J	Current density	A/m²
$A(s)$, S	Cross-sectional area	m²
K	The conductivity of the liquid	S/m
E	Electric field	V/m
L	Spinning distance	M
σ	Electric density	C/m²
P	Linear momentum	kg m/s, (N.S.)
F	Force	N
t	Time	S
m	Mass	Kg
ρ	Density	kg/m³
z	Jet axial position	m
p	Pressure	N/m²
τ_{zz}	Axial viscous normal stress	N/m²
γ	Surface tension	N/m
R'	Slope of the jet surface	–
t_t^e	Tangential electric force	N/m²
t_n^e	Normal electric force	N/m²

ε	Dielectric constant of the jet	–
$\bar{\varepsilon}$	Dielectric constant of the ambient air	–
We	Electric work	J
l	Distance	M
Ue	Electric potential energy	J
τ_{rr}	Radial normal stress	N/m²
$\rho P'$	Polarized charge density	C/m²
P'	Polarization	C/m²
$q, q_0 \& Q_b$	Charge	C
r	Distance between two charges	m
V	Electric potential	kV
τ	Shear stress	N/m²
K'	Flow consistency index	–
m	Flow behavior index	–
μ	Constant	–
$\dot{\gamma}$	Rate of strain tensor	s⁻¹
	Excess stress	N/m²
S_k	Configurational tensor	N/m²
C_k	Viscosity	P
η	Young's modulus	N/m²
E_k	Shear modulus	N/m²
μ_k	Constitutive mobility	m²/(V·s)
β_k	Number of beads per unit volume	m⁻³
N	Temperature	K
T	Elastic modulus	N/m²
G	Filament	m
ξ'	Lagrangian axial strain	–
b	Power exponent	–

Symbol	Description	Units
a	Fiber radius	m
T	Temperature	K
G	Elastic modulus	N/m²
ε'	Lagrangian axial strain	–
K	Coulomb constant	–
λ	Relaxation time	s
<QQ>	The suspension configuration tensor	N/m²
c	A spring constant	–
b_{max}	Maximum extensibility	m
k	Boltzmann's constant	–
n_d	Dumbbells density	Kg/m³
n	The number of fiber per unit area	Kg/m³
φ	Fiber orientation angle	Degree
$\psi(\varphi,l)$	Length distribution	m
$\bar{l}$	The mean segment length	m
I_m	Moment of inertia	Kgm²
ε_{xx}	Tensile strain	m
γ_{xy}	Shear strain	m
$\langle \varepsilon_{ij} \rangle, \langle \varepsilon_k \rangle$	Effective macroscopic strain	m
C_{ijkl}	Effective elasticity tensor	N/m²
$\langle \ \rangle$	Macroscopic properties	m/s
Ui	the velocity of the phase i	m/s
U_ζ	the velocity vector of the interface	N/m²
∇_ζ	surface divergence	–
Ti	Maxwell's constraint tensor	–
I	identity tensor	N/m²
subscripts O and I	the outer liquid and the inner liquid	–
Y_{eff}	effective value of the surface tension	–
α	measure of the liquid average viscosity	–
I_n and K_n	modified Bessel's functions	–
{u}	global displacement vector	V/m
{f}	global nodal force vector	m
$[K_e]$	global stiffness matrix	
Ei	electric field strength	
Di	dielectric displacement	

KEYWORDS

- **Coaxial**
- **Dielectric**
- **Displacement**
- **Displacement Vector**
- **Electrospinning**
- **Nanoparticles**
- **Nanotechnology**
- **Numerical Study**
- **Viscosity**

REFERENECES

1. Poole, C. P., Owens, F. J. (2003). Introduction to Nanotechnology, New Jersey, Hoboken: Wiley. 400.
2. Nalwa, H. S. (2001). Nanostructured Materials and Nanotechnology: Concise Edition: Gulf Professional Publishing. 324.
3. Gleiter, H. (1995). Nanostructured Materials: State of the Art and Perspectives. Nanostructured Materials. 6(1), 3–14.
4. Wong, Y., et al. (2006). Selected Applications of Nanotechnology in Textiles. AUTEX Research Journal. 6(1), 1–8.
5. Yu, B., Meyyappan, M. (2006). Nanotechnology: Role in Emerging Nanoelectronics. Solid-state electronics. 50(4), 536–544.
6. Farokhzad, O. C., Langer, R. (2009). Impact of Nanotechnology on Drug Delivery. ACS Nano. 3(1), 16–20.
7. Serrano, E., Rus, G., Garcia-Martinez, J. (2009). Nanotechnology for Sustainable Energy. Renewable and Sustainable Energy Reviews. 13(9), 2373–2384.
8. Dreher, K. L. (2004). Health and Environmental Impact of Nanotechnology: Toxicological Assessment of Manufactured Nanoparticles. Toxicological Sciences. 77(1), 3–5.
9. Bhushan, B., Introduction to Nanotechnology, in Springer Handbook of Nanotechnology(2010)., Springer. 1–13.
10. Ratner, D., Ratner, M. A. (2004). Nanotechnology and Homeland Security: New Weapons for New Wars: Prentice Hall Professional. 145.
11. Aricò, A. S., et al. (2005). Nanostructured Materials for Advanced Energy Conversion and Storage Devices. Nature Materials. 4(5), 366–377.
12. Wang, Z. L.,,, (2000). Nanomaterials for Nanoscience and Nanotechnology. Characterization of Nanophase Materials: 1–12.
13. Gleiter, H. (2000). Nanostructured Materials: Basic Concepts and Microstructure. Acta Materialia. 48(1), 1–29.
14. Wang, X., et al. (2005). A General Strategy for Nanocrystal Synthesis. Nature. 437(7055): 121–124.
15. Kelsall, R. W., et al. (2005). Nanoscale Science and Technology, New york: Wiley Online Library. 455.

16. Engel, E., et al. (2008). Nanotechnology in Regenerative Medicine: The Materials Side. Trends in Biotechnology. 26(1), 39–47.

17. Beachley, V., Wen, X. (2010). Polymer Nanofibrous Structures: Fabrication, Biofunctionalization, and Cell interactions. Progress in Polymer Science. 35(7), 868–892.

18. Gogotsi, Y., Nanomaterials Handbook(2006)., New york: CRC press. 779.

19. Li, C., Chou, T. A . (2003). Structural Mechanics Approach for the Analysis of Carbon Nanotubes. International Journal of Solids and Structures. 40(10), 2487–2499.

20. Delerue, C., M. Lannoo, Nanostructures: Theory and Modelling(2004). : Springer. 304.

21. Pokropivny, V., Skorokhod, V. (2007). Classification of Nanostructures by Dimensionality and Concept of Surface Forms Engineering in Nanomaterial Science. Materials Science and Engineering: C. 27(5), 990–993.

22. Balbuena, P., Seminario, J. M. (2006). Nanomaterials: Design and Simulation: Design and Simulation. Vol. 18, Elsevier. 523.

23. Kawaguchi, T., Matsukawa, H. (2002). Numerical Study of Nanoscale Lubrication and Friction at Solid Interfaces. Molecular Physics. 100(19), 3161–3166.

24. Ponomarev, S. Y., Thayer, K. M., Beveridge, D. L. (2004). Ion Motions in Molecular Dynamics Simulations on DNA. Proceedings of the National Academy of Sciences of the United States of America. 101(41), 14771–14775.

25. Loss, D., DiVincenzo, D. P. (1998). Quantum computation with quantum dots. Physical Review A. 57(1), 120–125.

26. Theodosiou, T. C., Saravanos, D. A. (2007). Molecular Mechanics Based Finite Element for Carbon Nanotube Modeling. Computer Modeling In Engineering And Sciences. 19(2), 19–24.

27. Pokropivny, V., Skorokhod, V. (2008). New Dimensionality Classifications of Nanostructures. Physica E: Low-dimensional Systems and Nanostructures. 40(7), 2521–2525.

28. Lieber, C. M. (1998). One-dimensional Nanostructures: Chemistry, Physics & Applications. Solid State Communications. 107(11), 607–616.

29. Emary, C. (2009). Theory of Nanostructures, New york: Wiley 141.

30. Edelstein, A. S., Cammaratra, R. C. (1998). Nanomaterials: Synthesis, Properties and Applications: CRC Press.

31. Grzelczak, M., et al., Directed Self-assembly of Nanoparticles. ACS Nano. (2010). 4(7), 3591–3605.

32. Hung, C., et al. (1999). Strain Directed Assembly of Nanoparticle Arrays within a Semiconductor. Journal of Nanoparticle Research. 1(3), 329–347.

33. Wang, L., Hong, R. (2001). Synthesis, Surface Modification and Characterization of Nanoparticles. Polymer Composites. 2, 13–51.

34. Lai, W., et al. (2012). Synthesis of Nanostructured Materials by Hot and Cold Plasma, in International Plasma Chemistry Society: Orleans, France. 5.

35. Petermann, N., et al. (2011). Plasma Synthesis of Nanostructures for Improved Thermoelectric Properties. Journal of Physics D: Applied Physics. 44(17), 174034.

36. Ye, Y., et al. (2001). RF plasma method, Google Patents: USA.

37. Hyeon, T. (2003). Chemical Synthesis of Magnetic Nanoparticles. Chemical Communications (8), 927–934.

38. Galvez, A., et al. (2002). Carbon Nanoparticles from Laser Pyrolysis. Carbon. 40(15), 2775–2789.

39. Porterat, D. (2012). Synthesis of Nanoparticles by Laser Pyrolysis., Google Patents: USA.

40. Tiwari, J. N., Tiwari, R. N., Kim, K. S. (2012). Zero-dimensional, One-dimensional, Two-dimensional and Three-dimensional Nanostructured Materials for Advanced Electrochemical Energy Devices. Progress in Materials Science. 57(4), 724–803.

41. Murray, P. T., et al. (2006). Nanomaterials Produced by Laser Ablation Techniques Part I: Synthesis and Passivation of Nanoparticles, in Nondestructive Evaulation for Health Monitoring and Diagnostics, International Society for Optics and Photonics. 61750–61750.

42. Dolgaev, S. I., et al. (2002). Nanoparticles Produced by Laser Ablation of Solids in Liquid Environment. Applied Surface Science. 186(1), 546–551.

43. Becker, M. F., et al. (1998). Metal Nanoparticles Generated by Laser Ablation. Nanostructured Materials. 10(5), 853–863.

44. Bonneau, F., et al. (2004). Numerical Simulations for Description of UV Laser Interaction with Gold Nanoparticles Embedded in Silica. Applied Physics B. 78(3–4), 447–452.

45. Chen, Y. H., Yeh, C. S. (2002). Laser Ablation Method: Use of Surfactants to Form the Dispersed Ag Nanoparticles. Colloids and Surfaces A: Physico-chemical and Engineering Aspects. 197(1), 133–139.

46. Andrady, A. L. (2008). Science and Technology of Polymer Nanofibers, Hoboken: John Wiley & Sons, Inc. 404.

47. Wang, H. S., Fu, G. D., Li, X. S. (2009). Functional Polymeric Nanofibers From Electrospinning. Recent Patents on Nanotechnology. 3(1), 21–31.

48. Ramakrishna, S. (2005). An Introduction to Electrospinning and Nanofibers: World Scientific Publishing Company. 396.

49. Reneker, D. H., Chun, I. (1996). Nanometer Diameter Fibres of Polymer, Produced by Electrospinning. Nanotechnology. 7(3), 216.

50. Doshi, J., Reneker, D. H. (1995). Electrospinning Process and Applications of Electrospun Fibers. Journal of Electrostatics. 35(2), 151–160.

51. Burger, C., Hsiao, B., Chu, B. (2006). Nanofibrous Materials and Their Applications. Annual Reviews Material Researchs. 36, 333–368.

52. Fang, J., et al. (2008). Applications of Electrospun Nanofibers. Chinese Science Bulletin. 53(15), 2265–2286.

53. Ondarcuhu, T., Joachim, C. (1998). Drawing a Single Nanofiber Over Hundreds of Microns. EPL (Europhysics Letters). 42(2), 215.

54. Nain, A. S., et al. (2006). Drawing Suspended Polymer Micro/Nanofibers Using Glass Micropipettes. Applied Physics Letters. 89(18), 183105–183105–3.

55. Bajakova, J., et al. (2011). "Drawing"-The Production of Individual Nanofibers by Experimental Method, in Nanoconference: Brno, Czech Republic, EU.

56. Feng, L., et al. (2002). Super Hydrophobic Surface of Aligned Polyacrylonitrile Nanofibers. Angewandte Chemie. 114(7), 1269–1271.

57. Delvaux, M., et al. (2000). Chemical and Electrochemical Synthesis of Polyaniline Micro- and Nano-tubules. Synthetic Metals. 113(3), 275–280.

58. Barnes, C. P., et al. (2007). Nanofiber Technology: Designing the Next Generation of Tissue Engineering Scaffolds. Advanced Drug Delivery Reviews. 59(14), 1413–1433.

59. Palmer, L. C., Stupp, S. I. (2008). Molecular Self-assembly into One-dimensional Nanostructures. Accounts of Chemical Research. 41(12), 1674–1684.

60. Hohman, M. M., et al. (2001). Electrospinning and Electrically Forced jets. I. Stability Theory. Physics of Fluids. 13, 2201–2220.

61. Hohman, M. M., et al. (2001). Electrospinning and Electrically Forced Jets. II. Applications. Physics of Fluids. 13, 2221.

62. Shin, Y. M., et al. (2001). Experimental Characterization of Electrospinning: The Electrically Forced Jet and Instabilities. Polymer. 42(25), 9955–9967.

63. Fridrikh, S. V., et al. (2003). Controlling the fiber diameter during electrospinning. Physical review letters. 90(14), 144502–144502.

64. Yarin, A. L., Koombhongse, S., Reneker, D. H. (2001). Taylor Cone and Jetting from Liquid droplets in Electrospinning of Nanofibers. Journal of Applied Physics. 90(9), 4836–4846.

65. Zeleny, J. (1914). The electrical discharge from liquid points, and a hydrostatic method of measuring the electric intensity at their surfaces. Physical Review. 3(2), 69–91.

66. Reneker, D. H., et al. (2000). Bending Instability of Electrically Charged Liquid Jets of Polymer Solutions in Electrospinning. Journal of Applied Physics. 87, 4531–4547.

67. Frenot, A., Chronakis, I. S. (2003). Polymer Nanofibers Assembled by Electrospinning. Current Opinion in Colloid & Interface Science. 8(1), 64–75.

68. Gilbert, W. (1958). De Magnete Transl. PF Mottelay, Dover, UK, New York: Dover Publications, Inc. 366.

69. Tucker, N., et al. (2012). The History of the Science and Technology of Electrospinning from 1600 to (1995). Journal of Engineered Fibers and Fabrics. 7, 63–73.

70. Hassounah, I. (2012). Melt electrospinning of thermoplastic polymers: Aachen : Hochschulbibliothek Rheinisch-Westfälische Technischen Hochschule Aachen. 650.

71. Taylor, G. I. (1971). The Scientific Papers of Sir Geoffrey Ingram Taylor. Mechanics of Fluids. 4.

72. Yeo, L. Y., Friend, J. R. (2006). Electrospinning Carbon Nanotube Polymer Composite Nanofibers. Journal of Experimental Nanoscience.,. 1(2), 177–209.

73. Bhardwaj, N., Kundu, S. C. (2010). Electrospinning: a Fascinating Fiber Fabrication Technique. Biotechnology Advances. 28(3), 325–347.

74. Huang, Z. M., et al. (2003). A review on polymer nanofibers by electrospinning and their applications in nanocomposites. Composites Science and Technology. 63(15), 2223–2253.

75. Haghi, A. K. (2011). Electrospinning of nanofibers in textiles, North Calorina: Apple Academic PressInc. 132.

76. Bhattacharjee, P., Clayton, V., Rutledge, A. G. (2011). Electrospinning and Polymer Nanofibers: Process Fundamentals, in Comprehensive Biomaterials., Elsevier. 497–512.

77. Garg, K., Bowlin, G. L. (2011). Electrospinning jets and nanofibrous structures. Biomicrofluidics. 5, 13403–13421.

78. Angammana, C. J., Jayaram, . S. H. (2011). A Theoretical Understanding of the Physical Mechanisms of Electrospinning, in Proc. ESA Annual Meeting on Electrostatics: Case Western Reserve University, Cleveland OH. 1–9.

79. Reneker, D. H., Yarin, A. L. (2008). Electrospinning Jets and Polymer Nanofibers. Polymer. 49(10), 2387–2425.

80. Deitzel, J., et al. (2001). The Effect of Processing Variables on the Morphology of Electrospun Nanofibers and Textiles. Polymer. 42(1), 261–272.

81. Rutledge, G. C., Fridrikh, S. V. (2007). Formation of Fibers by Electrospinning. Advanced Drug Delivery Reviews. 59(14), 1384–1391.

82. De Vrieze, S., et al. (2009). The Effect of Temperature and Humidity on Electrospinning. Journal of Materials Science. 44(5), 1357–1362.

83. Kumar, P. (2012). Effect of Colletor on Electrospinning to Fabricate Aligned Nanofiber, in Department of Biotechnology and Medical Engineering, National Institute of Technology Rourkela: Rourkela. 88.

84. Sanchez, C., Arribart, H., Guille, M. (2005). Biomimetism and bioinspiration as tools for the design of innovative materials and systems. Nature materials. 4(4), 277–288.

85. Ko, F., et al. (2003). Electrospinning of continuous carbon nanotube-filled nanofiber yarns. Advanced Materials. 15(14), 1161–1165.

86. Stuart, M., et al.,, (2010). Emerging applications of stimuli-responsive polymer materials. Nature materials. 9(2), 101–113.

87. Gao, W., Chan, J., Farokhzad, O. (2010). pH-responsive nanoparticles for drug delivery. Molecular pharmaceutics. 7(6), 1913–1920.
88. Li, Y., et al. (2010). Stimulus-responsive polymeric nanoparticles for biomedical applications. Science China Chemistry. 53(3), 447–457.
89. Tirelli, N. (Bio) Responsive nanoparticles. Current opinion in colloid & interface science. (2006). 11(4), 210–216.
90. Bonini, M., et al. (2002). A new way to prepare nanostructured materials: flame spraying of microemulsions. The Journal of Physical Chemistry B. 106(24), 6178–6183.
91. Thierry, B., et al. (2003). Nanocoatings onto arteries via layer-by-layer deposition: toward the in vivo repair of damaged blood vessels. Journal of the American Chemical Society. 125(25), 7494–7495.
92. Andrady, A. (2008). Science and technology of polymer nanofibers: Wiley. com.
93. Carroll, C. P., et al. (2008). Nanofibers From Electrically Driven Viscoelastic Jets: Modeling and Experiments. Korea-Australia Rheology Journal. 20(3), 153–164.
94. Zhao, Y., Jiang, L. (2009). Hollow micro/nanomaterials with multilevel interior structures. Advanced Materials. 21(36), 3621–3638.
95. Carroll, C. P. (2009). The development of a comprehensive simulation model for electrospinning. Vol. 70, Cornell University 300.
96. Song, Y. S., Youn, J. R. (2004). Modeling of Rheological Behavior of Nanocomposites by Brownian Dynamics Simulation. Korea-Australia Rheology Journal. 16(4), 201–212.
97. Dror, Y., et al. (2003). Carbon nanotubes embedded in oriented polymer nanofibers by electrospinning. Langmuir. 19(17), 7012–7020.
98. Gates, T., et al. (2005). Computational materials: multiscale modeling and simulation of nanostructured materials. Composites Science and Technology. 65(15), 2416–2434.
99. Agic, A. (2008). Multiscale mechanical phenomena in electrospun carbon nanotube composites. Journal of Applied Polymer Science. 108(2), 1191–1200.
100. Teo, W., Ramakrishna, S. (2009). Electrospun nanofibers as a platform for multifunctional, hierarchically organized nanocomposite. Composites Science and Technology. 69(11), 1804–1817.
101. Silling, S., Bobaru, F. (2005). Peridynamic modeling of membranes and fibers. International Journal of Non-Linear Mechanics. 40(2), 395–409.
102. Berhan, L., et al. (2004). Mechanical properties of nanotube sheets: Alterations in joint morphology and achievable moduli in manufacturable materials. Journal of applied physics. 95(8), 4335–4345.
103. Heyden, S. (2000). Network modeling for the evaluation of mechanical properties of cellulose fiber fluff, Lund University.
104. Collins, A. J., et al. (2011). The Value of Modeling and Simulation Standards, Virginia Modeling, Analysis and Simulation Center, Old Dominion University: Virginia. 1–8.
105. Kuwabara, S. (1959). The forces experienced by randomly distributed parallel circular cylinders or spheres in a viscous flow at small Reynolds numbers. Journal of the Physical Society of Japan. 14, 527.
106. Brown, R. (1993). Air filtration: an integrated approach to the theory and applications of fibrous filters, New York: Pergamon press New York.
107. Buysse, W. M. et al., (2008). A 2D model for the electrospinning process, in Department of Mechanical Engineering, Eindhoven University of Technology: Eindhoven 75.
108. Ante, A., Budimir, M. (2010). Design multifunctional product by nanostructures, Sciyo. com. 27.
109. Jackson, G., James, D. (1986). The permeability of fibrous porous media. The Canadian Journal of Chemical Engineering. 64(3), 364–374.

110. Sundmacher, K. (2010). Fuel cell engineering: toward the design of efficient electrochemical power plants. Industrial and Engineering Chemistry Research. 49(21), 10159–10182.

111. Kim, Y., et al. (2008). Electrospun bimetallic nanowires of PtRh and PtRu with compositional variation for methanol electrooxidation. Electrochemistry Communications. 10(7), 1016–1019.

112. Kim, H., et al. (2009). Pt and PtRh nanowire electrocatalysts for cyclohexane-fueled polymer electrolyte membrane fuel cell. Electrochemistry Communications. 11(2), 446–449.

113. Formo, E., et al. (2008). Functionalization of electrospun TiO2 nanofibers with Pt nanoparticles and nanowires for catalytic applications. Nano letters. 8(2), 668–672.

114. Xuyen, N., et al. (2009). Hydrolysis-induced immobilization of Pt (acac) 2 on polyimide-based carbon nanofiber mat and formation of Pt nanoparticles. Journal of Materials Chemistry. 19(9), 1283–1288.

115. Lee, K., et al. (2009). Nafion nanofiber membranes. ECS Transactions. 25(1), 1451–1458.

116. Qu, H., Wei, S., Guo, Z. (2013). Coaxial electrospun nanostructures and their applications. Journal of Material Chemistry A. 1(38), 11513–11528.

117. Thavasi, V., Singh, G., Ramakrishna, S. (2008). Electrospun nanofibers in energy and environmental applications. Energy and Environmental Science. 1(2), 205–221.

118. Dersch, R., et al. (2005). Nanoprocessing of polymers: applications in medicine, sensors, catalysis, photonics. Polymers for advanced technologies. 16(2-3), 276–282.

119. Yih, T., Al-Fandi, M. (2006). Engineered nanoparticles as precise drug delivery systems. Journal of cellular biochemistry. 97(6), 1184–1190.

120. Kenawy, E., et al. (2002). Release of tetracycline hydrochloride from electrospun poly (ethylene-covinylacetate), poly (lactic acid)., a blend. Journal of Controlled Release. 81(1), 57–64.

121. Verreck, G., et al. (2003). Incorporation of drugs in an amorphous state into electrospun nanofibers composed of a water-insoluble, nonbiodegradable polymer. Journal of Controlled Release. 92(3), 349–360.

122. Zeng, J., et al. (2003). Biodegradable electrospun fibers for drug delivery. Journal of Controlled Release. 92(3), 227–231.

123. Luu, Y., et al. (2003). Development of a nanostructured DNA delivery scaffold via electrospinning of PLGA and PLA–PEG block copolymers. Journal of Controlled Release. 89(2), 341–353.

124. Zong, X., et al. (2002). Structure and process relationship of electrospun bioabsorbable nanofiber membranes. Polymer. 43(16), 4403–4412.

125. Yu, D., et al. (2012). PVP nanofibers prepared using coaxial electrospinning with salt solution as sheath fluid. Materials Letters. 67(1), 78–80.

126. Verreck, G., et al. (2003). Preparation and characterization of nanofibers containing amorphous drug dispersions generated by electrostatic spinning. Pharmaceutical research. 20(5), 810–817.

127. Jiang, H., et al. (2004). Preparation and characterization of ibuprofen-loaded poly (lactide-coglycolide)/poly (ethylene glycol)-g-chitosan electrospun membranes. Journal of Biomaterials Science, Polymer Edition. 15(3), 279–296.

128. Yang, D., Li, Y., Nie, J. (2007). Preparation of gelatin/PVA nanofibers and their potential application in controlled release of drugs. Carbohydrate Polymers. 69(3), 538–543.

129. Kim, K., et al. (2004). Incorporation and controlled release of a hydrophilic antibiotic using poly (lactide-coglycolide)-based electrospun nanofibrous scaffolds. Journal of Controlled Release. 98(1), 47–56.

130. Xu, X., et al. (2005). Ultrafine medicated fibers electrospun from W/O emulsions. Journal of Controlled Release. 108(1), 33–42.

131. Zeng, J., et al., (2005). Poly(vinyl alcohol) nanofibers by electrospinning as a protein delivery system and the retardation of enzyme release by additional polymer coatings. Biomacromolecules. 6(3), 1484–1488.

132. Yun, J., et al. (2010). Effect of oxyfluorination on electromagnetic interference shielding behavior of MWCNT/PVA/PAAc composite microcapsules. European Polymer Journal. 46(5), 900–909.

133. Jiang, H., et al. (2005). A facile technique to prepare biodegradable coaxial electrospun nanofibers for controlled release of bioactive agents. Journal of Controlled Release. 108(2), 237–243.

134. He, C., Huang, Z., Han, X. (2009). Fabrication of drug-loaded electrospun aligned fibrous threads for suture applications. Journal of Biomedical Materials Research Part A. 89(1), 80–95.

135. Qi, R., et al. (2010). Electrospun poly (lactic-coglycolic acid)/halloysite nanotube composite nanofibers for drug encapsulation and sustained release. Journal of Materials Chemistry. 20(47), 10622–10629.

136. Reneker, D. H., et al. (2007). Electrospinning of Nanofibers from Polymer Solutions and Melts. Advances in Applied Mechanics. 41, 343–346.

137. Haghi, A. K., Zaikov, G. (2012). Advances in Nanofiber Research: Smithers Rapra Technology. 194.

138. Maghsoodloo, S., et al. (2012). A Detailed Review on Mathematical Modeling of Electrospun Nanofibers. Polymers Research Journal. 6, 361–379.

139. Fritzson, P. (2010). Principles of object-oriented modeling and simulation with Modelica 2.1, Wiley-IEEE Press.

140. Robinson, S. (2004). Simulation: the practice of model development and use: Wiley. 722.

141. Carson, I. I., John, S. (2004). Introduction to modeling and simulation, in Proceedings of the 36th conference on Winter simulation, Winter Simulation Conference: Washington, DC. 9–16.

142. Banks, J., Handbook of simulation(1998). : Wiley Online Library. 342.

143. Pritsker, A. B., Alan, B. (1998). Principles of Simulation Modeling, New York: Wiley. 426.

144. Yu, J. H., Fridrikh., S. V. Rutledge, G. C. (2006). The Role of Elasticity in the Formation of Electrospun Fibers. Polymer. 47(13), 4789–4797.

145. Han, T., Yarin, A. L., Reneker, D. H. (2008). Viscoelastic Electrospun Jets: Initial Stresses and Elongational Rheometry. Polymer. 49(6), 1651–1658.

146. Bhattacharjee, P. K., et al. (2003). Extensional Stress Growth and Stress Relaxation in Entangled Polymer Solutions. Journal of Rheology. 47, 269–290.

147. Paruchuri, S., Brenner, M. P. (2007). Splitting of A Liquid Jet. Physical Review Letters. 98(13), 134502–134504.

148. Ganan-Calvo, A. M. (1997). On the Theory of Electrohydrodynamically Driven Capillary Jets. Journal of Fluid Mechanics. 335, 165–188.

149. Liu, L., Dzenis, Y. A. (2011). Simulation of Electrospun Nanofiber Deposition on Stationary and Moving Substrates. Micro & Nano Letters. 6(6), 408–411.

150. Spivak, A. F., Dzenis, Y. A. (1998). Asymptotic decay of radius of a weakly conductive viscous jet in an external electric field. Applied Physics Letters. 73(21), 3067–3069.

151. Jaworek, A., Krupa, A. (1999). Classification of the Modes of EHD Spraying. Journal of Aerosol Science. 30(7), 873–893.

152. Senador, A. E., Shaw, M. T., Mather, P. T. (2000). Electrospinning of Polymeric Nanofibers: Analysis of jet formation, in Material research society., Cambridge Univ Press: California, USA. 11.

153. Feng, J. J. (2002). The stretching of an electrified non-Newtonian jet: A model for electrospinning. Physics of Fluids. 14(11), 3912–3927.

154. Feng, J. J. (2003). Stretching of a straight electrically charged viscoelastic jet. Journal of Non-Newtonian Fluid Mechanics. 116(1), 55–70.

155. Spivak, A. F., Dzenis, Y. A., Reneker, D. H., (2000). A Model of Steady State Jet in the Electrospinning Process. Mechanics Research Communications. 27(1), 37–42.

156. Yarin, A. L., Koombhongse, S.,. Reneker, D. H . (2001). Bending Instability in Electrospinning of Nanofibers. Journal of Applied Physics. 89, 3018.

157. Gradoń, L., Principles of Momentum, Mass and Energy Balances. Chemical Engineering and Chemical Process Technology. 1, 1–6.

158. Bird, R. B., Stewart, W. E., Lightfoot, E. N. (1960). Transport Phenomena. Vol. 2, New York: Wiley & Sons, Incorporated, John 808.

159. Peters, G. W. M., Hulsen, M. A., Solberg, R. H. M. (2007). A Model for Electrospinning Viscoelastic Fluids, in Department of Mechanical Engineering, Eindhoven University of Technology: Eindhoven 26.

160. Whitaker, R. D. (1975). An historical note on the conservation of mass. Journal of Chemical Education. 52(10), 658.

161. He, J. H., et al. (2007). Mathematical models for continuous electrospun nanofibers and electrospun nanoporous microspheres. Polymer International. 56(11), 1323–1329.

162. Xu, L., Liu, F., Faraz, N. (2012). Theoretical model for the electrospinning nanoporous materials process. Computers and Mathematics with Applications. 64(5), 1017–1021.

163. Heilbron, J. L. (1979). Electricity in the 17th and eighteenth century: A Study of Early Modern Physics: University of California Press. 437.

164. Orito, S., Yoshimura, M. (1985). Can the universe be charged? Physical Review Letters. 54(22), 2457–2460.

165. Karra, S. (2007). Modeling electrospinning process and a numerical scheme using Lattice Boltzmann method to simulate viscoelastic fluid flows, in Indian Institute of Technology, Texas A&M University: Chennai. 60.

166. Hou, S. H., Chan, C. K. (2011). Momentum Equation for Straight Electrically Charged Jet. Applied Mathematics and Mechanics. 32(12), 1515–1524.

167. Maxwell, J. C. (1878). Electrical Research of the Honorable Henry Cavendish, 426, in Cambridge University Press, Cambridge, Editor, Cambridge University Press, Cambridge, UK: UK.

168. Vught, R. V. (2010). Simulating the dynamical behavior of electrospinning processes, in Department of Mechanical Engineering, Eindhoven University of Technology: Eindhoven. 68.

169. Jeans, J. H. (1927). The Mathematical Theory of Electricity and Magnetism, London: Cambridge University Press. 536.

170. Truesdell, C., Noll, W. (2004). The nonlinear field theories of mechanics: Springer. 579.

171. Roylance, D. (2000). Constitutive equations, in Lecture Notes. Department of Materials Science and Engineering, Massachusetts Institute of Technology: Cambridge. 10.

172. He, J. H., Wu, Y., Pang, N. A. (2005). mathematical model for preparation by AC-electrospinning process. International Journal of Nonlinear Sciences and Numerical Simulation. 6(3), 243–248.

173. Little, R. W. (1999). Elasticity: Courier Dover Publications. 431.

174. Clauset, A., Shalizi C. R., Newman, M. E. J. (2009). Power-law Distributions in Empirical Data. SIAM Review. 51(4), 661–703.

175. Wan, Y., Guo, Q., Pan, N. (2004). Thermo-electro-hydrodynamic model for electrospinning process. International Journal of Nonlinear Sciences and Numerical Simulation. 5(1), 5–8.
176. Giesekus, H. (1966). Die elastizität von flüssigkeiten. Rheologica Acta. 5(1), 29–35.
177. Giesekus, H. (1973). The physical meaning of Weissenberg's hypothesis with regard to the second normal-stress difference, in The Karl Weissenberg 80th Birthday Celebration Essays, J. Harris and K. Weissenberg, Editors, East African Literature Bureau 103–112.
178. Wiest, J. M. (1989). A differential constitutive equation for polymer melts. Rheologica Acta. 28(1), 4–12.
179. Bird, R. B., J. M. Wiest, Constitutive Equations for Polymeric Liquids. Annual Review of Fluid Mechanics. (1995). 27(1), 169–193.
180. Giesekus, H. (1982). A simple constitutive equation for polymer fluids based on the concept of deformation-dependent tensorial mobility. Journal of Non-Newtonian Fluid Mechanics. 11(1), 69–109.
181. Oliveira, P. J. (2001). On the Numerical Implementation of Nonlinear Viscoelastic Models in a Finite-Volume Method. Numerical Heat Transfer: Part B: Fundamentals. 40(4), 283–301.
182. Simhambhatla, M., Leonov, A. I. (1995). On the Rheological Modeling of Viscoelastic Polymer Liquids with Stable Constitutive Equations. Rheologica Acta. 34(3), 259–273.
183. Giesekus, H. (1982). A unified approach to a variety of constitutive models for polymer fluids based on the concept of configuration-dependent molecular mobility. Rheologica Acta. 21(4–5), 366–375.
184. Eringen, A. C., Maugin, G. A. (1990). Electrohydrodynamics, in Electrodynamics of Continua II, Springer. 551–573.
185. Hutter, K. (1991). Electrodynamics of Continua (A. Cemal Eringen and Gerard A. Maugin). SIAM Review. 33(2), 315–320.
186. Kröger, M. (2004). Simple Models for Complex Nonequilibrium Fluids. Physics Reports. 390(6), 453–551.
187. Denn, M. M. (1990). Issues in Viscoelastic Fluid Mechanics. Annual Review of Fluid Mechanics. 22(1), 13–32.
188. Rossky, P. J., Doll, J. D., Friedman, H. L. (1978). Brownian Dynamics as Smart Monte Carlo Simulation. The Journal of Chemical Physics. 69, 4628–4633.
189. Chen, J. C., Kim, A. S. (2004). Brownian Dynamics, Molecular Dynamics, and Monte Carlo Modeling of Colloidal Systems. Advances in Colloid and Interface Science. 112(1), 159–173.
190. Pasini, P., Zannoni, C. (2005). Computer Simulations of Liquid Crystals and Polymers. Vol. 177, Erice: Springer. 380.
191. Zhang, H., Zhang, P. (2006). Local Existence for the FENE-dumbbell Model of Polymeric Fluids. Archive for Rational Mechanics and Analysis. 181(2), 373–400.
192. Isihara, A. (1951). Theory of High Polymer Solutions (The Dumbbell Model). The Journal of Chemical Physics. 19, 397–343.
193. Masmoudi, N. (2008). Well-posedness for the FENE Dumbbell Model of Polymeric Flows. Communications on Pure and Applied Mathematics. 61(12), 1685–1714.
194. Stockmayer, W. H., et al. (1970). Dynamic Properties of Solutions. Models for Chain Molecule Dynamics in Dilute Solution. Discussions of the Faraday Society. 49, 182–192.
195. Graham, R. S., et al. (2003). Microscopic Theory of Linear, Entangled Polymer Chains under Rapid Deformation Including Chain Stretch and Convective Constraint Release. Journal of Rheology. 47, 1171–1200.

196. Gupta, R. K., Kennel, E., Kim, K. S. (2010). Polymer nanocomposites handbook: CRC Press.
197. Marrucci, G. (1972). The free energy constitutive equation for polymer solutions from the dumbbell model. Journal of Rheology. 16, 321–331.
198. Reneker, D. H., et al. (2007). Electrospinning of Nanofibers from Polymer Solutions and Melts. Advances in Applied Mechanics. 41, 43–195.
199. Kowalewski, T. A., Barral, S., Kowalczyk, T. (2009). Modeling Electrospinning of Nanofibers, in IUTAM Symposium on Modelling Nanomaterials and Nanosystems, Springer: Aalborg, Denmark. 279–292.
200. Macosko, C. W., Rheology: Principles, Measurements, and Applications. Poughkeepsie, (1994)., Newyork: Wiley-VCH. 578.
201. Kowalewski, T. A., Blonski, S., Barral, S. (2005). Experiments and Modelling of Electrospinning Process. Technical Sciences. 53(4), 385–394.
202. Ma, W. K. A., et al. (2008). Rheological Modeling of Carbon Nanotube Aggregate Suspensions. Journal of Rheology. 52, 1311–1330.
203. Buysse, W. M. (2008). A 2D Model for the Electrospinning Process, in Department of Mechanical Engineering, Eindhoven University of Technology: Eindhoven. 71.
204. Silling, S. A., Bobaru, F. (2005). Peridynamic Modeling of Membranes and Fibers. International Journal of Non-Linear Mechanics. 40(2), 395–409.
205. Teo, W. E., Ramakrishna, S. (2009). Electrospun Nanofibers as a Platform for Multifunctional, Hierarchically Organized Nanocomposite. Composites Science and Technology. 69(11), 1804–1817.
206. Wu, X., Dzenis, Y. A. (2005). Elasticity of Planar Fiber Networks. Journal of Applied Physics. 98(9), 93501.
207. Tatlier, M., Berhan, L. (2009). Modelling the Negative Poisson's Ratio of Compressed Fused Fibre Networks. physica Status Solidi (b). 246(9), 2018–2024.
208. Kuipers, B. (1994). Qualitative reasoning: modeling and simulation with incomplete knowledge: the MIT press. 554.
209. West, B. J. (2004). Comments on the renormalization group, scaling and measures of complexity. Chaos, Solitons and Fractals. 20(1), 33–44.
210. De Gennes, P. G., Witten, T. A. (1980). Scaling Concepts in Polymer Physics. Vol. Cornell University Press. 324.
211. He, J. H., Liu, H. M. (2005). Variational approach to nonlinear problems and a review on mathematical model of electrospinning. Nonlinear Analysis. 63, e919-e929.
212. He, J. H., Wan, Y. Q., Yu, J. Y. (2004). Allometric scaling and instability in electrospinning. International Journal of Nonlinear Sciences and Numerical Simulation. 5(3), 243–252.
213. He, J. H., Wan, Y. Q., Yu, J. Y. (2004). Allometric Scaling and Instability in Electrospinning. International Journal of Nonlinear Sciences and Numerical Simulation. 5, 243–252.
214. He, J. H., Wan, Y. Q. Allometric scaling for voltage and current in electrospinning. Polymer. (2004). 45, 6731–6734.
215. He, J. H., Y. Q. Wan., J. Y. Yu, . (2005). Scaling law in electrospinning: relationship between electric current and solution flow rate. Polymer. 46, 2799–2801.
216. He, J. H., Wanc, Y. Q., Yuc, J. Y. (2004). Application of vibration technology to polymer electrospinning. International Journal of Nonlinear Sciences and Numerical Simulation. 5(3), 253–262.
217. Kessick, R., Fenn, J., Tepper, G. (2004). The use of AC potentials in electrospraying and electrospinning processes. Polymer. 45(9), 2981–2984.
218. Boucher, D. F., Alves, G. E. (1959). Dimensionless numbers, part 1 and 2.

219. Ipsen, D. C. (1960). Units Dimensions And Dimensionless Numbers, New York: McGraw Hill Book Company Inc. 466.

220. Langhaar, H. L. (1951). Dimensional analysis and theory of models. Vol. 2, New York: Wiley. 166

221. McKinley, G. H. (2005). Dimensionless groups for understanding free surface flows of complex fluids. Bulletin of the Society of Rheology(2005). : 6–9.

222. Carroll, C. P., et al. (2008). Nanofibers from Electrically Driven Viscoelastic Jets: Modeling and Experiments. Korea-Australia Rheology Journal. 20(3), 153–164.

223. Saville, D. (1997). .Electrohydrodynamics: the Taylor-Melcher leaky dielectric model. Annual review of fluid mechanics 29(1), 27–64.

224. Ramos, J. I. (1996). Force Fields on Inviscid, Slender, Annular Liquid. International Journal for Numerical Methods in Fluids. 23, 221–239.

225. Saville, D. A. (1997). Electrohydrodynamics: the Taylor-Melcher leaky dielectric model. Annual Review of Fluid Mechanics. 29(1), 27–64.

226. Senador, A. E., Shaw, M. T., Mather P. T. (2000). Electrospinning of Polymeric Nanofibers: Analysis of jet formation. in MRS Proceedings: Cambridge Univ Press.

227. Reneker, D. H., et al. (2000). Bending Instability of Electrically Charged Liquid Jets of Polymer Solutions in Electrospinning. Journal of Applied physics. 87, 4531.

228. Peters, G., Hulsen, M., Solberg, R. A Model for Electrospinning Viscoelastic Fluids.

229. Wan, Y., et al. (2012). Modeling and Simulation of the Electrospinning Jet with Archimedean Spiral. Advanced Science Letters. 10(1), 590–592.

230. Dasri, T. (2012). Mathematical Models of Bead-Spring Jets during Electrospinning for Fabrication of Nanofibers. Walailak Journal of Science and Technology. 9.

231. Solberg, R. H. M. (2007). Position-controlled deposition for electrospinning, Eindhoven University of Technology: Eindhoven. 75.

232. Holzmeister, A., Yarin, A. L., Wendorff, J. H. (2010). Barb Formation in Electrospinning: Experimental and Theoretical Investigations. Polymer. 51(12), 2769–2778.

233. Karra, S. (2012). Modeling electrospinning process and a numerical scheme using Lattice Boltzmann method to simulate viscoelastic fluid flows

234. Arinstein, A., et al. (2007). Effect of supramolecular structure on polymer nanofiber elasticity. Nature Nanotechnology, 2(1), 59–62.

235. Lu, C., et al. (2006). Computer Simulation of Electrospinning. Part I. Effect of Solvent in Electrospinning. Polymer, 47(3), 915–921.

236. Greenfeld, I., et al. (2011). Polymer dynamics in semidilute solution during electrospinning: A simple model and experimental observations. Physical Review 84(4), 41806–41815.

237. Ly, H. V., Tran, H. T. (2001). Modeling and Control of Physical Processes Using Proper Orthogonal Decomposition. Mathematical and Computer Modelling, 33(1), 223–236.

238. Peiró, J., Sherwin, S. (2005). Finite Difference, Finite Element and Finite Volume Methods for Partial Differential Equations, in Handbook of Materials Modeling. Springer: London. 2415–2446.

239. Kitano, H. (2002). Computational Systems Biology. Nature, 420(6912), 206–210.

240. Gerald, C. F., Wheatley, P. O. (2007). Applied Numerical Analysis, ed. 7th Addison-Wesley. 624.

241. Burden, R. L., Faires, J. D. (2005). Numerical Analysis. Vol. 8. Thomson Brooks/Cole. 850.

242. Lawrence, C. E. (2010). Partial Differential Equations. American Mathematical Society. 749.

243. Quarteroni, A., Quarteroni, A. M., Valli, A. (2008). Numerical Approximation of Partial Differential Equations. Vol. 23. Springer. 544.

244. Butcher, J. C. (1996). A History of Runge-Kutta Methods. Applied Numerical Mathematics, 20(3), 247–260.

245. Cartwright, J. H., Piro, E. O. (1992). The Dynamics of Runge–Kutta Methods. International Journal of Bifurcation and Chaos, 2(03), 427–449.

246. Zingg, D. W., Chisholm, T. T. (1999). Runge–Kutta Methods for Linear Ordinary Differential Equations. Applied Numerical Mathematics, 31(2), 227–238.

247. Butcher, J. C. (1987). The Numerical Analysis of Ordinary Differential Equations: Runge-Kutta and General Linear Methods. (1987), Wiley-Interscience. 512.

248. Reznik, S. N., et al. (2006). Evolution of a Compound Droplet Attached to a Core-shell Nozzle Under the Action of a Strong Electric Field. Physics of Fluids, 18(6), 062101–062101–13.

249. Reznik, S. N., et al. (2004). Transient and Steady Shapes of Droplets Attached to a Surface in a Strong Electric Field. Journal of Fluid Mechanics, 516, 349–377.

250. Donea, J., Huerta, A. (2003). Finite Element Methods for Flow Problems. Wiley. com. 362.

251. Zienkiewicz, O. C., Taylor, R. L. (2000). The Finite Element Method: Solid Mechanics. Vol. 2, Butterworth-heinemann. 459.

252. Brenner, S. C., Scott, L. R. (2008). The Mathematical Theory of Finite Element Methods. Vol. 15. Springer. 397.

253. Bathe, K. J. (1996). Finite Element Procedures. Vol. 2. Prentice hall Englewood Cliffs. 1037.

254. Reddy, J. N. (2006). An Introduction to the Finite Element Method. Vol. 2. McGraw-Hill New York. 912.

255. Ferziger, J. H., Perić, M. (1996). Computational Methods for Fluid Dynamics. Vol. 3. Springer Berlin. 423.

256. Baaijens, P. T., F. (1998). Mixed Finite Element Methods for Viscoelastic Flow Analysis: A Review. Journal of Non-Newtonian Fluid Mechanics, 79(2), 361–385.

257. Angammana, C. J., Jayaram, S. H., A Theoretical Understanding of the Physi-cal Mechanisms of Electrospinning. in Proceedings of the ESA Annual Meeting on Electrostatics. (2011).

258. Costabel, M. (1987). Principles of Boundary Element Methods. Computer Physics Reports, 6(1), 243–274.

259. Kurz, S., Fetzer, J., Lehner, G. (1995). An Improved Algorithm for the BEM-FEM-coupling Method Using Domain Decomposition. IEEE Transactions on Magnetics, 31(3), 1737–1740.

260. Mushtaq, M., Shah, N. A., Muhammad, G. (2010). Advantages and Disadvantages of Boundary Element Methods For Compressible Fluid Flow Problems. Journal of American Science, 6(1), 162–165.

261. Gaul, L., Kögl, M., Wagner, M. (2003). Boundary Element Methods for Engineers and Scientists. Springer. 488.

262. Kowalewski, T. A., Barral, S., Kowalczyk, T. (2009). Modeling Electrospinning of Nanofibers. in IUTAM Symposium on Modelling Nanomaterials and Nanosystems Springer.

263. Toro, E. F. (2009). Riemann Solvers and Numerical Methods for Fluid Dynamics: A Practical Introduction. Springer. 724.

264. Tonti, E. (2001). A Direct Discrete Formulation of Field Laws: The Cell Method. CMES-Computer Modeling in Engineering and Sciences, 2(2), 237–258.

265. Thomas, P. D., Lombard, C. K. (1979). Geometric Conservation Law and Its Application to Flow Computations on Moving Grids. American Institute of Aeronautics and Astronautics Journal, 17(10), 1030–1037.
266. Lyrintzis, A. S. (2003). Surface Integral Methods in Computational Aeroacoustics-From the (CFD) Near-field to the (Acoustic) Far-field. International Journal of Aeroacoustics, 2(2), 95–128.
267. Škerget, L., Hriberšek, M., Kuhn, G. (1999). Computational Fluid Dynamics by Boundary–domain Integral Method. International Journal for Numerical Methods in Engineering, 46(8), 1291–1311.
268. Rüberg, T., Cirak, F. (2011). An Immersed Finite Element Method with Integral Equation Correction. International Journal for Numerical Methods in Engineering, 86(1), 93–114.
269. Feng, J. J. (2002). The Stretching of an Electrified Non-Newtonian Jet: A Model for Electrospinning. Physics of Fluids, 14, 3912–3926.
270. Varga, R. S. (2009). Matrix Iterative Analysis. Vol. 27. Springer. 358.
271. Stoer, J., Bulirsch, R. (2002). Introduction to Numerical Analysis. Vol. 12. Springer. 744.
272. Bazaraa, M. S., Sherali, H. D., Shetty, C. M. (2006). Nonlinear Programming: Theory and Algorithms. John Wiley & Sons. 872.
273. Fox, L. (1947). Some Improvements in the Use of Relaxation Methods for the Solution of Ordinary and Partial Differential Equations. Proceedings of the Royal Society of London. Series A. Mathematical and Physical Sciences, 190(1020), 31–59.
274. Zauderer, E. (2011). Partial Differential Equations of Applied Mathematics. Vol. 71. Wiley. 968.
275. Fisher, M. L. (2004). The Lagrangian Relaxation Method for Solving Integer Programming Problems. Management Science, 50(12 supplement): 1861–1871.
276. Steger, T. M. (2005). Multi-Dimensional Transitional Dynamics: A Simple Numerical Procedure. Macroeconomic Dynamics, 12(3), 301–319.
277. Roozemond, P. C. (2007). A Model for Electrospinning Viscoelastic Fluids, in Department of Mechanical Engineering, Eindhoven University of Technology. 25.
278. Succi, S. (2001). The Lattice Boltzmann Equation: For Fluid Dynamics and Beyond Oxford University Press. 288.
279. Chen, S., Doolen, G. D. (1998). Lattice Boltzmann Method for Fluid Flows. Annual Review of Fluid Mechanics, 30(1), 329–364.
280. Aidun, C. K., Clausen, J. R. (2010), Lattice-Boltzmann Method for Complex Flows. Annual Review of Fluid Mechanics, 42, 439–472.
281. Guo, Z., Zhao, T. S. (2003). Explicit Finite-difference Lattice Boltzmann Method for Curvilinear Coordinates. Physical review E., 67(6), 066709-1-066709-12.
282. Albuquerque, P., et al. (2006). A Hybrid Lattice Boltzmann Finite Difference Scheme for the Diffusion Equation. International Journal for Multiscale Computational Engineering, 4(2), 209–219.
283. Tsutahara, M. (2012). The Finite-difference Lattice Boltzmann Method and Its Application in Computational Aero-acoustics. Fluid Dynamics Research, 44(4), 045507–1-045507–18.
284. Junk, M. (2001). A Finite Difference Interpretation of the Lattice Boltzmann Method. Numerical Methods for Partial Differential Equations, 17(4), 383–402.
285. So, R. M. C., Fu, S. C., Leung, R. C. K. (2010). Finite Difference Lattice Boltzmann Method for Compressible Thermal Fluids. American Institute of Aeronautics and Astronautics Journal, 48(6), 1059–1071.
286. Karra, S. (2007). Modeling Electrospinning Process and a Numerical Scheme Using Lattice Boltzmann Method to Simulate Viscoelastic Fluid Flows, in Mechanical Engineering. Indian Institute of Technology Madras. 60.

287. De Pascalis, R. (2010). The Semi-Inverse Method in Solid Mechanics: Theoretical Under-pinnings and Novel Applications, in Mathematics, Universite Pierre et Marie Curie and Universita del Salento. 140.

288. Nemenyi, P. F. (1951). Recent Developments in Inverse and Semi-inverse Methods in the Mechanics of Continua. Advances in Applied Mechanics, 2(11), 123–151.

289. Chen, J. T., Lee, Y. T., Shieh, S. C. (2009). Revisit of Two Classical Elasticity Problems by Using the Trefftz Method. Engineering Analysis with Boundary Elements, 33(6), 890–895.

290. Zhou, X. W. (2013). A Note on the Semi-Inverse Method and a Variational Principle for the Generalized KdV-mKdV Equation. in Abstract and Applied Analysis: Hindawi Publishing Corporation.

291. A. Narayan, S. P., Rajagopal, K. R. (2013). Unsteady Flows of a Class of Novel Gener-alizations of the Navier-Stokes Fluid. Applied Mathematics and Computation. 219(19), 9935–9946.

292. He, J. H. (2004). Variational Principles for Some Nonlinear Partial Differential Equations with Variable Coefficients. Chaos, Solitons & Fractals, 19(4), 847–851.

293. Tarantola, A. (2002). Inverse Problem Theory: Methods for Data Fitting and Model Pa-rameter Estimation: Elsevier Science. 613.

294. He, J. H., Liu, H. M., Pan, N. (2003). Variational Model for Ionomeric Polymer-metal Composite. Polymer, 44(26), 8195–8199.

295. He, J. H. (2001). Coupled Variational Principles of Piezoelectricity. International Journal of Engineering Science, 39(3), 323–341.

296. Starovoitov, É., Nağıyev, F. (2012). Foundations of the Theory of Elasticity, Plasticity, and Viscoelasticity: CRC Press. 320.

297. Bertero, M. (1986). Regularization Methods for Linear Inverse Problems, in Inverse Prob-lems. Springer. 52–112.

298. Wan, Y. Q., Guo, Q., Pan, N. (2004). Thermo-electro-hydrodynamic Model for Eectros-pinning Process. International Journal of Nonlinear Sciences and Numerical Simulation, 5(1), 5–8.

299. Mccann, J., Li, D., Xia, Y. (2005). Electrospinning of Nanofibers with Core-sheath, Hol-low, or Porous Structures. Journal of Materials Chemistry, 15(7), 735–738.

300. Srivastava, Y., et al. (2008). Electrospinning of hollow and core/sheath nanofibers using a microfluidic manifold. Microfluidics and Nanofluidics, 4(3), 245–250.

301. Sun, Z., et al. (2003). Compound Core-shell Polymer Nanofibers by Co-Electrospinning. Advanced Materials, 15(22), 1929–1932.

302. Li, D., Xia, Y. (2004). Direct fabrication of composite and ceramic hollow nanofibers by electrospinning. Nano Letters, 4(5), 933–938.

303. Moghe, A., Gupta, B. (2008). Co-axial Electrospinning for Nanofiber Structures: Prepara-tion and Applications. Polymer Reviews, 48(2), 353–377.

304. Sakuldao, S., Yoovidhya, T., Wongsasulak, S. (2011). Coaxial electrospinning and sus-tained release properties of gelatin-cellulose acetate core-shell ultrafine fibers. ScienceA-sia, 37(4), 335–343.

305. Yarin, A., et al. (2007). Material encapsulation and transport in core-shell micro/nanofi-bers, polymer and carbon nanotubes and micro/nanochannels. Journal of Materials Chem-istry, 17(25), 2585–2599.

306. Yarin, A. (2011). Coaxial electrospinning and emulsion electrospinning of core shell fi-bers. Polymers for Advanced Technologies, 22(3), 310–317.

307. Hufenus, R. (2008). Electrospun Core-Sheath Fibers for Soft Tissue Engineering, National Textile Center Annual Report. 18–31.

308. Mccann, J. M., Marquez, Y., Xia, (2006). Melt coaxial electrospinning: a versatile method for the encapsulation of solid materials and fabrication of phase change nanofibers. Nano Letters, 6(12), 2868–2872.
309. Loscertales, G., et al. (2004). Electrically Forced Coaxial Nanojets for One-step Hollow Nanofiber Design. Journal of the American Chemical Society, 126(17), 5376–5377.
310. Tan, S., Huang, X., Wu, B. (2007). Some Fascinating Phenomena in Electrospinning Processes and Applications of Electrospun Nanofibers. Polymer International, 56(11), 1330–1339.
311. Ganan-Calvo, A. M. (1997). Cone-jet analytical extension of Taylor's electrostatic solution and the asymptotic universal scaling laws in electrospraying. Physical Review Letters, 79(2), 217–220.
312. Chen, X., et al. (2005). Spraying modes in coaxial jet electrospray with outer driving liquid. Physics of Fluids, 17, 032101.
313. Gañán-Calvo, A. M. (1998). Generation of steady liquid microthreads and micron-sized monodisperse sprays in gas streams. Physical Review Letters, 80(2), 285.
314. Hwang, Y., Jeong, U., Cho, E. (2008). Production of uniform-sized polymer core-shell microcapsules by coaxial electrospraying. Langmuir, 24(6), 2446–2451.
315. Lopez-Herrera, J., et al. (2003). Coaxial jets generated from electrified Taylor cones. Scaling laws. Journal of aerosol Science, 34(5), 535–552.
316. Loscertales, I., et al. (2002). Micro/nano Encapsulation via Electrified Coaxial Liquid Jets. Science, 295(5560), 1695–1698.
317. Reznik, S., et al. (2006). Evolution of a compound droplet attached to a core-shell nozzle under the action of a strong electric field. Physics of Fluids, 18(6), 62101–62113.
318. Zussman, E., et al. (2006). Electrospun Polyaniline/Poly (methyl methacrylate)-Derived Turbostratic Carbon Micro-/Nanotubes. Advanced Materials, 18(3), 348–353.
319. La Mora, D., Fernandez, J. (1994). The current emitted by highly conducting Taylor cones. Journal of Fluid Mechanics, 260(1), 155–184.
320. Gamero-Castano, M., Hruby, V. (2002). Electric measurements of charged sprays emitted by cone-jets. Journal of Fluid Mechanics, 459(1), 245–276.
321. Higuera, F. (2003). Flow rate and electric current emitted by a Taylor cone. Journal of Fluid Mechanics, 484, 303–327.
322 Artana, G., Romat, H., Touchard, G. (1998). Theoretical analysis of linear stability of electrified jets flowing at high velocity inside a coaxial electrode. Journal of electrostatics, 43(2), 83–100.
323. Yu, J., Fridrikh, S., Rutledge, G. C. (2004). Production of submicrometer diameter fibers by two-fluid electrospinning. Advanced Materials, 16(17), 1562–1566.
324. Li, F., Yin, X., Yin, X. (2005). Linear instability analysis of an electrified coaxial jet. Physics of Fluids, 17, 77104.
325. Boubaker, K. (2012). A Confirmed Model to Polymer Core-Shell Structured Nanofibers Deposited via Coaxial Electrospinning. ISRN Polymer Science. (2012).
326. Zhang, L., et al. (2012). Coaxial electrospray of microparticles and nanoparticles for biomedical applications. Expert review of medical devices, 9(6), 595–612.
327. Mei, F., Chen, D. (2007). Investigation of compound jet electrospray: Particle encapsulation. Physics of Fluids, 19, 103303.
328. Batchelor, G. (2000) An introduction to fluid dynamicsCambridge: university press.
329. Castellanos, A. (1998). Electrohydrodynamics: Springer.
330. Gibbs, J., H., Bumstead, W., Longley, (1928). The collected works of J. Willard Gibbs. Vol. 1. Longmans, Green and Company.

331. Slattery, J. (1980). Interfacial Transport Phenomena Invited Review. Chemical Engineering Communications, 4(1–3), 149–166.
332. Castellanos, A., Gonzalez, A. (1996). Nonlinear waves and instabilities on electrified free surfaces. in Conduction and Breakdown in Dielectric Liquids, (1996), ICDL'96., 12th International Conference on.IEEE.
333. Drazin, P., Reid, W. (2004). Hydrodynamic stability, Cambridge Cambridge university press.
334. Levich, V., Spalding, D. (1962). Physico-chemical hydrodynamics. Vol. 689, Prentice-Hall Englewood Cliffs, N. J.
335. Taylor, G. (1969). Electrically driven jets. Proceedings of the Royal Society of London. A. Mathematical and Physical Sciences, 313(1515), 453–475.
336. Melcher, J., Waves, F. (1963). A Comparative Study of Surface-Coupled Electrohydrodynamic and Magnetohydrodynamic Systems. MIT Press, Cambridge, MA. 433.
337. Bailey, A. (1988). Electrostatic spraying of liquids. Research Studies Press Somerset, England.
338. Lin, S., Kang, D. (1987). Atomization of a liquid jet. Physics of Fluids, 30, (2000).
339. Reznik, S. N., et al. (2006). Evolution of a Compound Droplet Attached to a Core-shell Nozzle Under the Action of a Strong Electric Field. Physics of Fluids, 18(6), 62101–62113.
340. Hu, Y., Huang, Z. (2007). Numerical study on two-phase flow patterns in coaxial electrospinning. Journal of applied physics, 101(8), 084307–084307-7.
341. Sugiyama, H., et al. (2013). Simulation of Electrohydrodynamic Jet Flow in Dielectric Fluids. Journal of Applied Fluid Mechanics, 6(3).
342. Li, J., et al. (1991). Numerical study of laminar flow past one and two circular cylinders. Computers & Fluids, 19(2), 155–170.
343. Theron, S., Zussman, E., Yarin, A. (2004). Experimental investigation of the governing parameters in the electrospinning of polymer solutions. Polymer, 45(6), 2017–2030.
344. Marín, Á., et al. (2007). Simple and double emulsions via coaxial jet electrosprays. Physical Review Letters, 98(1), 14502.
345. Chen, X. et al., (2005). Spraying modes in coaxial jet electrospray with outer driving liquid. Physics of Fluids, 17, 32101.
346. Li, F., Yin, X., Yin, X. (2005). Linear instability analysis of an electrified coaxial jet. Physics of Fluids, 17, 077104.
347. Li, F., Yin, X., Yin, X. (2006). Linear instability of a coflowing jet under an axial electric field. Physical Review E., 74(3), 036304.
348. Higuera, F. (2007). Stationary coaxial electrified jet of a dielectric liquid surrounded by a conductive liquid. Physics of Fluids, 19, 012102.
349. Chandrasekhar, S. (1961). Hydrodynamic and hydromagnetic stability. International Series of Monographs on Physics, 1.
350. Shen, J., Li, X. (1996). Instability of an annular viscous liquid jet. Acta Mechanica, 114, 167–183.
351. Chen, J., Lin, S. (2002). Instability of an annular jet surrounded by a viscous gas in a pipe. Journal of Fluid Mechanics, 450, 235–258.
352. Turnbull, R. (1992). On the instability of an electrostatically sprayed liquid jet. Industry Applications, IEEE Transactions on, 28(6), 1432–1438.
353. Mestel, A. (1994). Electrohydrodynamic stability of a slightly viscous jet. Journal of Fluid Mechanics, 274, 93–114.
354. González, H., García, F. A. (2003). Castellanos, Stability analysis of conducting jets under ac radial electric fields for arbitrary viscosity. Physics of Fluids, 15, 395.

355. Huebner, A., Chu, H. (1971). Instability and breakup of charged liquid jets. J. Fluid Mech, 49(2), 361–372.
356. Son, P., Ohba, K. (1998). Theoretical and experimental investigations on instability of an electrically charged liquid jet. International journal of multiphase flow, 24(4), 605–615.
357. Lopez-Herrera, J., Ganan-Calvo, A. (2004). A note on charged capillary jet breakup of conducting liquids: experimental validation of a viscous one-dimensional model. Journal of Fluid Mechanics, 501, 303–326.
358. Zakaria, K. (2000). Nonlinear instability of a liquid jet in the presence of a uniform electric field. Fluid Dynamics Research, 26(6), 405.
359. Elhefnawy, A.,. Agoor., B Elcoot, A. (2001). Nonlinear electrohydrodynamic stability of a finitely conducting jet under an axial electric field. Physica A: Statistical Mechanics and its Applications, 297(3), 368–388.
360. Elhefnawy, A., Moatimid, G., Elcoot, A. (2004). Nonlinear electrohydrodynamic instability of a finitely conducting cylinder: Effect of interfacial surface charges. Zeitschrift für angewandte Mathematik und Physik ZAMP, 55(1), 63–91.
361. Moatimid, G. (2003). Non-linear electrorheological instability of two streaming cylindrical fluids. Journal of Physics A: Mathematical and General, 36(44), 11343.
362. Si, T., et al. (2009). Modes in flow focusing and instability of coaxial liquid-gas jets. Journal of Fluid Mechanics, 629(1), 1–23.
363. Si, T., et al. (2010). Spatial instability of coflowing liquid-gas jets in capillary flow focusing. Physics of Fluids, 22, 112105.
364. Lin, S. P. (2003). Breakup of Lliquid Sheets and Jets.Cambridge University Press New York.
365. Li, F., Yin, X., Yin, X. (2009). Axisymmetric and nonaxisymmetric instability of an electrified viscous coaxial jet. Journal of Fluid Mechanics, 632(1), 199–225.
366. Li, F., Yin, X., Yin, X., Linear instability of a coflowing jet under an axial electric field. Physical Review E, 74(3), 36304.
367. Si, T., Zhang, L., Li, G. (2012). Co-axial electrohydrodynamic atomization for multimodal imaging and image-guided therapy. in Proc. SPIE, Multimodal Biomedical Imaging VII, San Francisco, California, USA.
368. López-Herrera, J., RiescoChueca, P., Gañán-Calvo, A. (2005). Linear stability analysis of axisymmetric perturbations in imperfectly conducting liquid jets. Physics of Fluids, 17, 034106.
369. Lim, D., Redekopp, L. (1998). Absolute instability conditions for variable density, swirling jet flows. European Journal of Mechanics-B/Fluids, 17(2), 165–185.
370. Meyer, J., Weihs, D. (1987). Capillary instability of an annular liquid jet. Journal of Fluid Mechanics, 179, 531–45.
371. Chen, F., et al. (2003). On the axisymmetry of annular jet instabilities. Journal of Fluid Mechanics, 488, 355–367.
372. Liu, J., Xue, D. (2010). Hollow nanostructured anode materials for Li-ion batteries. Nanoscale research Letters, 5(10), 1525–1534.
373. Chen, J., Archer, L., Lou, X. (2011). SnO2 hollow structures and TiO2 nanosheets for lithium-ion batteries. Journal of Materials Chemistry, 21(27), 9912–9924.
374. Lee, B., et al. (2012). Fabrication of Si core/C shell nanofibers and their electrochemical performances as a lithium-ion battery anode. Journal of Power Sources, 206, 267–273.
375. Liu, B., et al. (2011). An enhanced stable-structure core-shell coaxial carbon nanofiber web as a direct anode material for lithium-based batteries. Electrochemistry Communications, 13(6), 558–561.

376. Hwang, T., et al. (2012). Electrospun Core-shell Fibers for Robust Silicon Nanoparticle-Based Lithium Ion Battery Anodes. Nano Letters, 12(2), 802–807.

377. Han, H., et al. (2011). Nitridated TiO_2 hollow nanofibers as an anode material for high power lithium ion batteries. Energy & Environmental Science, 4(11), 4532–4536.

378. Ueno, S., Fujihara, S. (2011). Effect of an Nb_2O_5 nanolayer coating on ZnO electrodes in dye-sensitized solar cells. Electrochimica Acta, 56(7), 2906–2913.

379. Greene, L., et al. (2007). ZnO-TiO_2 core-shell nanorod/P_3HT solar cells. The Journal of Physical Chemistry C, 111(50), 18451–18456.

380. Yang, H., Lightner, C., Dong, L. (2011). Light-emitting coaxial nanofibers. ACS nano, 6(1), 622–628.

381. Qin, X., Wang, S. (2006). Filtration properties of electrospinning nanofibers. Journal of Applied Polymer Science, 102(2), 1285–1290.

382. Anka, F., Balkus, Jr, K. (2013). Novel Nanofiltration Hollow Fiber Membrane Produced via Electrospinning. Industrial and Engineering Chemistry Research, 52(9), 3473–3480.

383. Sill, T., von Recum, H. (2008). Electrospinning: applications in drug delivery and tissue engineering. Biomaterials, 29(13), 1989–2006.

384. Kenawy, E., et al. (2009). Processing of polymer nanofibers through electrospinning as drug delivery systems. Materials Chemistry and Physics, 113(1), 296–302.

385. Huang, Z., et al. (2006). Encapsulating drugs in biodegradable ultrafine fibers through co-axial electrospinning. Journal of Biomedical Materials Research Part A, 77(1), 169–179.

386. Chakraborty, S., et al. (2009). Electrohydrodynamics: A facile technique to fabricate drug delivery systems. Advanced Drug Delivery Reviews, 61(12), 1043–1054.

387. Zhang, Y., et al. (2006). Coaxial electrospinning of (fluorescein isothiocyanate-conjugated bovine serum albumin)-encapsulated poly (ε-caprolactone) nanofibers for sustained release. Biomacromolecules, 7(4), 1049–1057.

388. Srikar, R., et al. (2008). Desorption-limited mechanism of release from polymer nanofibers. Langmuir, 24(3), 965–974.

389. Bognitzki, M., et al. (2000). Polymer, metal, hybrid nano-and mesotubes by coating degradable polymer template fibers (TUFT process). Advanced Materials, 12(9), 637–640.

INDEX

A

Acoustic waves, 84
Aeolotropic, 262
Aerospace, 85, 93, 210
Agglomerate, 87, 89, 192
Air drag force, 293
Algebraic, 134, 135, 140, 150, 289, 293
Algorithms, 118, 121, 123, 288, 293
Alkaline fuel cells, 238
Alkanethiol, 37, 38
Amorphous soft carbon, 353
Amphiphilic block copolymer, 242
Analytic dispersion equation, 307
Anisotropic coupling, 81
Anisotropic forces, 263
Anisotropy energy, 75, 79
  easy axes, 75
Annular viscous liquid jet, 367
Anode configuration, 353
Antenna properties, 49
Anti-fouling surface treatment, 89
Antimicrobial nanocoatings, 239
Aqueous solutions, 92
Arbitrary amplitude, 63
Arbitrary electric fields, 250, 258
Arc melting chamber, 205
Arc-melted ingot, 205
Area of catalysis, 203
Artificial systems, 103
  automobile, 93, 103
  space shuttle, 103
Astrophysics, 7, 115, 122
Asymmetric nanoelements, 174
Asymptotic regime, 245, 246
Atom mimicry, 4, 9
Atomic interaction, 172, 174, 178, 187
Atomic level, 122, 170, 204
Atomic masses, 7
Atomic scale, 11, 63, 88, 121, 129
Attrition, 18

Axial direction, 199, 286, 334
Axial distance, 274
Axial wave number, 329, 330
Axisymmetric droplet, 290
Azimuthal wave number, 329

B

Balance of momentum, 225, 226, 252, 267
Ball bearings, 91, 93
Ball milling, 21, 205
Ball-powder, 21
Band gap energy, 34
Base liquid, 142, 149, 150, 163
Battery capacity, 351
Battery system, 351
Battle suit, 94
Bead-spring model, 269, 271
Beam diameter, 67
Benchmark solutions, 300
Bentonite, 86
Bessel functions, 338, 364, 365
Bhatnagar-Gross-Krook, 161, 298
Binding energy, 82
Bioactive agents, 303, 357
Biochemistry, 122, 132
Bioengineering applications, 200, 220
Biological arrays, 39
Biological materials, 48, 218
Biological systems, 2, 39, 199, 218, 240
Biological weapons, 95
Biology, 5, 7, 35, 48, 132, 199, 288
Biomedical applications, 92, 221, 250
Biotechnology, 20, 199, 222, 240, 304
Blue shifts, 34
Boltzmann equation, 115, 161, 298, 299
Born-Oppenheimer, 121
Bottom-up, 2, 20, 28, 35, 36, 46, 47, 169, 200, 204
Boubaker Polynomials Expansion Scheme, 307

Boundary element method, 293, 295
Boundary integral equations, 293–295
Boundary-value problems, 300
Bovine serum albumin, 242
Brass rod, 64
Bridge method, 230
British Royal Society, 5
Brownian dynamics, 132, 133, 267, 268
Brownian particles, 149, 153, 163
Bubbly slug, 321
Bulk conduction, 245, 296, 334, 345

C

Cable-type structures, 360
Calorimetric, 172
Cantilever, 19, 66
Capillary force, 34, 348
Carbon fibers, 86, 353
Carbon nanotubes, 24, 27, 34, 48, 239
Carbon precursors, 352
Carbonization temperature, 352
Carcinogens, 89
Cartesian coordinate system, 251
Cathode material, 350
Cauchy stress tensor, 301
Celestial mechanics, 288
    galaxies, 288
    planets, 288
Cell biology, 35
Cell living microenvironment, 358
Cell scaffolding, 243
Center of mass, 137, 175, 186
Centerline bends, 247
Centerline equation, 233
Ceramic materials, 203
Charge storage, 219
Charge transportation, 344
Charge–discharge process, 351
Chemical engineering process, 252
Chemical homogeneity, 206
Chemical vapor reaction, 204
Chirality, 46, 48
Chlorinated hydrocarbons, 89
Chlorine, 89
Circuit theory, 74, 83
Circular discs, 353

Clay fillers, 87
Clean future energy resources, 355
Coarse fibers, 238
Coarse-grain method, 227, 230
Coarsening, 21, 25
Coaxial
    capillary tubes, 245
    co-electrospinning process, 242
    electerospinning process, 198
    electrospin, 321, 322, 324
    electrospray analysis, 329
    electrospray problem, 319
    electrospray process control, 329
    electrospun fibers, 355
    spinneret, 198, 303, 305
Coercive field, 77
Coherent synthetic environment, 244
Cold plasma methods, 205
Colloidal dispersions, 153, 154, 163, 220
Commercial graphite anode, 352, 354
Competing models, 306
Complex fluids, 122, 267, 269, 270
Complex system, 9, 104, 105, 112, 252, 267
    machine, 104
    U.S. economy, 104
Complex viscosity, 270
Comprehensive investigation, 301
Computational fluid dynamics, 236, 298, 321
Computational techniques, 201, 299, 324
Computer science, 48
Conductors, 26, 41, 74, 327–329
Cone-jet mode, 246, 308, 326
Conical meniscus, 245
Conical shaped structure, 214
Constitutive assumption, 262
Constitutive equation, 226, 261–264, 271, 279, 293, 301, 302
Constitutive models, 300
Continuum field, 71
Continuum mechanics, 56, 195, 200, 300
Continuum plane stress model, 231
Controlled release system, 241
Convection term, 299, 315
Conveyor belts, 103
Cool light, 356

Cooling rates, 22
Coordinate system, 175, 251, 264, 342, 364
Copper drum, 22
Copper oxide, 88
    doped zinc oxides, 88
Core mineral oil, 352
Core-shell nozzle, 290, 305
Corrosion resistance, 39, 222
Coulomb blockade, 34, 46, 50
Coulomb force, 258, 293
Coulomb's law, 257, 258, 297, 302
Coulombic attractive force, 81
Counter electrode, 291, 303
Crude representations, 270
Crystallography, 172
Cubic element, 59
Cubic microparticles, 189
Cylindrical coordinate system, 251, 364

D

Dalton's atom, 6
Dangled segments, 232
Dashpot element, 251
Degussa, 87
Delithiation process, 352
Dendrimers, 4, 48
Density functional theory, 121, 201
Derivative function, 290
Design engineer, 198, 220
Design principles, 350
Desorption model, 359
Deterministic system, 113, 128
Diffusion of dopant, 50
Dilatation, 60, 122, 314
Dimensional analysis, 274, 276, 277
Dimensionless quantities, 276
Dipole moment, 151, 225
Dirac delta, 149
Direct current, 28
Discharge capacity, 352, 353
Discrete materials, 67
Discrete node model, 251
Disk drives, 91
Distribution functions, 122
Downscaled machine, 19

Drag formula, 143
Driving force, 38, 361
Drug delivery capability, 357
Drug loading, 241, 361, 363
Drug nonelectrospinnability, 357
Dry milling, 21
Dual-orifice spinneret design, 228
Dumbbell particles, 142, 163
Dumbbells density, 227
Dynamic balance, 337
Dynamic Oblique Deposition method, 66
Dynamic viscosity, 308, 332, 334

E

Easy axes, 75
E-field equations, 283
Elastic dumbbell kinetic theory, 263
Elastic dumbbell model, 270
Elastic energy, 233
Elasticity theory, 300
Electric circuit theory, 74
Electric field equation, 225, 262
Electric heating evaporation method, 204
Electrical conduction, 74, 254
    current, 301
    Drude, 74
Electrical conductivity, 56, 71, 73, 201, 296, 308, 321, 328, 331
Electrical shear force, 215, 223, 256, 336
Electro spun fibers, 215
Electrochemical devices, 238
Electrochemistry, 39
Electrode volume expansion, 351, 353
Electrodes, 73, 90, 352, 353, 355
Electromagnetic wave, 64
Electromechanical probe, 26
Electromechanical stresses, 246
Electron beam heating, 204
Electron cloud, 74
Electron free motion, 73
Electron scattering, 71
Electron tunneling, 46, 50
Electron-beam, 36
Electron-hole pairs, 356
Electronic conduction, 71
Electronic devices, 48, 67, 356

Electronic theory, 6
Electrophoresis method, 28
Electrospinning dilation, 253
Electrospinning modeling equations, 198
Electrospun mat, 227, 233, 239
Electrospun sheets, 233
Electrostatic attraction, 40, 213
Electrostatic force, 185, 213, 214, 245, 249, 330
Electrostatic pulling force, 285
Elongational flow, 251, 263
Emitted light, 82
Energy storage devices, 350
Entropy, 145, 147
Environment problem, 249
Environmental impact, 89, 355
Environmental pollution, 249, 350
Epitaxial growth, 205
Ethanol, 38
Evacuated chamber, 205
Evaporation rate, 218, 362
Exciton, 11, 81, 82
Exfoliated, 87
Experimental methods, 198, 219
Explosive forces, 255
External circuit, 73, 350
External electric field, 293
External operation factors, 362

F

Fabricating electronic unit, 202
Facile technique, 35
Faraday capacitors, 75
Fast electron transportation, 352
Feed gasses, 21
Feed rate, 216, 218
Fermi energy, 72, 74
Fermi gas, 74
Ferroelectric, 93
Ferrofluid, 150, 267, 269
Ferromagnetic, 75, 76, 79
Fiber
    formation techniques, 213
    Knudson number, 236
    mats, 218, 356
    morphology, 216, 218, 241
    network, 212, 231–233
    orientation angle, 231
    surface hydrophobic, 242
Fiber-fiber crossing points, 233
Fibrous media, 236
Fibrous network, 230
Fickian drug diffusion, 359
Field of investigation, 201
Field strength, 56, 57, 60, 61, 311
Filtration applications, 243
Fine dimension, 362
Finite element
    beam mesh, 234
    interpolation, 234
    method, 233, 289, 292–296, 299
Finite volume method, 289, 299
Finite-dimensional subspace, 289
Finite-element code, 297
Fire-proof properties, 363
Flow behavior, 87, 263, 298
    concrete, 87
Flow field, 153–155, 236
Flow method, 111
Flow pattern maps, 321
Fluctuation regions, 182
Fluent code, 236
Fluid constitutive properties, 251
Fluid dynamics, 122, 236, 250, 252, 270, 287, 298, 301, 307, 321
Fluid mixtures, 122
Fluid velocity, 162
Fluorination treatment, 242
Fokker-Planck equation, 227
Fossil energy, 350, 355
Fractal geometry, 274
Fracture mechanics, 67, 295
Free charge density, 308
Free electrons, 70, 214
Friction coefficient, 149, 151, 156
Fridge magnets, 91
Fullerenes, 34, 132, 170
Functionalizing agents, 360
    antibacterial, 360
    biomolecules agents, 360

G

Galaxies, 35, 114, 119, 122, 288
Galaxy clusters, 114
Gas storage applications, 243
Gas-phase evaporation method, 204
Gaussian distribution, 172
Gaussian electrostatic system, 271
Gels, 21
General paints, 88
Generic sense, 344
Geometry, 16, 48, 78, 80, 222, 225, 250, 274, 292
    Euclidean, 16
    Lobachevski, 16
    Riemannian, 16
Gibbs' free energy, 80
Gold nanowires, 26
Graphite carbon, 351
Graphite layer, 351
Gravimetric specific energy, 350
Gravitational acceleration, 308, 329, 331, 335, 336
Gravitational force, 115, 261, 293

H

Halloysite nanotubes, 242
Hamiltonian, 12
Hard magnetic iron, 78
Hardened steel ball, 61
Hardness test, 61
Harmonic oscillation, 62, 63
Harmonic potential, 268
Hartree-Fock approach, 121
Hartree-Lock methods, 201
Hecrite, 88
Helium gas, 205
Hemispherical surface, 213
Heterogeneous particles arrays, 202
Heuristic method, 300
Hierarchical nanostructures, 202
High production rate, 249
Hole bound, 81
Hollow fibers, 350, 353, 354, 356
Hollow spheres, 202
Homocentric capillary tubes, 326
Homogeneous equations, 277

Homogeneous linear system, 365
Homogenization, 115, 223, 229, 230, 233
Hybrid approach, 46, 47
Hybrid materials, 87
Hydration, 34
    hydrophilic, 34, 38, 87, 92, 213, 241, 242
    hydrophobic, 34, 38, 87, 88, 92, 213, 241, 242
Hydrodynamic parameters, 331
Hydrodynamic stress tensor, 319, 333
Hydrogels, 92
Hydrogen bonds, 172
Hydrogen fuel cells, 90, 238
Hydrophilic
    corona, 92
    polymers, 241
    therapeutic agents, 92
Hydrophobic polymers, 92
Hydrostatic forces, 260
Hydrostatic pressure, 60, 324
Hydrothermal synthesis, 189
Hypocrystalline, 170

I

Improper languages,
    JavaSim, 128
    C++Sim, 128
    SimPy (thon), 128
Indigo carmine dye, 357
Inert gas atmosphere, 205
Infinite computational power, 237
Infinitesimal disturbances, 332
Information technology, 199
Infrared, 83, 84, 89
Ingenious solutions, 218
Initial drug burst, 242
Inner liquid cylinder, 331
Inorganic fillers, 86
    glass fiber, 85, 86
    kaolin, 86
    talcum, 86
In-plane deformation, 67
Input energy sources, 356
Input parameter, 107, 108, 124, 125, 128, 233, 244

Instability
    analysis, 307, 312, 327–330, 335, 367
    theory, 329
Insulating liquid jets, 327
Insulators, 71, 73, 74
Intelligent catalysis, 222
Interaction force, 35, 170, 172, 173, 182, 183, 188, 195, 225
Interatomic bonds, 22
Internal combustion engine, 350
Internet, 91
Inverse investigations, 300
Inviscid liquid, 349
Ionization, 50, 82, 355
Isaac Newton, 7, 258, 276
    laws of motion, 7, 255
Isotropic distribution, 169
Isotropic electric contribution, 323
Isotropic mobility, 265
Iterative methods, 297

J

Java, 126, 128
    computer language, 126
JavaSim, 128
Jet breakup process, 347
Jet instability, 327–329, 341, 343, 344, 346–348
Jetting region, 246

K

Kinematic interface, 319
Kinetic energy, 12
Kinetic theory derivations, 268
Kutta-Merson method, 290, 291, 319

L

Ladder structure, 352
Lagrange multipliers, 301
Lagrangian axial strain, 272
Laminar basic flow, 329
Landau–Lifshitz–Gilbert equation, 80
Langevin equation, 149
    dynamics, 201
Laplace equation, 310, 311, 316, 325, 333, 364, 365

Lasers, 5, 49, 70, 202
    laser beam, 23, 207, 208
    laser heating, 204
Lattice Boltzmann methods, 132, 133, 298
Lattice cell, 159
Lattice vibration waves, 70
Leaky dielectric model, 327–329, 343, 349
Leapfrog method, 136
Light emitting diodes, 49, 202
Linear dispersion relation, 329
Linear jet portion, 251
Linear momentum balance, 297
Linear momentum, 236, 255, 317
Linear stability analysis, 313
Linearized kinematic boundary, 365
Liquid filament, 214
Liquid filtration, 236
Liquid phase method, 204
Lithium alloys, 351
Lithium ion batteries, 350–355, 362
Lithographic methods, 78
Lithography, 36, 39, 78
Log-conformation techniques, 293
Lorentz force law, 266
Lower packing density, 218
Low-storage methods, 290
Lubricants, 91
    inorganic materials, 91
Luminescence, 356
    chemiluminescence, 356
    electroluminescence, 356
    triboluminescence, 356
Luminescence, 82, 207, 350, 355, 356

M

Magnetic force, 150
Magnetic induction, 308
Magnetic intensity field, 308
Magnetic material, 75, 203
Magnetic nanocomposites, 363
Magnetic properties, 4, 22, 75–77, 92, 151
Magnetization energy, 75, 76
Magneto-optical sensitivity, 78
Man-made materials, 218
Markov chains, 288
Mass conservation, 252, 302, 310

Mass emission, 325
Mass production, 36, 40, 249, 250, 361
Mass transfer, 309, 310, 361, 363
Mass transport control, 221
Materials science, 48, 115, 204
Mathematic physical model, 321
Mathematical model, 106–110, 114, 116,
  117, 123, 168, 170, 198, 214, 279, 287,
  296
Max Planck, 7
Maximal packing density, 185
Maxwell equations, 83, 333
Maxwell's stress, 247
Mean segment length, 231
Mechanical oscillators, 64
  piano string, 64
  violin, 64
Mechanical resonance frequencies, 62, 96
Medical devices, 198, 219
Medical imaging, 92, 95
Medical technology, 86
Melt spinning, 22
Melting temperature, 22, 68, 69
Mesoscopic model, 154, 230
Mesoscopic scale, 79
Mesoscopic transport, 201
Methylene carbons, 38
Metropolis method, 146
Mica, 87
Micelles, 92
Microfluidics, 198, 303
Micromagnetic model, 79
  Brown, 79
Micromechanical elements, 226, 267
  beads, 226
  platelet, 226
Micron, 3, 84, 142, 204, 210, 213
Microprocessors, 8, 70
Microscopic interactions, 298
Microscopic models, 48, 226, 227
Microscopic scale, 115, 122
Microsensors, 19, 92
Microvalves, 19
Military applications, 239
Millimolar solution, 38
Milling method, 204
Minimal packing density, 185

Modern computational, 198, 219
Molecular anions, 206
Molecular bonding, 63
Molecular reorientation, 200
Molecular simulation method, 132, 159,
  267
Molecular theory, 6
Monodisperse films, 90
Montmorillonite, 87, 88
Morphological network model, 235
Morse potential, 187
Multifilament thread, 215
Multilevel organization, 228
Multiple material scale, 224
Multi-scale materials design, 219
Mutual charge repulsion, 213

N

Nanoassemblers, 48
Nanoceramics, 93
Nanoclays, 86
Nanoclusters, 4, 16, 24, 208
Nanocoatings, 14, 70, 222, 223, 239
Nanocrystalline, 14, 22, 23, 34, 36, 37, 71,
  73, 76, 92, 195
Nanocrystals, 34, 36, 44, 45, 48, 200, 203
Nano-electro-mechanical systems, 220
Nanoelectronics, 18, 28, 198, 199, 303
Nanoemulsions, 44
Nanolayered film, 45
Nanolubricants, 91
Nanomagnets, 77–80, 96
  isolated circular, 78, 96
  pentagonal, 78, 79, 96
  square, 78
  triangular, 78, 79, 96
Nanometer fabrication techniques, 77
Nanometer scale, 2, 6, 11, 18, 21, 36, 65,
  77
Nanometric scale, 2, 8
Nanoneedles, 45
Nano-Newton, 66
Nanoparticulate oxides, 88
Nanopatterned film, 45
Nanoplatelet, 199
Nanoporous film, 45

Nanopowders, 21
Nanorods, 14, 21, 45, 202, 210
Nanoscale element, 34, 36
Nanoscale lubrication, 201
Nanoscience, 2, 3, 5, 6, 9, 47, 99
Nanosprings, 67
Nanotem plate, 26
Nanowires, 14, 15, 26, 40, 48, 70, 96, 202, 239, 291
National security, 199
Natural biological systems, 218
Natural drying, 354
Natural systems, 103
    river, 103
    Universe, 103
Navier-Stokes equations, 154, 155, 287, 298, 301, 308
Newton's second law, 12, 62, 65, 119, 260
Newtonian flow, 227, 297
Nitrate, 18, 354
Nitrogen atmosphere, 352, 353
Noble metals, 37
Nonuniform grids, 299
Nonwoven fibrous sheet, 224
Notch root, 67
Novel multifunctional nanomaterials, 198
Novel properties, 3, 4, 6, 43, 200
Nuclear acids, 122
Nucleation, 170, 207
Nuclei, 121
Nucleic acids, 224
Numerical linear algebra, 288
Numerical procedure, 297
Numerical relaxation method, 285, 298
Nylon 6, 87
    synthetic polymer, 87

O

Oblique angle, 66
Ohm's law, 74
Ohmic contacts, 74, 96
One-dimensional nanomaterials, 14, 198
Optical emission, 82
Optical signals, 83
Optics, 48, 204, 267, 360
Optoelectronic, 5, 35, 86, 88, 202, 356

Ordinary differential equations, 119, 120, 283, 285, 288, 289, 298
Organic nanotubes, 48, 305
Oscillation, 63, 64, 78, 247
Osmosis separation, 356
Outer-driving state, 307

P

Parametric analysis, 283
Parasinuous mode, 330, 343, 347
Paravaricose mode, 330
Pegden, 126
Pendulum, 62, 63
Peridynamic modeling, 230
Permeate flow, 94
Perth, 21
Photocatalytic environmental remediation, 350
Photoelectric effect, 9
Photoelectric properties, 249
Photolithographic approaches, 35
Photon energy, 82
Photosynthesis, 90
Photothermal therapy, 82
Photovoltaic, 90, 243
Physical model, 106, 115, 244, 321, 329
Physical phenomena, 132, 136, 150, 201, 267, 287–289
Physical vapor deposition, 66
Picoscale particles, 6
Piezoelectric, 93
Planar gate length, 49
Planar technology, 49
Planetary system, 115
Plasma arcing, 21
Plasma heating, 204
Plasma proteins, 92
Plasmon, 82–84
Platinum-group metals, 90
Poisonous solvents, 249
Polarized charge density, 258
Polymer architecture, 87
Polymer chemist, 270
Polymer concentration, 212, 362
Polymer electrolyte membrane fuel cell, 239

Polymer kinetic theory, 263
Polymer shell, 242, 358
    barrier diffusion, 358
Polymer tensile force, 286
Polystyrene spheres, 36
Polyurethane film, 27
Poor-water soluble drugs, 241
Poor-water soluble, 242
Porous fibrous structures, 219
Portable fuel cells, 198, 219
Post-treatment methods, 363
Potential applications, 202, 220
Potential energy, 12, 63, 122, 179, 181,
    187–189, 257
Potential future energy, 356
Power exponent, 274
Power law equation, 358
Premature fusion, 87
Primary candidate, 350
Pritsker, 126
Process parameters, 170, 216, 224, 302,
    320, 361, 363
    applied electric field, 224
    flow rate, 224
Processors, 118
Programmable medical nanomachines, 219
Protein binding, 39
Proton exchange mat, 238, 239
Pyramidal diamond, 61

Q

Quantum computer, 7, 8
Quantum confinement, 34, 49, 70, 71, 73,
    96
Quantum dots, 5, 201, 202
Quantum effects, 5
Quantum mechanics, 6, 7, 11, 114, 255
Quantum number, 63, 73
Quantum physics, 2, 7, 8
    wave mechanics, 7
Quasicontinuum, 230
Quiescent fluid, 142, 143

R

Radial electric field, 326–329, 331, 335,
    336, 339, 341, 345, 347

Radial polymer stress, 286
Radiation chemical synthesis, 204
Radio-frequency, 205
Radius vector, 186
Random point field, 233
Rarefied gases, 115
Rayleigh-Benard instability, 122
Rayleigh-Taylor instability, 122
Red shift, 82
Regenerated energy sectors, 350
Relative motion, 156
Relaxation algorithm, 298
Relaxation process, 176, 179–182
Repulsion, 170, 188, 213, 214, 247, 254,
    281
Repulsive force, 145, 157, 260, 271
Rheological parameters, 224
    molecular weight, 224
    surface tension, 224
    viscosity, 224
Ribbons, 22
Rotary diffusivity, 227
Rotary movement, 170
Rotating collector, 218
Royal Academy, 5
Runge-Kutta method, 177, 289

S

Saddle-point stability, 298
Safety issues, 351
Salt, 18, 217
    matrix, 21
Sandwich, 75
Saturation magnetization, 75–77, 80
Scalar mobility constants, 264
Scaling laws, 274, 306, 325, 326
Scanning probe, 36, 132
Scattering length, 201
Schrödinger equation, 11, 12, 121
Second-order accuracy, 299
Seismic waves, 84
Self-replication, 198, 219
Semectic clays, 87
    montmorillonite, 87, 88
Semiconductor lasers, 49, 70
Semiconductor technology, 199

Semiconductor, 34, 40, 48, 49, 50, 71, 74, 81, 203, 204, 363
    insulators, 71
Sensors, 48, 49
Shear stress, 59, 60, 121, 286, 325, 336
Shear-lag model, 230
Signaling biomolecules, 221
Silanes, 40
Silicon carbide, 93
Silicon nitride, 93
Simple joint morphology, 235
Simple mesh collector, 215
Simulation language, 126, 127
    arena, 126
    GPSS, 126
    SIMSCRIPT, 126
    SIMULINK (Matlab), 126
Simulation models, 105, 111, 119
    computational models, 105
Simulation packages, 110, 111, 114, 126, 127
    arena, 111
    extend, 111
    pro model, 111
    witness, 111
Slippage, 86
Slug-annular, 321
Small electrical relaxation time limit, 345
Smart clothing, 198, 219
Sodium chloride solutions, 357
Solar cells, 23, 202, 208, 350, 355, 362
Solar photovoltaic cells, 90
Solid fibers, 249, 250, 354, 361
Solid oxide fuel cells, 90, 238
Solid phase method, 204
    solid-state reaction, 204
    spark discharge, 204
    thermal decomposition, 204
Solid state physics, 7, 74, 121, 262
Solid-fluid interaction force, 225
Solution filtration, 356
Solvent evaporation pyrolysis, 204
Solvent thermal method, 204
Source of inspiration, 218
Sparse droplets, 309
Spatial-temporal behavior, 267
Spatiotemporal distribution, 360

Spectral method, 307
Spherical coordinate system, 251
Spheroid particle, 142
Spraying process, 320
Sputtering method, 204
Stain-resistant fabrics, 88
Stainless steel alloys, 93
Static system, 104, 271
    bridge, 104
    building, 104
Stimuli-responsive capsules, 221
Stochastic fiber network, 233
Stochastic law, 144, 163
Stochastic system, 113, 128
Stock method, 111
Stoichiometries, 4, 7
Strain energy, 232, 233
Strain-hardening fluids, 246
Stretching process, 223, 297
Styrene-coacrylonitrile, 352
Submicron dimensions, 199
Sulfur binding, 38
Sun blockers, 94
Sunflower oil, 325, 344
Superficial energy, 190, 191
Supramolecular chemistry, 18, 28
Surface tension forces, 213, 215, 247
Surface-to-volume ratio, 16, 28, 200

T

Tactoids, 88
Taylor cone, 213–215, 217, 245, 296, 303, 306, 307, 326, 362
Taylor series expansion, 134, 139, 142, 289
Tensile
    force, 286, 302
    strain, 60, 61
    strength, 57, 86
    stress, 58, 60, 169
Terminal stages, 274
Tetracycline hydrochloride, 242
Textile fiber technology, 213
Therapeutic drugs, 242
Thermal energy, 75
Thermal expansion coefficient, 250

Thermal lattice Boltzmann methods, 298
Thermodynamic equilibrium, 133, 144, 145, 157, 159, 163
Thermosetting coatings, 88
Thiol chains, 38
Tissue engineering applications, 358
Tissue engineering scaffolds, 243
Tissue engineering, 211, 243, 358, 363
Titanium dioxide, 88, 94
Titanium precursor, 354
Toxic reagents, 206
Toyota, 86
Traction force, 68, 183, 184
Transactional-flow approach, 111
Transistors, 40, 49, 202
Tributyltin, 89
Tubular structure, 352
Tunable band gap, 34
Tunable porosity, 243
Tungsten filament, 356
Turbostatic arrangement, 352

U

Ultrafine fibers, 243, 360
Ultrafine particles, 205
Ultraviolet radiation, 356
Uniaxial structure, 222
Unique core-sheath structures, 198, 303
Units of measurement, 74
    farad, 74
    hertz, 74
    ohm, 74
Unspinnable polymers, 360

V

Vacuum deposition, 204
Vacuum gauge, 206
Valency, 4, 7
Vander Waals forces, 34, 38, 172
Varistor, 93
Vector computers, 118
Velocities, 119, 135–137, 141–144, 155, 157–159, 162, 171, 172, 209, 255, 299, 329, 342
Ventilation system, 215
Verlet method, 135, 136

Vesicles, 92
Vibrating atoms, 201
Vinyl acetates, 88
Viruses, 90, 240
Viscoelastic
    elements, 199
    force, 293
    medium, 271
    models, 251
Viscoplastic, 86
Viscous coaxial jet, 329, 335
Viscous coflowing jet, 339
Volatile solvent, 245
Voltage generator, 214
Volume-integral method, 297

W

Washing machines, 24
Waste water purification, 356
Water-filled channels, 358
Water-insoluble polymer, 242
Wave function, 11, 12, 71, 72
Weber number, 321, 330, 339, 347, 348
Whipping jet, 246, 247, 251
Whipping region, 215, 246
Wireless data transmission, 239
Wound healing, 211, 243

X

X-rays regimes, 84

Y

Young's modulus, 64, 66, 67, 230, 371

Z

Zeeman energy, 76, 78
Zeolites, 122
Zero-order kinetics, 241, 359
Zigzags, 66
Zinc oxides, 88
Zirconia, 93
Zirconium oxide, 93